Lecture Notes in Computer Science 7748

Commenced Publication in 1973
Founding and Former Series Editors:
Gerhard Goos, Juris Hartmanis, and Jan van Leeuwen

Editorial Board

Subir Kumar Ghosh   Takeshi Tokuyama (Eds.)

WALCOM: Algorithms and Computation

7th International Workshop, WALCOM 2013
Kharagpur, India, February 14-16, 2013
Proceedings

 Springer

Volume Editors

Subir Kumar Ghosh
Tata Institute of Fundamental Research
School of Technology and Computer Science
Homi Bhabha Road, Mumbai 400005, India
E-mail: ghosh@tifr.res.in

Takeshi Tokuyama
Tohoku University
Graduate School of Information Sciences (GSIS)
Aobayama Campus, GSIS Building, Sendai 980-8578, Japan
E-mail: tokuyama@dais.is.tohoku.ac.jp

ISSN 0302-9743   e-ISSN 1611-3349
ISBN 978-3-642-36064-0   e-ISBN 978-3-642-36065-7
DOI 10.1007/978-3-642-36065-7
Springer Heidelberg Dordrecht London New York

Library of Congress Control Number: 2012955562

CR Subject Classification (1998): F.2, G.2.1-2, G.4, I.1, I.3.5, E.1

LNCS Sublibrary: SL 1 – Theoretical Computer Science and General Issues

Typesetting: Camera-ready by author, data conversion by Scientific Publishing Services, Chennai, India

Printed on acid-free paper

Springer is part of Springer Science+Business Media (www.springer.com)

Preface

The 7th International Workshop on Algorithms and Computation (WALCOM 2013) was held during February 14–16, 2013, at the Indian Institute of Technology, Kharagpur, India, and was organized by the Department of Computer Science and Engineering, Indian Institute of Technology, Kharagpur. The workshop covered a diverse range of topics on algorithms and computations including computational geometry, approximation algorithms, graph algorithms, parallel and distributed computing, graph drawing, and computational complexity.

This volume contains 29 contributed papers presented at WALCOM 2013. The workshop received 86 submissions from 25 countries. Submissions were rigorously refereed by the Program Committee members with the help of the external reviewers. Abstracts of two invited talks delivered at WALCOM 2013 are also included in the volume.

We would like to thank the authors for contributing high-quality research papers to the workshop. We express our heartfelt thanks to the Program Committee members and the external referees for their active participation in reviewing the papers. We are grateful to Nicola Santro and Rina Panigrahy for delivering excellent invited talks. We thank the Organizing Committee, chaired by Partha Bhowmick and Sudebkumar Prasant Pal, for the smooth functioning of the workshop. We thank Springer for publishing the proceedings in the reputed *Lecture Notes in Computer Science* series. We thank our sponsors for their support. Finally, we remark that EasyChair conference management system was very efficient for handling the reviewing process.

February 2013
Subir Kumar Ghosh
Takeshi Tokuyama

WALCOM 2013 Organization

Steering Committee

Kyung-Yong Chwa	KAIST, Korea
Costas S. Iliopoulos	King's College London, UK
M. Kaykobad	BUET, Bangladesh
Petra Mutzel	Unversity of Dortmund, Germany
Shin-ichi Nakano	Gunma University, Japan
Subhas Chandra Nandy	Indian Statistical Institute, Kolkata, India
Takao Nishizeki	Tohoku University, Japan
C. Pandu Rangan	Indian Institute of Technology, Madras, India
Md. Saidur Rahman	BUET, Bangladesh

Program Committee

Joseph Cheriyan	University of Waterloo, Cannada
Sandip Das	Indian Statistical Institute, Kolkata, India
Sumit Ganguly	Indian Institute of Technology, Kanpur, India
Daya Gaur	University of Lethbridge, Canada
Subir Kumar Ghosh (Co-Chair)	Tata Institute of Fundamental Research, India
Sathish Govindarajan	Indian Institute of Science, Bangalore, India
Soekhee Hong	University of Sydney, Australia
Jesper Jansson	Kyoto University, Japan
Matya Katz	Ben-Gurion University of the Negev, Israel
Akinori Kawachi	Tokyo Institute of Technology, Japan
Shuji Kijima	Kyushu University, Japan
Andrzej Lingas	Lund University, Sweden
Anil Maheshwari	Carleton University, Canada
David Mount	University of Maryland, USA
Krishnendu Mukhopadhyaya	Indian Statistical Institute, Kolkata, India
Shin-Ichi Nakano	Gunma University, Japan
N. S. Narayanaswamy	Indian Institute of Technology, Madras, India
Janos Pach	EPFL, Lausanne, Switzerland
Md Saidur Rahman	BUET, Bangladesh
Sohel Rahman	BUET, Bangladesh
Venkatesh Raman	The Institute of Mathematical Sciences, India
Takeshi Tokuyama (Co-Chair)	Tohoku University, Japan
Peter Widmayer	ETH Zurich, Switzerland
Chee Yap	New York University, USA
Xiao Zhou	Tohoku University, Japan

Organizing Committee

Partha Bhowmick (Co-Chair) Indian Institute of Technology, Kharagpur
Arijit Bishnu   Indian Statistical Institute, Kolkata
Arindam Biswas   Bengal Engineering and Science University,
                               Shibpur
Partha Pratim Goswami   Calcutta University, Kolkata
Shovonlal Kundu   Jadavpur University, Kolkata
Sudebkumar P. Pal (Co-Chair) Indian Institute of Technology, Kharagpur
Sasanka Roy   Chennai Mathmetical Institute, Chennai

Additional Reviewers

Abu-Affash, A. Karim
Anonymous, Anon
Anonymouse, Anon
Anup Joshi, Anup
Aschner, Rom
Augustine, John
Auluck, Nitin
Averbakh, Igor
Basavaraju, Manu
Beauquier, Joffroy
Benkoczi, Robert
Bishnu, Arijit
Carmi, Paz
Chakraborty, Sourav
CS, Rahul
Di Giacomo, Emilio
Dolev, Shlomi
Dujmovic, Vida
Durand de Gevigney, Olivier
Dutta, Kunal
Fekete, Sandor
Floderus, Peter
Fraser, Robert
Frati, Fabrizio
Fukasawa, R.
Gagarin, Andrei
Ghosh, Arijit
Ghosh, Sasthi Charan
Hasan, Masud
Hasan, Md. Mahbubul
Hossain, Shahadat
Hruz, Tomas

Imai, Tatsuya
Inkulu, R.
Iranmanesh, Ehsan
Ito, Takehiro
Izumi, Taisuke
Kamiyama, Naoyuki
Kar, Purushottam
Karim, Md. Rezaul
Klein, Karsten
Kolay, Sudeshna
Korwar, Arpita
Kusakari, Yoshiyuki
Levcopoulos, Christos
Levin, Asaf
Lundell, Eva-Marta
Mandal, Partha Sarathi
Manuch, Jan
Mihalak, Matus
Miura, Kazuyuki
Mizuki, Takaaki
Mondal, Debajyoti
Morgenstern, Gila
Mudgal, Apurva
Nandakumar, Satyadev
Nandy, Subhas
Narasimhan, Sadagopan
Nekrich, Yakov
Nilsson, Bengt J.
Nishat, Rahnuma Islam
Okamoto, Yoshio
Pajor, Thomas
Pal, Sudebkumar

Penna, Paolo
Persson, Mia
Rajendraprasad, Deepak
Ramakrishna, Gadhamsetty
Ray, Saurabh
Roy, Sasanka
Satti, Srinivasa Rao
Sau, Buddhadeb
Saurabh, Saket
Schoengens, Marcel
Seto, Kazuhisa

Shibuya, Tetsuo
Sikdar, Somnath
Singla, Sahil
Sledneu, Dzmitry
Sur-Kolay, Susmita
Suzuki, Akira
Thachuk, Chris
Watanabe, Osamu
Yamamoto, Masaki
Yen, Hsu-Chun

Table of Contents

Invited Talks

Mobility and Computations: Some Open Research Directions
(Abstract) 1
    Nicola Santoro

Adversarial Prediction: Lossless Predictors and Fractal Like Adversaries
(Abstract) 4
    Rina Panigrahy

Computational Geometry

A Novel Efficient Approach for Solving the Art Gallery Problem 5
    Alexander Kröller, Mahdi Moeini, and Christiane Schmidt

Fixed-Orientation Equilateral Triangle Matching of Point Sets 17
    Jasine Babu, Ahmad Biniaz, Anil Maheshwari, and Michiel Smid

Online Exploration and Triangulation in Orthogonal Polygonal
Regions.. 29
    Sándor P. Fekete, Sophia Rex, and Christiane Schmidt

A Competitive Strategy for Distance-Aware Online Shape Allocation ... 41
    Sándor P. Fekete, Nils Schweer, and Jan-Marc Reinhardt

Base Location Problems for Base-Monotone Regions................... 53
    Jinhee Chun, Takashi Horiyama, Takehiro Ito,
    Natsuda Kaothanthong, Hirotaka Ono, Yota Otachi,
    Takeshi Tokuyama, Ryuhei Uehara, and Takeaki Uno

Counting Maximal Points in a Query Orthogonal Rectangle 65
    Ananda Swarup Das, Prosenjit Gupta, and Kannan Srinathan

Voronoi Game on Graphs 77
    Sayan Bandyapadhyay, Aritra Banik, Sandip Das, and Hirak Sarkar

Approximation and Randomized Algorithms

Approximation Schemes for Covering and Packing 89
    Rom Aschner, Matthew J. Katz, Gila Morgenstern, and
    Yelena Yuditsky

A Randomised Approximation Algorithm for the Hitting Set
Problem 101
    Mourad El Ouali, Helena Fohlin, and Anand Srivastav

Exact and Approximation Algorithms for Densest k-Subgraph
(Extended Abstract) 114
    *Nicolas Bourgeois, Aristotelis Giannakos, Giorgio Lucarelli,
    Ioannis Milis, and Vangelis Th. Paschos*

Linear-Time Constant-Ratio Approximation Algorithm and Tight
Bounds for the Contiguity of Cographs 126
    Christophe Crespelle and Philippe Gambette

Approximation Algorithms for the Partition Vertex Cover Problem 137
    *Suman Kalyan Bera, Shalmoli Gupta, Amit Kumar, and
    Sambuddha Roy*

Parallel and Distributed Computing

Daemon Conversions in Distributed Self-stabilizing Algorithms 146
    Wayne Goddard and Pradip K. Srimani

Broadcasting in Conflict-Aware Multi-channel Networks 158
    *Francisco Claude, Reza Dorrigiv, Shahin Kamali,
    Alejandro López-Ortiz, Paweł Prałat, Jazmín Romero,
    Alejandro Salinger, and Diego Seco*

Shared-Memory Parallel Frontier-Based Search 170
    *Shogo Takeuchi, Jun Kawahara, Akihiro Kishimoto, and
    Shin-ichi Minato*

Graph Algorithms

Smoothed Analysis of Belief Propagation for Minimum-Cost Flow
and Matching 182
    *Tobias Brunsch, Kamiel Cornelissen, Bodo Manthey, and
    Heiko Röglin*

Triangle-Partitioning Edges of Planar Graphs, Toroidal Graphs
and k-Planar Graphs 194
    Jiawei Gao, Ton Kloks, and Sheung-Hung Poon

Alliances and Bisection Width for Planar Graphs 206
    Martin Olsen and Morten Revsbæk

The Cyclical Scheduling Problem 217
    *Binay Bhattacharya, Soudipta Chakraborty, Ehsan Iranmanesh, and
    Ramesh Krishnamurti*

Complexity and Bounds

Generalized Rainbow Connectivity of Graphs 233
  Kei Uchizawa, Takanori Aoki, Takehiro Ito, and Xiao Zhou

Fixed-Parameter Tractability of Error Correction in Graphical Linear
Systems 245
  Peter Damaschke, Ömer Eğecioğlu, and Leonid Molokov

Lower Bounds for Ramsey Numbers for Complete Bipartite
and 3-Uniform Tripartite Subgraphs 257
  Tapas Kumar Mishra and Sudebkumar Prasant Pal

Improved Fixed-Parameter Algorithm for the Minimum Weight 3-SAT
Problem 265
  Venkatesh Raman and Bal Sri Shankar

On Directed Tree Realizations of Degree Sets 274
  Prasun Kumar, Jayalal Sarma M.N., and Saurabh Sawlani

An FPT Algorithm for TREE DELETION SET 286
  Venkatesh Raman, Saket Saurabh, and Ondřej Suchý

Graph Drawing

Circular Graph Drawings with Large Crossing Angles 298
  *Hooman Reisi Dehkordi, Quan Nguyen, Peter Eades, and
  Seok-Hee Hong*

On Graphs That Are Not PCGs 310
  Stephane Durocher, Debajyoti Mondal, and Md. Saidur Rahman

On Embedding of Certain Recursive Trees and Stars into Hypercube ... 322
  Indhumathi Raman

Box-Rectangular Drawings of Planar Graphs (Extended Abstract) 334
  Md. Manzurul Hasan, Md. Saidur Rahman, and Md. Rezaul Karim

Author Index 347

Mobility and Computations:
Some Open Research Directions
(Abstract)

Nicola Santoro

School of Computer Science, Carleton University, Canada
santoro@scs.carleton.ca

Distributed computing is the study of the computational and complexity issues arising in systems of autonomous computational entities interacting with each other (e.g., to solve a problem, to perform a task). The focus is on the autonomy of the entities (i.e., absence of control(lers) external to the system) and decentralization (i.e., absence of pre-defined controllers within the system). Traditionally the entities have been assumed to be stationary. However, there is a large and varied class of distributed environments where the interacting entities, autonomous and decentralized, are *mobile*. This class comprises of very different environments, including for example: software mobile agents in communication networks, mobile sensors networks, robotic networks, etc. These systems have been long the subject of intensive investigations in fields as diverse as AI, robotics, and software engineering. In recent years, an increasing number of algorithmic investigations in distributed computing have started to examine these settings, creating the research area of *distributed computing by mobile entities*.

In the effort to understand the algorithmic limitations of distributed computing by mobile entities (e.g, swarms of robots, systems of software agents, mobile sensor networks), the theoretical research has focused on the minimal what minimal hypotheses (e.g., capabilities of the entities, restrictions on the environment) allow a given problem to be solved.

Starting from the obvious fact that mobility, to take place, needs a place, the research has distinguished two basic settings for the spatial universe U in which the autonomous mobile entities operate and move. The first setting, called *discrete universe* or *graph world*, is when U is a simple graph (e.g., mobile agents in communication networks). The second setting, called *geometric* or *continuous universe*, is when U is a connected region of $\mathcal{R}^2$ or $\mathcal{R}^3$ (e.g., autonomous robots moving on a terrain or in space). In both settings, the entities have computational capabilities (i.e., storage and processing), can move in U (their movements are constrained by the nature of U), exhibit the same behavior (i.e., execute the same protocol), and are autonomous in their actions (e.g., they are not directed by an external controller); depending on the context, the entities are also called *agents* or *robots*.

The two settings are clearly different with quite distinct applications; however, although the different nature of the universes imposes the use of very different mathematical tools (e.g., geometry and calculus in one, graph theory and combinatorics in the other), in my experience the algorithmic mind-frame required is the same.

S.K. Ghosh and T. Tokuyama (Eds.): WALCOM 2013, LNCS 7748, pp. 1–3, 2013.
© Springer-Verlag Berlin Heidelberg 2013

The problems range from electing a leader (a temporary coordinator), to creating maps of the environment; from locating black holes (harmful hosts that destroys incoming agents), to decontaminating a network being infected by a mobile intruder (e.g. a virus); from gathering autonomous mobile vehicles (e.g., tanks) dispersed on a terrain, to scattering mobile sensors on a terrain so to obtain a uniformly coverage of the space (e.g., for surveillance).

In spite of the differences, the same fundamental principles occur in all these contexts. The discovery of these principles is a new important research challenge. The research efforts within distributed computing has been rapidly increasing, becoming intensive over the last decade; however, still little is known.

This talk highlights some open research directions (touching some of the references below), which in my opinion are important, interesting, and intriguing.

References

1. Agmon, N., Peleg, D.: Fault-tolerant gathering algorithms for autonomous mobile robots. SIAM J. Comput. 36(1), 56–82 (2006)
2. Balamohan, B., Dobrev, S., Flocchini, P., Santoro, N.: Asynchronous Exploration of an Unknown Anonymous Dangerous Graph with $O(1)$ Pebbles. In: Even, G., Halldórsson, M.M. (eds.) SIROCCO 2012. LNCS, vol. 7355, pp. 279–290. Springer, Heidelberg (2012)
3. Barrière, L., Flocchini, P., Fomin, F.V., Fraigniaud, P., Nisse, N., Santoro, N., Thilikos, D.: Connected graph searching. Information and Computation (to appear, 2012)
4. Barrière, L., Flocchini, P., Fraigniaud, P., Santoro, N.: Rendezvous and election of mobile agents: Impact of sense of direction. Theory of Computing Systems 40(2), 143–162 (2007)
5. Cao, J., Das, S. (eds.): Mobile Agents in Networking and Distributed Computing. Wiley (2012)
6. Casteigts, A., Flocchini, P., Mans, B., Santoro, N.: Measuring temporal lags in delay-tolerant networks. IEEE Transactions on Computers (to appear, 2012)
7. Casteigts, A., Flocchini, P., Quattrociocchi, W., Santoro, N.: Time-varying graphs and dynamic networks. Int. Journal of Parallel, Emergent and Distributed Systems 27, 346–359 (2012)
8. Chalopin, J., Flocchini, P., Mans, B., Santoro, N.: Network Exploration by Silent and Oblivious Robots. In: Thilikos, D.M. (ed.) WG 2010. LNCS, vol. 6410, pp. 208–219. Springer, Heidelberg (2010)
9. Cielibak, M., Flocchini, P., Prencipe, G., Santoro, N.: Distributed computing by mobile robots: Gathering. SIAM Journal on Computing (to appear, 2012)
10. Clementi, A.E.F., Silvestri, R., Trevisan, L.: Information spreading in dynamic graph. In: 31st ACM Symposium on Principles of Distributed Computing, PODC, pp. 37–46 (2012)
11. Cohen, R., Peleg, D.: Convergence of Autonomous Mobile Robots with Inaccurate Sensors and Movements. In: Durand, B., Thomas, W. (eds.) STACS 2006. LNCS, vol. 3884, pp. 549–560. Springer, Heidelberg (2006)
12. Czyzowicz, J., Dobrev, S., Královič, R., Miklík, S., Pardubská, D.: Black Hole Search in Directed Graphs. In: Kutten, S., Žerovnik, J. (eds.) SIROCCO 2009. LNCS, vol. 5869, pp. 182–194. Springer, Heidelberg (2010)
13. Das, S., Flocchini, P., Kutten, S., Nayak, A., Santoro, N.: Map construction of unknown graphs by multiple agents. Theoretical Computer Science 385(1-3), 34–48 (2007)

14. Das, S., Flocchini, P., Prencipe, G., Santoro, N., Yamashita, M.: Synchronizing asynchronous robots using visible bits. In: 32nd Int. Conf. Distributed Computing Systems, ICDCS (2012)
15. Das, S., Flocchini, P., Santoro, N., Yamashita, M.: On the computational power of oblivious robots: forming a series of geometric pattern. In: 29th ACM Symp. on Principles of Distributed Computing, PODC, pp. 267–276 (2010)
16. Défago, X., Gradinariu, M., Messika, S., Raipin-Parvédy, P.: Fault-Tolerant and Self-stabilizing Mobile Robots Gathering. In: Dolev, S. (ed.) DISC 2006. LNCS, vol. 4167, pp. 46–60. Springer, Heidelberg (2006)
17. Dobrev, S., Flocchini, P., Kralovic, R., Santoro, N.: Exploring an unknown dangerous graph using tokens. Theoretical Computer Science (to appear, 2012)
18. Dobrev, S., Flocchini, P., Prencipe, G., Santoro, N.: Searching for a black hole in arbitrary networks: optimal mobile agents protocol. Distributed Computing 19(1), 1–19 (2006)
19. Dobrev, S., Flocchini, P., Prencipe, G., Santoro, N.: Mobile search for a black hole in an anonymous ring. Algorithmica 48, 67–90 (2007)
20. Dobrev, S., Santoro, N., Shi, W.: Using scattered mobile agents to locate a black hole in a unoriented ring with token. Int. J. Foundations of Computer Science 19(6), 1355–1372 (2008)
21. Flocchini, P., Mans, B., Santoro, N.: On the exploration of time-varying networks. Theoretical Computer Science (to appear, 2012)
22. Flocchini, P., Prencipe, G., Santoro, N.: Self-deployment of mobile sensors on a ring. Theoretical Computer Science 402(1), 67–80 (2008)
23. Flocchini, P., Prencipe, G., Santoro, N.: Distributed Computing by Oblivious Mobile Robots. Morgan & Claypool (2012)
24. Flocchini, P., Prencipe, G., Santoro, N., Widmayer, P.: Arbitrary pattern formation by asynchronous oblivious robots. Theoretical Computer Science 407(1-3), 412–447 (2008)
25. Flocchini, P., Santoro: Distributed Security Algorithms for Mobile Agents. In: Cao, J., Das, S. (eds.) Mobile Agents in Networking and Distributed Computing. Wiley (2012)
26. Fraigniaud, P., Gasieniec, L., Kowalski, D., Pelc, A.: Collective tree exploration. Networks 48, 166–177 (2006)
27. Fujinaga, N., Yamauchi, Y., Kijima, S., Yamashita, M.: Asynchronous Pattern Formation by Anonymous Oblivious Mobile Robots. In: Aguilera, M.K. (ed.) DISC 2012. LNCS, vol. 7611, pp. 312–325. Springer, Heidelberg (2012)
28. Ilcinkas, D., Flocchini, P., Santoro, N.: Ping pong in dangerous graphs: optimal black hole search with pebbles. Algorithmica 62(3-4), 1006–1033 (2012)
29. Izumi, T., Souissi, S., Katayama, Y., Inuzuka, N., Defago, X., Wada, K., Yamashita, M.: The gathering problem for two oblivious robots with unreliable compasses. SIAM J. Computing 41(1), 26–46 (2012)
30. Kuhn, F., Lynch, N., Oshman, R.: Distributed computation in dynamic networks. In: 42nd Symposium on Theory of Computing, STOC, pp. 513–522 (2010)
31. Masuzawa, T., Tixeuil, S.: Quiescence of self-stabilizing gossiping among mobile agents in graphs. Theoretical Computer Science 411(14-15), 1567–1582 (2010)
32. Sudo, Y., Baba, D., Nakamura, J., Ooshita, F., Kakugawa, H., Masuzawa, T.: An agent exploration in unknown undirected graphs with whiteboards. In: 3rd Workshop on Reliability, Availability, and Security, WRAS (2010)
33. Suzuki, I., Yamashita, M.: Distributed anonymous mobile robots: formation of geometric patterns. SIAM Journal of Computing 28(4), 1347–1363 (1999)

Adversarial Prediction: Lossless Predictors and Fractal Like Adversaries

(Abstract)

Rina Panigrahy

Microsoft, Mountain View, CA
rina@microsoft.com

Abstract. In this talk we will look at the classical prediction game where the adversary (or nature) is producing a sequence of bits and a prediction algorithm is trying to predict the future bit(s) from the past bits. This is like gambling on the future bits which involves the risk of making mistakes while shooting for profit from right predictions. Say the algorithm gets a payoff of 1 on a right prediction and -1 on wrong predictions (and is also make fractional bets $c \leq 1$ in which case its payoff is $+c$ or $-c$). We will see an algorithm [1] that has a good performance while almost never taking a risk of having a net loss where loss is said to happen when the number of wrong predictions exceeds the number of right predictions. Our algorithm gets no more than an exponentially small loss $e^{-\Omega(\epsilon^2 T)}$ over T bits on any sequence (where ϵ is a constant parameter). Further as compared to the payoff that would have been achieved by predicting the majority bit (in hindsight) our algorithms payoff is not lower by more than $O(\epsilon T)$ (which is commonly known as regret). We will also see experimental results on how these algorithms perform on stock data. Our algorithms build upon several classical works on the experts problem [2–4]

We will also see what kind of sequences are best from the adversary's perspective. We will show that under a certain formulation of predictive payoff it is best for the adversary to generate a "fractal like" sequence [5].

References

1. Kapralov, M., Panigrahy, R.: Prediction strategies without loss. In: Proceedings of NIPS (2011)
2. Even-Dar, E., Kearns, M., Mansour, Y., Wortman, J.: Regret to the best vs. regret to the average. Machine Learning 72, 21–37 (2008)
3. Cover, T.: Behaviour of sequential predictors of binary sequences. In: Transactions of the Fourth Prague Conference on Information Theory, Statistical Decision Functions, Random Processes (1965)
4. Littlestone, N., Warmuth, M.: The weighted majority algorithm. In: FOCS (1989)
5. Poppat, P., Panigrahy, R.: Fractal structures in adversarial prediction (manuscript, 2012)

S.K. Ghosh and T. Tokuyama (Eds.): WALCOM 2013, LNCS 7748, p. 4, 2013.
© Springer-Verlag Berlin Heidelberg 2013

A Novel Efficient Approach for Solving
the Art Gallery Problem

Alexander Kröller*, Mahdi Moeini, and Christiane Schmidt

Braunschweig University of Technology, IBR, Algorithms Group,
Mühlenpfordtstr. 23, 38106 Braunschweig, Germany
{a.kroeller,m.moeini,c.schmidt}@tu-bs.de

Abstract. In this paper, we consider the Art Gallery Problem (AGP)
that asks for the minimum number of guards placed in a polygon to over-
see the whole polygon. The AGP is known to be NP-hard even for very
restricted special cases. This paper describes a primal-dual algorithm
based on continuous optimization techniques for solving large-scale instances of the Art Gallery Problem. More precisely, the algorithm is a
combination of methods from computational geometry, linear programming (LP), and Difference of Convex functions (DC) programming. The
structure of the algorithm permits to provide lower and upper bounds
on the minimum number of guards. In order to evaluate the algorithm,
we measure its performance by solving some standard test instances including some non-orthogonal polygons with holes.

Keywords: Art Gallery Problem, Linear Programming, DC Programming, Duality, Separation, Visibility.

1 Introduction

The classical Art Gallery Problem (AGP) is one of the most famous problems
in computational geometry. It asks for the *minimum* number of guards, $G(P)$,
placed inside of a polygon P, that are sufficient to cover the entire polygon.
The problem has been studied for several classes of polygons and variants on
the placement of guards and it has been shown to be NP-hard (see for example
Lee and Lin [9]). Not only theoretical interest in the algorithmic challenges, but
numerous applications of the Art Gallery Problem (such as robotics, telecommunications, etc.) motivate the development of efficient methods for solving it.

The problem originated from a question posed by Klee. The first result was
given by Chvátal [3] who proved that for a polygon P of n vertices $\lfloor \frac{n}{3} \rfloor$ guards
are sometimes neccessary and always sufficient to monitor the entire polygon.
In other words, $g(n) \leq \lfloor \frac{n}{3} \rfloor$, where $g(n)$ is the maximum $G(P)$ over all polygons
of n vertices. Later a simpler proof for this statement was provided by Fisk
[8]. For orthogonal polygons Kahn et al. [12] showed $g(n) \leq \lfloor \frac{n}{4} \rfloor$. These results
refer to polygons without holes. Similar results can be found for polygons with

---

* Corresponding author.

S.K. Ghosh and T. Tokuyama (Eds.): WALCOM 2013, LNCS 7748, pp. 5–16, 2013.
© Springer-Verlag Berlin Heidelberg 2013

holes. Particularly, $\lfloor \frac{n+h}{3} \rfloor$ guards are always sufficient and sometimes necessary for polygon with h holes, see[14].

Variants have been studied that allow for vertex guards, edge guards or diagonal guards (guards restricted to be located on vertices, edges or diagonals, respectively). For an overview we refer to the textbook by O'Rourke [14].

While the above results refer to the number of guards that is always sufficient and sometimes neccessary to cover any polygon with n vertices, i.e., $g(n)$, we are interested in the minimum number of guards, $G(P)$, for a particular polygon P. Eidenbenz et al. [17] gave a lower bound of $\Omega(\log n)$ for the achievable approximation ratio for polygons with holes. Previous works on bounds mostly proposed heuristics [1] or studied restricted special cases of the Art Gallery Problem (see [2] and [5-7]). In this paper, we are interested in finding good upper bounds on the number of guards for the General Art Gallery Problem: the polygon may contain holes and there is no restriction on the guards' location. In particular, we are interested in solving the problem when the size of the polygon (i.e., the number of vertices n) is large. In a predecessor paper. [13], we developed a primal-dual approach for the general art gallery problem. Our method computes a sequence of lower and upper bounds on the optimal number of guards until—in case of convergence and integrality—eventually an optimal solution is reached. The algorithm solves the problem using a cutting plane and column generation approach. The computational time and integrality of the solutions provided by this approach are the main issues. In this paper, we propose a new approach by improving the method of [13] in order to find integer solutions more often and in shorter time. The key point is the use of Difference of Convex function (DC) programming techniques (see [10, 11, 15, 16] and references therein) and DC Algorithms (DCA) that are used to solve the DC programming problems. The choice of DC programming and DCA is based on their efficiency in solving large-scale binary programming problems.

The rest of the paper is organized as follows. In the following Section 2, we provide notations. In Section 3, we describe the mathematical formulation of the art gallery problem using linear programming. Section 4 is dedicated to a short introduction to Difference of Convex function (DC) programming; in fact, we use the DC programming techniques for developing our algorithm. We present our algorithm in Section 5. Section 6 discusses implementation aspects and the performance evaluation of our algorithm. Finally, Section 7 contains some concluding remarks and future research plans.

2   Definitions and Notations

We consider a polygon P that may contain holes. The set of all points of P that are visible from a point $p \in P$ is the *visibility polygon* $\mathcal{V}(p)$ (it is star-shaped).

The original Art Gallery Problem (AGP) is defined as follows: a *guard set* $G \subseteq P$ *covers* P iff $\cup_{g \in G} \mathcal{V}(g) = P$. We ask for a set G of minimum cardinality. Note that visibility is symmetric, i.e., $p \in \mathcal{V}(q) \iff q \in \mathcal{V}(p)$ and the inverse $\mathcal{V}(\cdot)$ simply denotes the set of points that can see a given point.

Points of P are used for two purposes: we select points to place guards on them, we denote these points as *guard positions*. In addition, each point of the polygon must be located in the visibility polygon of at least one guatrd. Thus, an uncovered points certifies that the current solution is not feasible, it "witnesses" this state. Hence, we refer to points that we monitor for being covered as *witnesses*.

The *separation problem* for an instance of a linear programming problem, e.g., $max\{c^T x | Ax \leq b\}$, asks for a point y whether y belongs to the polyhedron $\{x | Ax \leq b\}$ or not, and in the latter case, for a violated constraint. The separation problem for the primal and dual problem is refered to as the *primal separation problem* or *dual separation problem*, respectively.

3   The Art Gallery Problem: Mathematical Programming Formulation and LP-Based Solution Procedure

AGP can be formulated as an integer linear program with infinitely (actually uncountably) many binary variables and inequalities:

$$\min \quad \sum_{g \in P} x_g \tag{1}$$

$$\text{s.t.} \quad \sum_{g \in \mathcal{V}(w)} x_g \geq 1 \;\; \forall w \in P \tag{2}$$

$$x_g \in \{0, 1\} \quad \forall g \in P \tag{3}$$

Since the size of the above formulation is infinite, it is impossible to solve the problem directly. Instead, we restrict the guard positions to be from a finite set $G \subset P$, and only require a finite set $W \subset P$ of "witnesses" to be covered. We denote this problem by AGP(G, W):

$$\min \quad \sum_{g \in G} x_g \tag{4}$$

$$\text{s.t.} \quad \sum_{g \in G \cap \mathcal{V}(w)} x_g \geq 1 \;\; \forall w \in W \tag{5}$$

$$x_g \in \{0, 1\} \quad \forall g \in G \tag{6}$$

When the number of elements in G and W increases, it becomes quite difficult to solve this problem. Instead of solving this problem directly, we consider the LP relaxation AGR(G, W). It is obtained by relaxing the integrality constraint (6) to read

$$0 \leq x_g \leq 1 \;\; \forall g \in G \tag{7}$$

instead. Therefore, AGR(P, P) is the LP relaxation of AGP, and AGR(G, W) is a relaxation of AGR(P, P). In a predecessor paper [13], we have shown that AGR(P, P), can be solved efficiently for many problem instances. The procedure uses primal and dual separation (i.e., cutting planes and column generation) to connect $AGR(G, W)$ and $AGR(P, P)$. We start with restricted sets G and W, and solve $AGR(G, W)$ using the simplex method. This produces an optimal primal solution x^* and dual solution y^* with objective value z^*. Now there are three cases:

1. If there exists a point $w \in P \setminus W$ with $x^*(G \cap \mathcal{V}(w)) < 1$, then w corresponds to an inequality of $AGR(P, P)$ that is violated by x^*. The new witness w is added to W and the LP is re-solved. If such a w cannot be found, then x^* is optimal for $AGR(G, P)$, and z^* is an upper bound for $AGR(P, P)$.
2. If there exists a point $g \in P \setminus G$ with $y^*(W \cap \mathcal{V}(g)) > 1$, then it corresponds to a violated dual inequality of $AGR(P, P)$. We create the LP column for g and re-solve the LP. If such a g does not exist, y^* is an optimal dual solution for $AGR(P, W)$ and z^* is a lower bound for $AGR(P, P)$.
3. If neither of the previous cases produces a new point, then x^* is optimal for $AGR(P, P)$. The algorithm terminates.

The first two cases above reflect the two separation problems, asking for the existence of a w (primal) resp. g (dual). We have shown how to solve these efficiently, using the geometric interpretation. Consider the overlay of the visibility polygons of all points $g \in G$ with $x_g^* > 0$. This decomposes P into a planar arrangement of bounded complexity. The coverage function $c(p) := \sum_{g \in G \cap \mathcal{V}(p)} x^*(p)$ is constant over every face and edge of the arrangement. Consequently, an algorithm for the primal separation problem simply has to iterate over all faces, edges, and vertices of the arrangement to identify one point where the coverage value is less than 1. Due to the symmetry of visibility, the dual separation problem can be solved in the same manner as the primal (looking for coverage higher than 1).

4   DC Programming: A Short Introduction

In this section, we review some of the main definitions and properties of DC programming and DC Algorithms (DCA); where, "DC" stands for "difference of convex functions".

Consider the following primal DC program

$$(P_{dc}) \qquad \beta_p := \inf\{F(x) := g(x) - h(x) \ : \ x \in \mathbb{R}^n\},$$

where g and h are convex and differentiable functions. F is a *DC function*, g and h are *DC components* of F, and $g - h$ is called a *DC decomposition* of F.

Let C be a nonempty closed convex set and χ_C be the indicator function of C, i.e., $\chi_C(x) = 0$ if $x \in C$ and $+\infty$ otherwise. Then, by using χ_C, one can transform the constrained problem

$$\inf\{g(x) - h(x) \ : \ x \in C\}, \tag{8}$$

into the following unconstrained DC program

$$\inf\{f(x) := \phi(x) - h(x) \ : \ x \in \mathbb{R}^n\}, \tag{9}$$

where $\phi(x)$ is a convex function defined by $\phi(x) := g(x) + \chi_C(x)$.

Hence, without loss of generality, we can suppose that the primal DC program is unconstrained and in the form of (P_{dc}).

For any convex function g, its conjugate is defined by $g^*(y) := \sup\{\langle x, y \rangle - g(x) : x \in \mathbb{R}^n\}$ and the dual program of (P_{dc}) is defined as follows

$$(D_{dc}) \qquad \beta_d := \inf\{h^*(y) - g^*(y) : y \in \mathbb{R}^n\}. \tag{10}$$

One can prove that $\beta_p = \beta_d$ [16].

For a convex function θ and $x_0 \in \operatorname{dom} \theta := \{x \in \mathbb{R}^n | \theta(x_0) < +\infty\}$, the subdifferential of θ at x_0 is denoted by $\partial\theta(x_0)$ and is defined by

$$\partial\theta(x_0) := \{y \in \mathbb{R}^n : \theta(x) \geq \theta(x_0) + \langle x - x_0, y \rangle, \forall x \in \mathbb{R}^n\}. \tag{11}$$

We note that $\partial\theta(x_0)$ is a closed convex set in $\mathbb{R}^n$ and is a generalization of the concept of derivative.

For the primal DC program (P_{dc}) and $x^* \in \mathbb{R}^n$, the necessary local optimality condition is described as follows

$$\partial h(x^*) \subset \partial g(x^*). \tag{12}$$

This condition is also sufficient for many important classes of DC programs, for example, for the polyhedral DC programs [15] (in order to have a *polyhedral* DC program, at least one of the functions g and h must be a polyhedral convex function; i.e., the point-wise supremum of a finite collection of affine functions).

We are now ready to present the main scheme of the DC Algorithms (DCA) [15, 16] that are used for solving the DC programming problems. The DC Algorithms (DCA) are based on local optimality conditions and duality in DC programming, and consist of constructing two sequences $\{x^l\}$ and $\{y^l\}$. The elements of these sequences are trial solutions for the primal and dual programs, respectively. In fact, x^{l+1} and y^{l+1} are solutions of the following convex primal program (P_l) and dual program (D_{l+1}), respectively:

$$(P_l) \qquad \inf\{g(x) - h(x^l) - \langle x - x^l, y^l \rangle : x \in \mathbb{R}^n\}, \tag{13}$$
$$(D_{l+1}) \quad \inf\{h^*(y) - g^*(y^l) - \langle y - y^l, x^{l+1} \rangle : y \in \mathbb{R}^n\}. \tag{14}$$

One must note that, (P_l) and (D_{l+1}) are convexifications of (P_{dc}) and (D_{dc}), respectively, in which h and g^* are replaced by their corresponding affine minorizations. By using this approach, the solution sets of (P_{dc}) and (D_{dc}) are $\partial g^*(y^l)$ and $\partial h(x^{l+1})$, respectively. To sum up, in an iterative scheme, DCA takes the following simple form

$$y^l \in \partial h(x^l); \quad x^{l+1} \in \partial g^*(y^l). \tag{15}$$

One can prove that the sequences $\{g(x^l) - h(x^l)\}$ and $\{h^*(y^l) - g^*(y^l)\}$ are decreasing, and $\{x^l\}$ (respectively, $\{y^l\}$) converges to a primal feasible solution (respectively, a dual feasible solution) satisfying the local optimality conditions. More details, on convergence properties and theoretical basis of the DCA, can be found in [15, 16].

5   DC Programming for Solving the Art Gallery Problem

Our previously published LP-based procedure [13], as described in Section 3, focuses on finding good lower bounds. It may find upper bounds, corresponding to feasible (i.e., integer) solutions of $AGP(G, W)$. At each iteration, we solve continuous linear programs using IBM CPLEX, and find integer solutions only by chance. Due to this fact, there is no guarantee to get integer solutions.

To overcome this problem, one may use an integer program (IP) solver such as the IP solver of IBM CPLEX. Although this approach provides integer solutions and, consequently, upper bounds for the AGP, it is not an efficient way for solving the AGP for polygons with large number of vertices. Indeed, by increasing the number of integer variables (in other words, the number of elements in G and W) the integer programming problem becomes more and more difficult to solve, and it may become computationally expensive to solve such a problem.

Since using IP solvers for finding global solutions can be computationally expensive, we propose an alternative to integer solvers. It is able to find good integer solutions in short time. To this aim, we use DC programming techniques. Starting from the fractional solution, the objective is to find good local *integer* solutions. In this way, we expect to find high quality solutions in short time to cover the same set of witnesses by the same set of guards as the relaxed LP problem. Such a method is quite suitable to apply to large-scale AGP problems.

Algorithm 1 shows the resulting algorithm. Note that it is equal to the previous approach, with the exception of the new DC technique in Step 2.

Algorithm 1. LP-DC procedure

Step 0. Generate initial $G \subset P$, $W \subset P$.
Step 1. Solve relaxed $AGR(G, W)$, get optimal solution x^*, dual y^*, and objective value z^*;
Step 2. If x^* is fractional, call DC Algorithm to find an integer solution;
    Step 2.1. Verify the quality of the integer solution, if it is an improving solution, update the upper bound.
Step 3. Run primal separation
    Step 3.1. If primal separation produces a point w, **then** $W \leftarrow W \cup \{w\}$,
    Step 3.2. else
        – Output "z^* is an upper bound to relaxed $AGR(P, P)$",
        – **if** x^* is integral **then** Output "x^* is feasible for AGP, z^* is an upper bound for AGP".
Step 4. Run dual separation
    Step 4.1. If dual separation produces a point g, **then** $G \leftarrow G \cup \{g\}$,
    Step 4.2. else
        – Output "z^* is a lower bound to the relaxed $AGR(P, P)$ and to AGP".
Step 4. If both primal and dual separation failed or lower and upper bounds meet, **stop**, otherwise go to **Step 1**.

In the following, we describe the details of Step 2.

At each iteration of *Algorithm 1* the relaxed problem $AGR(G, W)$ is solved for $G \subset P$, $W \subset P$. If the optimal solution x^* is integral then it is feasible for $AGP(G, W)$, but this is rarely the case and in general x^* is a fractional solution. In order to overcome this problem and to find integer solutions for $AGP(G, W)$, we use DC programming techniques. First, let us define the following penalty function

$$\alpha(x) = \sum_{g \in G} x_g (1 - x_g),$$

and set $\mathcal{C} \subset \mathbb{R}^n$ as the set of all vectors satisfying the constraints (5) and (7). $\alpha(x)$ is a concave function that is nonnegative on $\mathcal{C}$; according to an exact penalty result presented in [10], there is a sufficiently large positive number θ_0 such that for all $\theta \geq \theta_0$ the following program is equivalent to $AGP(G, W)$

$$\min \ \sum_{g \in G} x_g + \theta \sum_{g \in G} x_g (1 - x_g) \tag{16}$$

s.t.

$$\sum_{g \in G \cap \mathcal{V}(w)} x_g \geq 1 \quad : \forall w \in W, \tag{17}$$

$$0 \leq x_g \leq 1 \quad : \forall g \in G. \tag{18}$$

We denote this program by $AGP_{DC}(G, W)$. As $AGP(G, W)$ and $AGP_{DC}(G, W)$ are equivalent, one can solve $AGP_{DC}(G, W)$ in place of $AGP(G, W)$. $AGP_{DC}(G, W)$ is a continuous optimization problem. Since (16) is a concave function, it is a DC function and we can write $AGP_{DC}(G, W)$ as follows

$$\min \left\{ F(x) := \sum_{g \in G} x_g - \theta \sum_{g \in G} x_g (x_g - 1) + \chi_{\mathcal{C}} : x \in \mathbb{R}^{|G|} \right\}.$$

In other words, $AGP_{DC}(G, W)$ is a DC program and a natural DC decomposition of $F(x)$ is

$$F(x) = f_1(x) - f_2(x),$$

where

$$f_1(x) := \sum_{g \in G} x_g + \chi_{\mathcal{C}},$$

and

$$f_2(x) := \theta \sum_{g \in G} x_g (x_g - 1).$$

Needless to say that $f_1(x)$ and $f_2(x)$ are convex functions.

The DC Algorithm for solving $AGP_{DC}(G, W)$, will construct two sequences $\{x^k\}$ and $\{u^k\}$ such that

$$u^k \in \partial f_2(x^k),$$

and

$$x^{k+1} \in \partial f_1^*(u^k).$$

To this aim, we compute u^k by using the following formula:

$$u_g^k = \theta(2x_g^k - 1) \quad : \forall g \in G, \tag{19}$$

and in order to compute x^{k+1}, it is sufficient to solve the following linear program

$$min \left\{ \sum_{g \in G} (1 - u_g^k)x_g : x \in C \right\}. \tag{20}$$

Any standard LP solver, such as IBM CPLEX, can solve (20) efficiently.
We are now ready to present the DC Algorithm for solving $AGP_{DC}(G, W)$:

Algorithm 2. DC Algorithm (DCA) for solving $AGP_{DC}(G, W)$

Step 1. (Initialization) Set $\epsilon > 0$ as the expected precision of the solutions. Set $k = 0$ and choose x^* (the fractional solution of $AGR(G, W)$) as the starting point (i.e., x^0) for DCA (see **Step 2.** of the *Algorithm 1*).
Step 2. Compute the vector u^k by means of (19).
Step 3. Solve (20) to obtain x^{k+1}.
Step 4. If $\|x^{k+1} - x^k\| \leq \epsilon$, **stop** and take x^{k+1} as the optimal solution, otherwise set $k \longleftarrow k + 1$ and go to **Step 2**.

As $AGP_{DC}(G, W)$ is equivalent to $AGP(G, W)$ and any deviation of x_g from 0 and 1 is penalized by a factor of θ (see the objective function of $AGP_{DC}(G, W)$), *Algorithm 2* will seek for optimal binary solutions that minimize the objective function of $AGP(G, W)$. Since the starting point of the algorithm is the global solution of $AGR(G, W)$, we expect to find a *good integer* solution for $AGP(G, W)$ in a *short time*.

In a similar way to the papers [15, 16], one can prove that

Theorem 1. *(Convergence properties of **Algorithm 2**)*

(i) *Algorithm 2 generates a sequence $\{x^k\}$ that is contained in the vertex set of C, in a way that the sequence $\{F(x^k)\}$ is decreasing.*
(ii) *The sequence $\{x^k\}$ converges to x^*, after a finite number of iterations, where x^* is a critical point of $AGP_{DC}(G, W)$.*
(iii) *There is a nonnegative number θ_1 such that for every $\theta > \theta_1$ the sequence $\{\alpha(x^k)\}$ is decreasing. In particular, if x^r is a feasible solution of the $AGP(G, W)$, so is x^k, for $k > r$.*
(iv) *x^* is almost always a local minimizer of $AGP_{DC}(G, W)$.*

6   Experimental Results

In this section, we describe implementation aspects and numerical results of our algorithms.

We implemented the algorithms in C++. The standard solver IBM CPLEX has been used to solve the LP problems generated at each iteration of the LP-based and LP-DC procedures. Furthermore, we employ the Computational Geometry Algorithms Library [4] for visibility queries and the separation problems. We make heavy use of the `Arrangement_2` package for both.

We tested our implementation on a variety of different classes of polygons, including randomly generated non-orthogonal polygons, von Koch polygons, and Spike polygons (see Figure 1), compare [13] for a more detailed description of the polygons used. In this section, we analyze how the LP-DC procedure performs on these instances and we compare it with the LP-based procedures using an LP/IP solver.

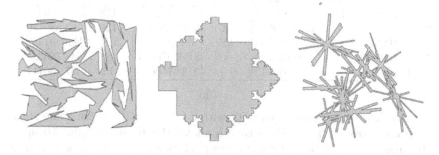

Fig. 1. Examples of randomly generated non-orthogonal, von Koch, and Spike polygons

The tolerance ϵ for the DCA is set to 10^{-6}. We use a time limit of 10 minutes on the run time of the algorithms. The parameter θ_0 is an input to the DCA. In general, it is not evident to compute an exact value of θ_0. In practice, we choose a sufficiently large value of θ_0 to make sure that for any $\theta \geq \theta_0$, the DC program (AGP$_{\mathrm{DC}}(G, W)$) is equivalent to the optimization problem (AGP(G, W)). For this purpose, we take θ_0 such that the solution provided by DCA is integer, i.e., it is feasible for the problem $AGP(G, W)$. Thus, according to the Theorem of exact penalty in [10], we have the equivalence between AGP$_{\mathrm{DC}}(G, W)$ and AGP(G, W). In our computational experiments, we set $\theta = 2n$ where n is the dimension of x^* (starting point of the DCA).

Numerical Results. The results of our experiments are shown in Table 1. The performance of the LP-DC procedure versus purely LP-based procedures is shown. We give results for LP-based procedures using both an *LP solver* and an *IP solver*. The tests have been carried out on polygons with different number of vertices: 200, 500, and 1000 vertices. We compare the performance of the different algorithms using

- the best upper bound. This bound represents the best integer solutions found by each procedure. They are reported in the column *Best UB*.

Table 1. Results of the experiments carried out over randomly generated polygons

		LP solver		IP solver		LP-DC	
Instance	Vertices	Best UB	CPU	Best UB	CPU	Best UB	CPU
1	200	30	578.519221	30	573.369291	29	3.239406
2	200	—	592.544785	28	578.832172	26	5.018052
3	200	26	594.126563	26	6.314784	26	6.838053
4	200	7	6.902821	7	6.014137	7	6.408391
5	200	7	8.059715	7	7.861637	7	6.390596
6	500	—	599.462758	75	594.722171	72	19.499690
7	500	—	596.492954	75	597.166775	73	23.046765
8	500	—	594.190658	71	589.272784	68	19.259176
9	500	11	598.982780	10	74.281563	10	69.564039
10	500	10	269.002230	10	203.192250	10	139.007263
11	1000	175	592.929514	154	595.962218	153	566.239325
12	1000	—	596.982976	127	594.172367	127	563.889150
13	1000	—	597.497764	135	597.285079	134	563.891582
14	1000	—	590.638740	—	2192.976454	14	558.789729
15	1000	—	593.640413	—	2433.750670	12	507.200717

- the run time (in seconds) of each algorithm (column *CPU*). For some of the large instances, we relaxed the time limit for the *IP solver* from **10** up to **40** minutes, hoping to find some integer solutions. This relaxation on time limit was not used for the two other procedures.

Integer Solutions. The LP-based procedure fails to find integer solutions (consequently, upper bounds) for most of the instances. This is not surprising because the procedure gives no guarantee on finding integer solutions (i.e., upper bounds). In particular, we observe this for polygons with a large number of vertices.

When we replace the LP solver by an IP solver, integer solutions are found more frequently. The drawback of this option being the decline of the performance of any IP solver when we try to solve the AGP for polygons with large number of vertices. In particular, we see that, in a similar way to the LP solver, the IP solver fails to find integer solutions for instances number 14 and 15. Even after a relaxation of the time limit up to 40 minutes, the IP solver fails to provide integer solutions. This fact is not surprising, because it is computationally expensive to an exact (Branch-and-Bound) method to solve a large-scale IP problem.

According to our experiments, the LP-DC procedure has a good performance in solving the AGP. Apart a few number of instances, we could solve the AGP problem by using the LP-DC procedure until global optimality and the LP-DC procedure provides integer solutions for *all* of the tested instances. As the LP-DC procedure provides good integer solutions and consequently good upper bounds, the procedure converges to the global solution in a shorter time than the other procedures (i.e., LP-based procedures with LP/IP solver).

The LP-DC procedure improves on the integer solution quality and the run time of the LP-based procedure using an IP solver. In particular, this is true for the test instances number 14 and 15 for which the IP solver fails to provide integer solutions.

No matter the class of polygons, the running time of the algorithms increases with problem size. A particular observation concerns the Spike polygons for which, a huge amount of computational time is needed for geometric evaluations (such as computing visibility polygons, etc.) and the time limit is over after a few number of iterations.

7 Conclusion

In this paper, we presented a novel approach, based on linear programming and DC programming techniques, for solving the Art Gallery Problem (AGP). In order to evaluate the proposed algorithm, we measured its performance by solving some standard test instances. According to the numerical results, we could solve the AGP for a majority of the test instances and for the remain of them, we could provide some integer bounds. This work improved our previous work [13] by providing upper bounds and solving some of the unsolved instances. For future work, we have a plan to do more experiments on different classes of polygons, particularly we are interested in examining our approach for solving the AGP for polygons with larger number of vertices.

Acknowledgments. This work was supported by the Deutsche Forschungs-gemeinschaft (DFG) under contract number KR 3133/1-1 (Kunst!).

References

1. Amit, Y., Mitchell, J.S.B., Packer, E.: Locating Guards for Visibility Coverage of Polygons. In: ALENEX, pp. 120–134 (2007)
2. Bottino, A., Laurentini, A.: A nearly optimal sensor placement algorithm for boundary coverage. Pattern Recognition 41(11), 3343–3355 (2008)
3. Chvátal, V.: A Combinatorial Theorem in Plane Geometry. Journal of Combinatorial Theory (B) 18, 39–41 (1975)
4. Computational Geometry Algorithms Library, http://www.cgal.org
5. Couto, M.C., de Rezende, P.J., de Souza, C.C.: An Exact and Efficient Algorithm for the Orthogonal Art Gallery Problem. In: SIBGRAPI 2007: Proceedings of the XX Brazilian Symposium on Computer Graphics and Image Processing, pp. 87–94. IEEE Computer Society, Washington, DC (2007)
6. Couto, M.C., de Souza, C.C., de Rezende, P.J.: Experimental Evaluation of an Exact Algorithm for the Orthogonal Art Gallery Problem. In: McGeoch, C.C. (ed.) WEA 2008. LNCS, vol. 5038, pp. 101–113. Springer, Heidelberg (2008)
7. Couto, M.C., de Rezende, P.J., de Souza, C.C.: An Exact Algorithm for an Art Gallery Problem. Technical report, Institute of Computing, University of Campinas (November 2009)

8. Fisk, S.: A Short Proof of Chvátal's Watchman Theorem. Journal of Combinatorial Theory (B) 24, 374–375 (1978)
9. Lee, D.T., Lin, A.K.: Computational Complexity of art gallery problems. IEEE Transactions on Information Theory 32(2), 276–282 (1986)
10. Le Thi, H.A., Pham Dinh, T., Muu, L.D.: Exact penalty in DC programming. Vietnam Journal of Mathematics 27, 169–178 (1999)
11. Le Thi, H.A., Moeini, M., Pham Dinh, T.: Portfolio Selection under Downside Risk Measures and Cardinality Constraints based on DC Programming and DCA. Computational Management Science 6(4), 477–501 (2009)
12. Klawe, M., Kleitman, D.: Traditional art galleries require fewer watchmen. SIAM Journal on Algebraic and Discrete Methods 4(2), 194–206 (1983)
13. Kröller, A., Baumgartner, T., Fekete, S., Schmidt, C.: Exact Solutions and Bounds for General Art Gallery Problems. To appear in Journal of Experimental Algorithms
14. O'Rourke, J.: Art Gallery Theorems and Algorithms. Oxford University Press, New York (1987)
15. Pham Dinh, T., Le Thi, H.A.: Convex analysis approach to d.c. programming: Theory, Algorithms and Applications. Acta Mathematica Vietnamica. Dedicated to Professor Hoang Tuy on the Occasion of his 70th Birthday 22(1), 289–355 (1997)
16. Pham Dinh, T., Le Thi, H.A.: DC optimization algorithms for solving the trust region subproblem. SIAM Journal on Optimization 8, 476–505 (1998)
17. Eidenbenz, S., Stamm, C., Widmayer, P.: Inapproximability results for guarding polygons and terrains. Algorithmica 31, 79–113 (2001)

Fixed-Orientation Equilateral Triangle Matching of Point Sets

Jasine Babu[1], Ahmad Biniaz[2], Anil Maheshwari[2], and Michiel Smid[2]

[1] Department of Computer Science and Automation,
Indian Institute of Science, Bangalore, India
[2] School of Computer Science, Carleton University, Ottawa, Canada
jasine@csa.iisc.ernet.in, ahmad.biniaz@gmail.com,
{anil,michiel}@scs.carleton.ca

Abstract. Given a point set P and a class $\mathcal{C}$ of geometric objects, $G_{\mathcal{C}}(P)$ is a geometric graph with vertex set P such that any two vertices p and q are adjacent if and only if there is some $C \in \mathcal{C}$ containing both p and q but no other points from P. We study $G_{\bigtriangledown}(P)$ graphs where $\bigtriangledown$ is the class of downward equilateral triangles (ie. equilateral triangles with one of their sides parallel to the x-axis and the corner opposite to this side below that side). For point sets in general position, these graphs have been shown to be equivalent to half-Θ_6 graphs and TD-Delaunay graphs.

The main result in our paper is that for point sets P in general position, $G_{\bigtriangledown}(P)$ always contains a matching of size at least $\left\lceil \frac{n-2}{3} \right\rceil$ and this bound cannot be improved above $\left\lceil \frac{n-1}{3} \right\rceil$.

We also give some structural properties of $G_{\bigcirc}(P)$ graphs, where $\bigcirc$ is the class which contains both upward and downward equilateral triangles. We show that for point sets in general position, the block cut point graph of $G_{\bigcirc}(P)$ is simply a path. Through the equivalence of $G_{\bigcirc}(P)$ graphs with Θ_6 graphs, we also deduce that any Θ_6 graph can have at most $5n - 11$ edges, for point sets in general position.

Keywords: Geometric graphs, Delaunay graphs, Matchings.

1 Introduction

In this work, we study the structural properties of some special geometric graphs defined on a set P of n points on the plane. An equilateral triangle with one side parallel to the x-axis and the corner opposite to this side below (resp. above) that side as in $\bigtriangledown$ (resp. $\triangle$) will be called a down (resp. up)-triangle. A point set P is said to be in general position, if the line passing through any two points from P does not make angles $0°$, $60°$ or $120°$ with the horizontal [4,13]. In this paper, we consider only point sets that are in general position and our results assume this pre-condition.

Given a point set P, $G_{\bigtriangledown}(P)$ (resp. $G_{\triangle}(P)$) is defined as the graph whose vertex set is P and that has an edge between any two vertices p and q if and only if there is a down-(resp. up-)triangle containing both points p and q but no

S.K. Ghosh and T. Tokuyama (Eds.): WALCOM 2013, LNCS 7748, pp. 17–28, 2013.

other points from P. (See Fig. 1.) We also define another graph $G_{\diamondsuit}(P)$ as the graph whose vertex set is P and that has an edge between any two vertices p and q if and only if there is a down-triangle or an up-triangle containing both points p and q but no other points from P. In Section 2 we will see that, for any point set P in general position, its $G_{\bigtriangledown}(P)$ graph is the same as the well known Triangle Distance Delaunay (TD-Delaunay) graph of P and the half-Θ_6 graph of P on so-called negative cones. Moreover, $G_{\diamondsuit}(P)$ is the same as the Θ_6 graph of P [4,6].

Given a point set P and a class $\mathcal{C}$ of geometric objects, the maximum $\mathcal{C}$-matching problem is to compute a subclass $\mathcal{C}'$ of $\mathcal{C}$ of maximum cardinality such that no point from P belongs to more than one element of $\mathcal{C}'$ and for each $C \in \mathcal{C}'$, there are exactly two points from P which lie inside C. Dillencourt [9] proved that every point set admits a perfect circle-matching. Ábrego et al. [1] studied the isothetic square matching problem. Bereg et al. concentrated on matching points using axis-aligned squares and rectangles [3].

A matching in a graph G is a subset M of the edge set of G such that no two edges in M share a common end-point. A matching of maximum cardinality is called a maximum matching in G. If all vertices of G appear as end-points of some edge in the matching, then it is called a perfect matching. It is not difficult to see that for a class $\mathcal{C}$ of geometric objects, computing the maximum $\mathcal{C}$-matching of a point set P is equivalent to computing the maximum matching in the graph $G_{\mathcal{C}}(P)$ [1].

The maximum $\triangle$-matching problem, which is the same as the maximum matching problem on $G_{\triangle}(P)$, was previously studied by Panahi et al. [13]. It was claimed that, for any point set P of n points in general position, any maximum matching of $G_{\triangle}(P)$ (and $G_{\bigtriangledown}(P)$) will match at least $\lfloor \frac{2n}{3} \rfloor$ vertices. But we found that their proof of Lemma 7, which is very crucial for their result, has gaps. By a completely different approach, we show that for any point set P in general position, $G_{\bigtriangledown}(P)$ (and by symmetric arguments, $G_{\triangle}(P)$) will have a maximum matching of size at least $\lceil \frac{n-2}{3} \rceil$; i.e, at least $2(\lceil \frac{n-2}{3} \rceil)$ vertices are matched. We also give examples where our bound is tight, in all cases except when $|P|$ is one less than a multiple of three.

We also prove some structural and geometric properties of the graphs $G_{\bigtriangledown}(P)$ (and by symmetric arguments, $G_{\triangle}(P)$) and $G_{\diamondsuit}(P)$. It will follow that for point sets in general position, Θ_6 graphs can have at most $5n - 11$ edges and their block cut point graph is a simple path.

2   Preliminaries

Our notations are similar to those in [4], with minor modifications. A *cone* is the region in the plane between two rays that emanate from the same point, its apex. Consider the rays obtained by a counter-clockwise rotation of the positive x-axis by angles of $\frac{i\pi}{3}$ with $i = 1, \ldots, 6$ around a point p. Each pair of successive rays, $\frac{(i-1)\pi}{3}$ and $\frac{i\pi}{3}$, defines a cone, denoted by $A_i(p)$, whose apex is p. For $i \in \{1, \ldots, 6\}$, when i is odd, we denote $A_i(p)$ using $C_{\frac{i+1}{2}}(p)$ and the cone opposite to $C_i(p)$ using $\overline{C}_i(p)$. We call $C_i(p)$ a positive cone around p and $\overline{C}_i(p)$ a negative cone around p. For each cone $\overline{C}_i(p)$ (resp. $C_i(p)$), let $\ell_{\overline{C}_i(p)}$ (resp. $\ell_{C_i(p)}$) be its bisector. If $p' \in \overline{C}_i(p)$, then let $\overline{c}_i(p, p')$ denote the distance between p and the orthogonal projection of p' onto $\ell_{\overline{C}_i(p)}$. Similarly, if $p' \in C_i(p)$, then let $c_i(p, p')$ denote the distance between p and the orthogonal projection of p' onto $\ell_{C_i(p)}$. For $1 \leq i \leq 3$, let $V_i(p) = \{p' \in P \mid p' \in C_i(p), p' \neq p\}$ and $\overline{V}_i(p) = \{p' \in P \mid p' \in \overline{C}_i(p), p' \neq p\}$. For any two points p and q, the smallest down-triangle containing p and q is denoted by $\bigtriangledown pq$ and the smallest up-triangle containing p and q is denoted by $\bigtriangleup pq$. If G_1 and G_2 are graphs on the same vertex set, $G_1 \cap G_2$ (resp. $G_1 \cup G_2$) denotes the graph on the same vertex set whose edge set is the intersection (resp. union) of the edge sets of G_1 and G_2.

The class of down-triangles (and up-triangles) admits a shrinkability property [1]: each triangle object in this class that contains two points p and q, can be shrunk such that p and q lie on its boundary. It is also clear that we can continue the shrinking process—from the edge that does not contain neither p or q—until at least one of the points, p or q, becomes a triangle vertex and the other point lies on the edge opposite to this vertex. After this, if we shrink the triangle further, it cannot contain p and q together. Therefore, for any pair of points p and q, $\bigtriangledown pq$ ($\bigtriangleup pq$) has one of the points p or q at a vertex of $\bigtriangledown pq$ ($\bigtriangleup pq$) and the other point lies on the edge opposite to this vertex (see Fig. 1, triangles are shown after shrinking) By the shrinkability property, for the $\bigtriangledown$-matching problem, it is enough to consider the smallest down-triangle for every pair of points (p, q) from P. Thus, $G_{\bigtriangledown}(P)$ is equivalent to the graph whose vertex set is P and that has an edge between any two vertices p and q if and only if $\bigtriangledown pq$ contains no other points from P. Notice that if $\bigtriangledown pq$ has p as one of its vertices, then $q \in \overline{C}_1(p) \cup \overline{C}_2(p) \cup \overline{C}_3(p)$. The following two properties are simple, but useful. Their proofs are easy and is included in the full version [2].

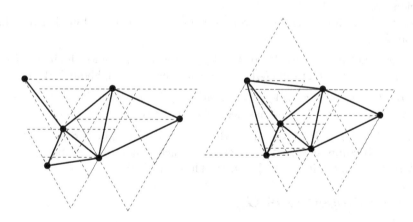

Fig. 1. A point set P and its (a) $G_{\bigtriangledown}(P)$ and (b) $G_{\lozenge}(P)$

Property 1. Let p and p' be two points in the plane. Let $i \in \{1,2,3\}$. The point p is in the cone $C_i(p')$ if and only if the point p' is in the cone $\overline{C}_i(p)$. Moreover, if p is in the cone $C_i(p')$, then $c_i(p', p) = \overline{c}_i(p, p')$.

Property 2. Let P be a point set, $p \in P$ and $i \in \{1,2,3\}$. If $\overline{V}_i(p)$ is non-empty, then, in $G_\triangledown(P)$, the vertex p' corresponding to the point in $\overline{V}_i(p)$ with the minimum value of $\overline{c}_i(p, p')$ is the unique neighbour of vertex p in $\overline{V}_i(p)$.

Consider a point set P and let $p, q \in P$ be two distinct points. By Property 1, $\exists i \in \{1,2,3\}$ such that $p \in \overline{C}_i(q)$ or $q \in \overline{C}_i(p)$; by the general position assumption, both conditions cannot hold simultaneously. Since $\triangledown pq$ has either p or q as a vertex, Property 2 implies that we can construct $G_\triangledown(P)$ as follows. For every point $p \in P$, and for each of the three cones, $\overline{C}_i$, for $i \in \{1,2,3\}$, add an edge from p to the point p' in $\overline{V}_i(p)$ with the minimum value of $\overline{c}_i(p, p')$, if $\overline{V}_i(p) \neq \emptyset$. This definition of $G_\triangledown(P)$ is the same as the definition of the half-Θ_6-graph on negative cones $(\overline{C}_i)$, given by Bonichon et al. [4]. We can similarly define the graph $G_\triangledown(P)$ using the cones C_i instead of $\overline{C}_i$, for $i \in \{1,2,3\}$, and show that it is equivalent to the half-Θ_6 graph on positive cones (C_i), given by Bonichon et al. [4]. In Bonichon et al. [4], it was shown that for point sets in general position, the half-Θ_6-graph, the *triangular distance-Delaunay graph* (TD-Del) [6], which are 2-spanners, and the *geodesic embedding* of P, are all equivalent.

The Θ_k-graphs discovered by Clarkson [7] and Keil [10] in the late 80's, are also used as spanners [11]. In these graphs, adjacency is defined as follows: the space around each point p is decomposed into $k \geqslant 2$ regular cones, each with apex p, and a point q of a given cone C is linked to p if, from p, the orthogonal projection of q onto C's bisector is the nearest point in C. In Bonichon et al. [4], it was shown that every Θ_6-graph is the union of two half-Θ_6-graphs, defined by C_i and $\overline{C}_i$ cones. In our notation this is same as the graph $G_\triangledown(P) \cup G_\triangle(P)$, which by definition, is equivalent to $G_{\triangle\kern-0.5em\triangledown}(P)$. Thus, for a point set in general position, $\Theta_6(P) = G_{\triangle\kern-0.5em\triangledown}(P)$.

Proofs of the following easy properties of $G_\triangledown(P)$ can be found in the full version [2].

Property 3. Let $p \in P$ with $V_i(p) \neq \emptyset$, $\overline{V}_j(p) = \emptyset$, $\overline{V}_k(p) = \emptyset$ for distinct $i, j, k \in \{1,2,3\}$. Then, in the graph $G_\triangledown(P)$, p has at least one neighbour in $V_i(p)$.

Property 4. Let $p \in P$ with $V_i(p) \neq \emptyset$ and $\overline{V}_i(p) \neq \emptyset$ for some $i \in \{1,2,3\}$. Then the vertex corresponding to p has degree at least two in $G_\triangledown(P)$.

Property 5. Let $p \in P$ be such that the vertex corresponding to p is of degree one in $G_\triangledown(p)$. Suppose $\exists i, j \in \{1,2,3\}, i \neq j$, such that $V_i(p) \neq \emptyset$ and $V_j(p) \neq \emptyset$. Let k be the element in $\{1,2,3\} \setminus \{i,j\}$. Then, $\overline{V}_k(p) \neq \emptyset$.

3   Some Properties of $G_\triangledown(P)$

Planarity: Planarity of $G_\triangledown(P)$ was proven in [5,6]. We have included a direct proof of the following lemma in the full version [2]. We use $G_\triangledown(P)$ to represent both the abstract graph and its planar embedding described in Lemma 1.

Lemma 1. *For a point set P, its $G_\nabla(P)$ is a plane graph, where its edges are straight line segments between the corresponding end-points.*

Connectivity: As stated in the following lemma, between every pair of vertices, there exist a path with a special structure. See the full version [2] for the proof.

Lemma 2. *Let P be a point set with $p, q \in P$. Then, in $G_\nabla(P)$, there is a path between p and q which lies fully in ∇pq and hence $G_\nabla(P)$ is connected.*

Number of Degree-One Vertices: The following fact is important for our proof of the lower bound of the cardinality of a maximum matching in $G_\nabla(P)$.

Lemma 3. *For a point set P, its $G_\nabla(P)$ has at most three degree-one vertices.*

Proof. (By contradiction.) Let p_1, p_2, p_3 and p_4 be four points such that the vertices corresponding to them are of degree one in $G_\nabla(p)$. Since the points are in general position, without loss of generality, we can assume that these points are given in the bottom to top order of their y co-ordinates. We analyse different relative positionings of p_2 and p_3 with respect to p_1 and prove that in none of these cases, we can properly place all the four points consistently. Since p_1 is below p_2 and p_3, the relative positioning of p_2 and p_3 should be one of the following : (1) $p_2 \in \overline{V_3}(p_1)$, (2) $p_2 \notin \overline{V_3}(p_1)$ but $p_3 \in \overline{V_3}(p_1)$, (3) $p_2, p_3 \in V_1(p_1)$ or $p_2, p_3 \in V_2(p_1)$, (4) $p_2 \in V_1(p_1)$, $p_3 \in V_2(p_1)$ or $p_2 \in V_2(p_1)$, $p_3 \in V_1(p_1)$.

Case 1. Since $p_2 \in \overline{V_3}(p_1)$, we have $p_1 \in V_3(p_2)$. Since p_2 is of degree one, by Property 4, $\overline{V_3}(p_2) = \emptyset$. Since p_3 and p_4 are above p_2, and p_4 is above p_3, we have only the following sub-cases to consider: (1a) $p_3, p_4 \in V_i(p_2)$ and $p_4 \in V_i(p_3)$, where $i \in \{1, 2\}$, (1b) $p_3, p_4 \in V_i(p_2)$, where $i \in \{1, 2\}$, and $p_4 \in \overline{V_3}(p_3)$, (1c) $p_3, p_4 \in V_i(p_2)$ and $p_4 \in V_j(p_3)$, where $i, j \in \{1, 2\}$ and $i \neq j$, (1d) $p_3 \in V_i(p_2)$, $p_4 \in V_j(p_2)$, where $i, j \in \{1, 2\}, i \neq j$. (See Fig. 2.) Without loss of generality, assume that $i = 2$ and $j = 1$.
Case 1a : We have $p_3, p_4 \in V_2(p_2)$, implying that $p_2 \in \overline{V_2}(p_3)$ and $p_2 \in \overline{V_2}(p_4)$. Since $p_4 \in V_2(p_3)$ and $p_2 \in \overline{V_2}(p_3)$, by Property 4, the degree of p_3 is at least two. This is a contradiction.
Case 1b : We have $p_3, p_4 \in V_2(p_2)$. This implies that $p_2 \in \overline{V_2}(p_3)$ and $p_2 \in \overline{V_2}(p_4)$. Since $p_4 \in \overline{V_3}(p_3)$ and $p_2 \in \overline{V_2}(p_3)$, by Property 2, the degree of p_3 is at least two. This is a contradiction.
Case 1c : We have $p_3, p_4 \in V_2(p_2)$. This implies that $p_2 \in \overline{V_2}(p_3)$ and $p_2 \in \overline{V_2}(p_4)$. Since $p_4 \in V_1(p_3)$, we have $p_3 \in \overline{V_1}(p_4)$. Since we already had $p_2 \in \overline{V_2}(p_4)$, by Property 2, the degree of p_4 is at least two, which is a contradiction.
Case 1d : Since $p_3 \in V_2(p_2)$ and $p_4 \in V_1(p_2)$, by Property 5, we get $\overline{V_3}(p_2) \neq \emptyset$. This contradicts the property $\overline{V_3}(p_2) = \emptyset$, that we observed at the beginning of the analysis of Case 1.

Cases 2, 3 and 4 also lead to similar contradictions. (See the full version [2] for details.) $\square$

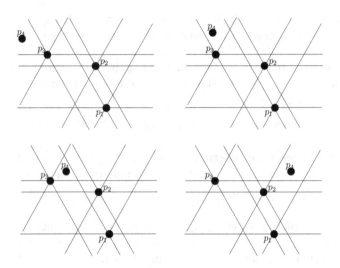

Fig. 2. Sub-cases of Case 1

Internal Triangulation: It can be shown that for a point set P, the plane graph $G_\triangledown(P)$ is internally triangulated. This property will be used in Section 4 to derive the lower bound for the cardinality of maximum matchings in $G_\triangledown(P)$. The proof of the following lemma is included in the full version [2].

Lemma 4. *For a point set P, all the internal faces of $G_\triangledown(P)$ are triangles and hence all the cut vertices of $G_\triangledown(P)$ lie on its outer face.*

4   Maximum Matching in $G_\triangledown(P)$

In this section, we show that for any point set P of n points, $G_\triangledown(P)$ contains a matching of size $\lceil \frac{n-2}{3} \rceil$; i.e, at least $2(\lceil \frac{n-2}{3} \rceil)$ vertices are matched. Consider a point set P containing n points. If we have only two points in P, then the graph contains a perfect matching. Hence, we assume that $|P| \geq 3$.

We construct a graph G' such that it is a 2-connected planar graph of minimum degree at least 3 and then make use of the following theorem of Nishizeki [12] to get a lower bound on the size of a maximum matching of G'. Using this, we will then derive a lower bound on the size of a maximum matching of $G_\triangledown(P)$.

Theorem 1 ([12]). *Let G be a connected planar graph with n vertices having minimum degree at least 3 and let M be a maximum matching in G. Then,*

$$|M| \geq \begin{cases} \lceil \frac{n+2}{3} \rceil & \text{when } n \geq 10 \text{ and } G \text{ is not 2-connected} \\ \lceil \frac{n+4}{3} \rceil & \text{when } n \geq 14 \text{ and } G \text{ is 2-connected} \\ \lfloor \frac{n}{2} \rfloor & \text{otherwise} \end{cases}$$

Initialize G' to be the same as $G_\triangledown(P)$. Consider a simple closed curve $\mathcal{C}$ in the plane such that (1) the entire graph $G_\triangledown(P)$ (all vertices and edges) lies inside the bounded region enclosed by $\mathcal{C}$, (2) the vertices of $G_\triangledown(P)$ which lie on $\mathcal{C}$ are precisely the degree-one vertices of $G_\triangledown(P)$, (3) except for the end points, every edge of $G_\triangledown(P)$ lies in the interior of the bounded region enclosed by $\mathcal{C}$.

Let the degree-one vertices of $G_\triangledown(P)$ be denoted by $p_0, p_1, \ldots, p_{k-1}$. In the previous section, we proved that $k \leq 3$.

If $k \geq 2$, the region of the outer face of $G_\triangledown(P)$ bounded by the curve $\mathcal{C}$ can be divided into k regions $R_0, \ldots, R_{k-1}$ where R_i is the region bounded by the edge at p_i, the edge at $p_{(i+1) \bmod k}$, the boundary of the outer face of $G_\triangledown(P)$ and the curve $\mathcal{C}$. See Fig. 3. (Here onwards, in this section we assume that indices of vertices and regions are taken modulo k.) Notice that every vertex on the outer-face of $G_\triangledown(P)$ lies on at least one of these regions and p_i lies on the regions R_i and R_{i-1}, for $0 \leq i \leq k-1$. We insert k new vertices $x_0, \ldots, x_{k-1}$ into G'. (To visualize the abstract graph G', vertex x_i may be assumed to lie on the boundary of the region R_i, a point distinct from p_i and p_{i+1}.) New edges are added between x_i and x_{i+1}, for $0 \leq i \leq k-1$. We also insert new edges into G' between each x_i and all the vertices of $G_\triangledown(P)$ which lie on the region R_i, for $0 \leq i \leq k-1$. This transformation maintains planarity. (Edges between new vertices and old vertices can be drawn inside the corresponding region R_i. The edges among the new vertices can be drawn outside these regions, except at their end points.) Each degree-one vertex p_i, $0 \leq i \leq k-1$, of $G_\triangledown(P)$ lies on

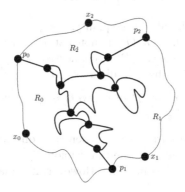

Fig. 3. Regions on the outer face of $G_\triangledown(P)$

two regions R_i and R_{i-1}, in G' it gets two new edges; one to x_i and the other to x_{i-1}. Thus the degree of p_i becomes 3. All other vertices on the outer face of $G_\triangledown(P)$ were of degree at least two. Since they belong to at least one of the regions $R_0, \ldots, R_{k-1}$, they get at least one new edge in G' and their degree is at least three in G'. Since $G_\triangledown(P)$ is an internally triangulated planar graph, we know that all vertices except those on the outer face had degree at least 3. These vertices maintain the same degrees in G' as in G. The degree of x_i, $0 \leq i \leq k-1$, is also at least 3 in G', because it is adjacent to p_i, p_{i+1} and at least one more vertex on the outer face of $G_\triangledown(P)$. Thus, G' has minimum degree at least three.

If $k = 0$ or 1, the modification of G' is similar. We insert a new vertex x in the outer face of G' and add edges between x and all other vertices in the outer face of $G_\triangledown(P)$. This transformation maintains planarity. As earlier, all vertices in G' except the vertex p_0 (present only when $k = 1$) have degree ≥ 3.

If $k = 1$, the degree of p_0 has become two in G' at this stage. In this case, let f be a face of the current graph G' which contains p_0 and x. Modify G' by inserting a new vertex y inside f and adding edges from this new vertex to all other vertices belonging to f. As earlier, this transformation maintains planarity. Now, the degree of p_0 becomes 3.

Claim 1. G' is 2-connected in all the above cases.

Proof. See the full version [2].

Thus, the graph G' is a 2-connected planar graph of minimum degree at least 3, having at most $n + 3$ vertices. Let $n' = n + k$ be the number of vertices of G'. By Theorem 1, the cardinality of a maximum matching M' in G' is at least $\left\lceil \frac{n'+4}{3} \right\rceil$ when $n' \geq 14$ and $|M'| \geq \lfloor \frac{n}{2} \rfloor$, otherwise. Since $G_\triangledown(P)$ is an induced subgraph of G', if we delete the edges in M' which have at least one end point which is not in P, we get a matching M of $G_\triangledown(P)$. We have $|M| \geq |M'| - k$, where $k = n' - n$ with $k \in \{0, 1, 2, 3\}$. From this, we get,

$$|M| \geq \begin{cases} \left\lceil \frac{n+4-2k}{3} \right\rceil & \text{when } n \geq 14 - k \\ \lfloor \frac{n-k}{2} \rfloor & \text{otherwise} \end{cases}$$

Whenever $n \geq 7$, from the above inequality, we get $|M| \geq \left\lceil \frac{n-2}{3} \right\rceil \geq 2$. When $n \geq 5$, Lemma 3 implies that $G_\triangledown(P)$ is not a star with $n - 1$ leaves and hence $|M| \geq 2$. When $n > 1$, since $G_\triangledown(P)$ is connected, we get $|M| \geq 1$. From this discussion, we can conclude that, in all cases, $|M| \geq \left\lceil \frac{n-2}{3} \right\rceil$.

Theorem 2. *For any point set P of n points in general position, $G_\triangledown(P)$ contains a matching of size $\left\lceil \frac{n-2}{3} \right\rceil$.*

Our bound is tight in all cases except when n is one less than a multiple of three. To find some examples, see the full version [2]. From this, it is clear that no bound better than $\left\lceil \frac{n-1}{3} \right\rceil$ is possible. However, it remains open whether our bound can be improved to $\left\lceil \frac{n-1}{3} \right\rceil$.

3-Connected Down Triangle Graphs without Perfect Matching: The example given by Panahi et al. [13], for a point set P for which $G_\triangledown(P)$ has a maximum matching of size $\left\lceil \frac{n-2}{3} \right\rceil$, contained many cut vertices. However, for general planar graphs, we get a better lower bound for the size of a maximum matching, when the connectivity of the graph increases. By Theorem 1, we know that any 3-connected planar graph on n vertices has a matching of size $\left\lceil \frac{n+4}{3} \right\rceil$, if $n \geq 14$ and has a matching of size $\lfloor \frac{n}{2} \rfloor$ if $n < 14$ or it is 4-connected. Hence, it was interesting to see whether there exist a point set P in general position, with an even number of points, such that $G_\triangledown(P)$ is 3-connected but does not contain a perfect matching. The answer is positive. In the full version [2], we have given examples of 3-connected down triangle graphs corresponding to point sets in general position, for which the size of their maximum matching is only $\lceil \frac{n+5}{3} \rceil$. Since the known lower bound for the size of maximum matching in 3-connected planar graphs is $\lceil \frac{n+4}{3} \rceil$, it remains open whether the bound for 3-connected down triangle graphs can be improved to $\lceil \frac{n+5}{3} \rceil$.

5   Some Properties of $G_{\diamondsuit}(P)$

In this section, we prove that for a point set P, the 2-connectivity structure of $G_{\diamondsuit}(P)$ is simple and $G_{\diamondsuit}(P)$ can have at most $5n - 11$ edges.

Block Cut Point Graph: Let $G(V, E)$ be a graph. A block of G is a maximal connected subgraph having no cut vertex. The block cut point graph of G is a bipartite graph $B(G)$ whose vertices are cut-vertices of G and blocks of G, with a cut-vertex x adjacent to a block X if x is a vertex of block X. The block cut point graph of G gives information about the 2-connectivity structure of G.

Since $G_{\diamondsuit}(P)$ is the union of two connected graphs $G_{\triangledown}(P)$ and $G_{\triangle}(P)$ (Lemma 2), it is connected and hence its block-cut point graph is a tree [8]. We will show that the block cut point graph of $G_{\diamondsuit}(P)$ is a simple path. We use the following lemma in our proof.

Lemma 5. *Let P be a point set and $p \in P$ be a cut vertex of $G_{\diamondsuit}(P)$. Then, there exists an $i \in \{1, 2, 3\}$ such that $V_i(p) \neq \emptyset$, $\overline{V_i}(p) \neq \emptyset$ and for all $j \in \{1, 2, 3\} \setminus \{i\}$, $V_j(p) = \emptyset$ and $\overline{V_j}(p) = \emptyset$. Moreover, $G_{\diamondsuit}(P) \setminus p$ has exactly two connected components, one containing all vertices in $V_i(p)$ and the other containing all vertices of $\overline{V_i}(p)$.*

Proof. Since p is a cut vertex of $G_{\diamondsuit}(P)$, we know that there exist $v_1, v_2 \in P$ that are in different components of $G_{\diamondsuit}(P) \setminus p$. We will show that v_1 and v_2 should be in opposite cones with reference to the apex point p.

Without loss of generality, assume that $v_1 \in A_1(p) \cap P \setminus \{p\}$. If $v_2 \in (A_1(p) \cup A_2(p) \cup A_6(p)) \cap (P \setminus \{p\})$, then, $p \notin \triangledown v_1 v_2$ and hence by Lemma 2, there is a path in $G_{\triangledown}(P)$ between v_1 and v_2 that does not pass through p, which is not possible. Similarly, if $v_2 \in (A_3(p) \cup A_5(p)) \cap (P \setminus \{p\})$, then, $p \notin \triangle v_1 v_2$ and there is a path in $G_{\triangle}(P)$ between v_1 and v_2 that does not pass through p, which is not possible. Therefore, $v_2 \in A_4(p)$, the cone which is opposite to $A_1(p)$ which contains v_1. Thus any two points v_1 and v_2 which are in different connected components of $G_{\diamondsuit}(P) \setminus p$, are in opposite cones around p.

Let C_1 and C_2 be two connected components of $G_{\diamondsuit}(P) \setminus p$ with $v_1 \in C_1$ and $v_2 \in C_2$. Without loss of generality, assume that such $v_1 \in V_1(p)$ and $v_2 \in \overline{V_1}(p)$. From the paragraph above, we know that every vertex of $G_{\diamondsuit}(P) \setminus p$ which is not in C_1 is in $\overline{V_1}(p)$ and every vertex of $G_{\diamondsuit}(P) \setminus p$ which is not in C_2 is in $V_1(p)$. This implies that for all $j \in \{2, 3\}$, $V_j(p) = \emptyset$ and $\overline{V_j}(p) = \emptyset$. This proves the first part of our lemma.

For any $v_1, v_2 \in \overline{V_i}(p)$, we have $p \notin \triangledown v_1 v_2$ and hence by Lemma 2, there is a path in $G_{\triangledown}(P)$ between v_1 and v_2 that does not pass through p. Similarly, for any

$v_1, v_2 \in V_i(p)$, $p \notin \triangle v_1 v_2$ and there is a path in $G_\triangle(P)$ between v_1 and v_2 that does not pass through p. Therefore, there are exactly two connected components in $G_✡(P) \setminus p$, one containing all vertices in $V_i(p)$ and the other containing all vertices of $\overline{V}_i(p)$. □

Theorem 3. *Let P be a point set in general position and let k be the number of blocks of $G_✡(P)$. Then, the blocks of $G_✡(P)$ can be arranged linearly as $B_1, B_2, \ldots B_k$ such that, for $i > j$, $B_i \cap B_j$ contains a single (cut) vertex p_i when $j = i + 1$ and $B_i \cap B_j$ is an empty graph otherwise. That is, the block cut point graph of $G_✡(P)$ is a path.*

Proof. Since $G_✡(P)$ is a connected graph, its block cut point graph is a tree. Any two blocks can have at most one vertex in common and the common vertex is a cut vertex. If the node corresponding to block B_i is a leaf node in the block cut point graph of $G_✡(P)$, then B_i is adjacent to only one another block and they share a single (cut) vertex.

If the node corresponding to the block B_i is not a leaf node of the block cut point graph of $G_✡(P)$, then using Lemma 5, we can show that exactly two vertices in B_i are cut vertices of $G_✡(P)$. Since no three blocks can share a common vertex by Lemma 5, we are done. A detailed proof is given in the full version [2]. □

Number of Edges of $G_✡(P)$: Since $G_\triangledown(P)$ and $G_\triangle(P)$ are planar graphs and $G_✡(P) = G_\triangledown(P) \cup G_\triangle(P)$, it is obvious that $G_✡(P)$ has at most $2 \times (3n - 6) = 6n - 12$ edges, where $n = |P|$ [8]. In this section, we show that for any point set P, its $G_✡(P)$ has a spanning tree of a special structure, which will imply that $G_✡(P)$ can have at most $5n - 11$ edges.

Lemma 6. *For a point set P, the intersection of $G_\triangledown(P)$ and $G_\triangle(P)$ is a connected graph.*

Proof. We will prove this algorithmically. At any point of execution of this algorithm, we maintain a partition of P into two sets S and $P \setminus S$ such that the induced subgraph of $G_\triangledown(P) \cap G_\triangle(P)$ on S is connected. When the algorithm terminates, we will have $S = P$, which will prove the lemma.

We start by adding any arbitrary point $p_1 \in P$ to S. The induced subgraph of $G_\triangledown(P) \cap G_\triangle(P)$ on S is trivially connected now.

At any intermediate step of the algorithm, let $S = \{p_1, p_2, \ldots, p_k\} \neq P$, such that the invariant is true. We will show that we can add a point p_{k+1} from $P \setminus S$ into S, and still maintain the invariant.

For any point $p \in S$, let $d_1(p) = \min\limits_{i \in \{1,2,3\}, p' \in V_i(p) \cap P \setminus S} c_i(p, p')$, $d_2(p) = \min\limits_{i \in \{1,2,3\}, p' \in \overline{V}_i(p) \cap P \setminus S} \overline{c}_i(p, p')$ and $d(p) = \min(d_1(p), d_2(p))$. Since $|P \setminus S| \geq 1$, $d(p) < \infty$. Let $d = \min\limits_{p \in S} d(p)$.

Consider $p \in S$ such that $d(p) = d$. By definition of d, such a point exists. Consider the area enclosed by the hexagon around p which is defined by

$$H_p = \bigcup_{i=1}^{3} \{p' \in C_i(p) \mid c_i(p,p') \le d\} \cup \bigcup_{i=1}^{3} \{p' \in \overline{C}_i(p) \mid \overline{c}_i(p,p') \le d\}. \text{ (See Fig. 4}$$

(a).) We know that there exists a point $q \in P \setminus S$ such that q is on the boundary of H_p. We claim that pq is an edge in $G_\nabla(P) \cap G_\triangle(P)$.

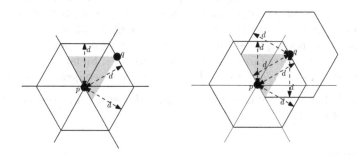

Fig. 4. (a) Closest point to p. (b) Hexagons around closest pairs.

Let $H_q = \bigcup_{i=1}^{3} \{p' \in C_i(q) \mid c_i(q,p') \le d\} \cup \bigcup_{i=1}^{3} \{p' \in \overline{C}_i(q) \mid \overline{c}_i(q,p') \le d\}$,

which is a hexagonal area around q. (See Fig. 4 (b).) Without loss of generality, assume that $q \in C_1(p)$. Note that, by Property 1, $c_1(p,q) = \overline{c}_1(q,p) = d$ and hence, $\nabla pq \cup \triangle pq \subseteq H_p \cap H_q$.

If there exists a point $q' \in (P \setminus \{q\}) \setminus S$ such that q' lies in the interior of H_p, then $d(p) < d$, which is a contradiction. Similarly, if there exists a point $p' \in (P \setminus \{p\}) \cap S$ such that p' lies in the interior of H_q, then $d(p) < d$. This is also a contradiction. Therefore, $H_p \cap H_q \cap (P \setminus \{p,q\}) = \emptyset$. Since, $\nabla pq \cup \triangle pq \subseteq H_p \cap H_q$, this implies that $\nabla pq \cap (P \setminus \{p,q\}) = \emptyset$ and $\triangle pq \cap (P \setminus \{p,q\}) = \emptyset$. This implies that pq is an edge in $G_\nabla(P)$ as well as in $G_\triangle(P)$.

Since pq is an edge in $G_\nabla(P) \cap G_\triangle(P)$, we can add $p_{k+1} = q$ to the set S, thus increasing the cardinality of S by one, and still maintaining the invariant that the induced subgraph of $G_\nabla(P) \cap G_\triangle(P)$ on S is connected. Since we can keep on doing this until $S = P$, we conclude that $G_\nabla(P) \cap G_\triangle(P)$ is connected. □

Theorem 4. *For a set P of n points in general position, $G_{\not\triangle}(P)$ has at most $5n - 11$ edges and hence its average degree is less than 10.*

Proof. Since $G_\nabla(P)$ and $G_\triangle(P)$ are both planar graphs we know that each of them can have at most $3n - 6$ edges. From Lemma 6, we know that the intersection of $G_\nabla(P)$ and $G_\triangle(P)$ contains a spanning tree and hence they have at least $n - 1$ edges in common. From this, we conclude that the number of edges in $G_{\not\triangle}(P) = G_\nabla(P) \cup G_\triangle(P)$ is at most $(3n - 6) + (3n - 6) - (n - 1) = 5n - 11$. Hence, the average degree of $G_{\not\triangle}(P)$ is less than 10. □

Corollary 1. *For a set P of n points in general position, its Θ_6 graph has at most $5n - 11$ edges.*

6   Conclusions

We have shown that for any set P of n points in general position, any maximum $\triangledown$ (resp. $\triangle$) matching of P will match at least $2\left(\left\lceil\frac{n-2}{3}\right\rceil\right)$ points. This also implies that any half-Θ_6 graph for point sets in general position has a matching of size at least $\left\lceil\frac{n-2}{3}\right\rceil$. This bound is tight except when $|P|$ is one less than a multiple of three. We also proved that when P is in general position, the block cut point graph of its Θ_6 graph is a simple path and that the Θ_6 graph has at most $5n-11$ edges. It is an interesting question to see whether for every point set in general position, its Θ_6 graph contains a matching of size $\left\lfloor\frac{n}{2}\right\rfloor$. So far, we were not able to get any counter examples for this claim and hence we conjecture the following.

Conjecture 1. For every set of n points in general position, its Θ_6 graph contains a matching of size $\left\lfloor\frac{n}{2}\right\rfloor$.

References

1. Ábrego, B.M., Arkin, E., Fernández-Merchant, S., Hurtado, F., Kano, M., Mitchell, J., Urrutia, J.: Matching points with squares. Discrete and Computational Geometry 41, 77–95 (2009)
2. Babu, J., Biniaz, A., Maheshwari, A., Smid, M.: Fixed-orientation equilateral triangle matching of point sets. CoRR abs/1211.2734 (2012), http://arxiv.org/abs/1211.2734
3. Bereg, S., Mutsanas, N., Wolff, A.: Matching points with rectangles and squares. Comput. Geom. Theory Appl. 42(2), 93–108 (2009)
4. Bonichon, N., Gavoille, C., Hanusse, N., Ilcinkas, D.: Connections between Theta-Graphs, Delaunay Triangulations, and Orthogonal Surfaces. In: Thilikos, D.M. (ed.) WG 2010. LNCS, vol. 6410, pp. 266–278. Springer, Heidelberg (2010)
5. Bose, P., Carmi, P., Collette, S., Smid, M.: On the stretch factor of convex Delaunay graphs. Journal of Computational Geometry 1, 41–56 (2010)
6. Chew, L.P.: There are planar graphs almost as good as the complete graph. Journal of Computer and System Sciences 39(2), 205–219 (1989)
7. Clarkson, K.: Approximation algorithms for shortest path motion planning. In: Proceedings of the Nineteenth Annual ACM Symposium on Theory of Computing, STOC 1987, pp. 56–65. ACM (1987)
8. Diestel, R.: Graph Theory, 4th edn. Springer (2010)
9. Dillencourt, M.: Toughness and Delaunay triangulations. In: Proceedings of the Third Annual Symposium on Computational Geometry, SCG 1987, pp. 186–194. ACM (1987)
10. Keil, J.M.: Approximating the Complete Euclidean Graph. In: Karlsson, R., Lingas, A. (eds.) SWAT 1988. LNCS, vol. 318, pp. 208–213. Springer, Heidelberg (1988)
11. Narasimhan, G., Smid, M.: Geometric Spanner Networks. Cambridge University Press (2007)
12. Nishizeki, T.: Lower bounds on the cardinality of the maximum matchings of planar graphs. Discrete Mathematics 28, 255–267 (1979)
13. Panahi, F., Mohades, A., Davoodi, M., Eskandari, M.: Weak matching points with triangles. In: Proceedings of the 23rd Annual Canadian Conference on Computational Geometry (2011)

Online Exploration and Triangulation in Orthogonal Polygonal Regions

Sándor P. Fekete, Sophia Rex, and Christiane Schmidt

Department of Computer Science, TU Braunschweig,
D–38116 Braunschweig, Germany
{s.fekete,c.schmidt}@tu-bs.de, mail.s.rex@gmail.com

Abstract. We consider the problem of exploring and triangulating a region with a swarm of robots with limited vision and communication range. For an unknown polygonal region P, the Online Minimum Relay Triangulation Problem (OMRTP) asks for an exploration strategy that maintains a triangulation with limited edge length and achieves a minimum number of robots (relays), such that the triangulation covers P; for a given number n of robots, the Online Maximum Area Triangulation Problem (OMATP) asks for maximizing the triangulated area. Both problems have been studied before, with a competitive factor of 3 for the OMRTP in general polygons, and an unbounded competitive factor for the OMATP; the latter holds for polygons with very narrow corridors.

In this paper, we study the OMRTP for polygons without such bottlenecks: polyominoes, i.e., orthogonal polygons with integer edge lengths. Based on optimal solutions for small squares, we establish a competitive factor of $\frac{17\sqrt{3}}{16+\sqrt{3}} \approx 1.661$ for polyominoes with and $\frac{19\sqrt{3}}{20+\sqrt{3}} \approx 1.514$ for polyominoes without holes. We also give a lower bound of $\frac{38}{37} \approx 1.027$ for any deterministic strategy for the OMRTP in polyominoes. For the OMATP, we establish a competitive factor of $\frac{2}{3\sqrt{3}} \approx 0.3849$, and argue that this is asymptotically optimal.

1 Introduction

Consider a swarm of robots that has to explore a region P. Each robot has limited capabilities: both vision and communication are restricted in range. Incrementally, the swarm has to build a rigid, stable formation that covers all of P. This gives rise to the Minimum Relay Triangulation Problem (MRTP): find a triangulation T with limited edge length, such that P is fully covered by T and the number of relays is minimized. In the online version of this problem (OMRTP), the polygon is unknown in advance. Closely related is the Maximum Area Triangulation Problem (MATP), and its online version OMATP, in which the number n of available robots is fixed, and the enclosed area needs to be maximized. Note that considering the problem as a domain decomposition we give solutions for mesh generation with triangle elements with a bounded edge length, such that the number of Steiner points is minimized.

S.K. Ghosh and T. Tokuyama (Eds.): WALCOM 2013, LNCS 7748, pp. 29–40, 2013.

Both of these problems have been considered before for the case of arbitrary polygonal regions that need to be explored. As Fekete et al. [6] showed, there is a 3-competitive strategy for the OMRTP, while the OMATP does not allow any constant competitive factor. Limiting factors for both of these results are sharp turns along the boundary, as well as tight bottlenecks; however, when exploring buildings, we are typically faced with orthogonal walls, as well as corridors of reasonable dimensions that are multiples of some underlying size. This makes it natural to consider orthogonal polygons with integer dimensions, i.e., *polyominoes*. In this paper, we provide a number of refined results for this class of environments. Our results are as follows.

- We give a strategy with a competitive factor of $\frac{17\sqrt{3}}{16+\sqrt{3}} \approx 1.661$ for polyominoes with holes.
- We give a strategy with a competitive factor of $\frac{7}{8}\sqrt{3} \approx 1.516$ and sketch one of $\frac{19\sqrt{3}}{20+\sqrt{3}} \approx 1.514$ for polyominoes without holes.
- We establish a lower bound of $\frac{38}{37} \approx 1.027$ for any deterministic algorithm for the OMRTP in polyominoes.
- We show that the OMATP in polyominoes does allow a constant competitive ratio of $\frac{2}{3\sqrt{3}} \approx 0.3849$.
- We argue that the value of $\frac{2}{3\sqrt{3}} \approx 0.3849$ is asymptotically optimal.

Related Work. There exists a broad spectrum of work on **triangulations**, both in theory and in practical applications. Here we just mention the work by Bern and Eppstein [1], who investigated triangulations with certain characteristics, e.g., Steiner points, bounds on angles, and even minimizing the sum of edge lengths. For **online robot exploration**, Hoffman et al. [11] presented a 26.5-competitive algorithm for exploring a simple polygon with a single robot with continuous vision. For discrete vision, the problem was studied by Fekete and Schmidt [9]. Another variant with discrete vision and limited visibily range was investigated by Fekete, Mitchell and Schmidt [8]. A related problem on polyominoes was considered by Icking et al. [13].

Strongly related to exploration problems is the task of **deploying a robot swarm into an unknown environment**. Hsiang et al. [12] studied the problem for polyominoes, but they placed one robot per unit square instead of triangulating the area. Practical aspects of deploying strategies were investigated by McLurkin and Smith [15]. Brunner et al. [3] examined the minimum set of abilities a robot needs to perform a certain task, see also Suri et al. [17] for exploration and triangulation algorithms with such robots.

Offline **relay placement** has also been studied, especially in the context of network properties. Bredin et al. [2] considered deploying a minimal number of sensors with limited communication range in an outdoor area such that they form a network with k-connectivity. Differences to our scenario include the absence of defined boundary and holes as well as the requirement that sensors need to stay connected during deployment. Moreover, their aim is to guarantee the connectivity and not to form a triangulation, so some faces of the final network graph may not be triangles. They also considered adding sensors to an existing network with lower connectivity in order to achieve k-connectivity. Kashyap et al. [14] studied the problem for $k = 2$ in higher dimensions. For $k = 1$ and two dimensions, a similar problem was studied by Degener et al. [4] with the added difficulty of finding shortest paths from the entry point to the final locations of the robots carrying the sensors. Moreover, the robots only perceive their current environment and have to make decisions based on this local information without knowing the global situation. Another variant studied by Efrat et al. [5] is to equip sensors with short-range communication devices and then place a minimum number of relays with long-range communication devices in the sensor network to establish connectivity.

The MRTP and OMRTP discussed in this paper were first considered by Fekete et al. [7] and a competitive ratio of 6 on polygons was achieved for the OMRTP. The results were refined by Schmidt [16] and Fekete et al. [6]; they include an NP-hardness proof and a PTAS for the MRTP and the MATP, as well as a lower bound of 1.2 and a 3-competitive algorithm for the OMRTP on polygons. Moreover, they showed that no competitive online algorithm for the online MATP can exist.

Finally, Friedman [10] has applied results similar to our optimality results for unit squares and two-by-two squares in his proofs of lower bounds on *packing* unit squares into squares. However, their optimality was neither mentioned nor proved. Note, however, that these (offline) packing problems are notoriously difficult, and there are still major gaps in what is known.

2   Preliminaries

A *triangulation* T is a set of relays and edges that subdivide a given polygon P into triangles. All edges must lie in P, i.e., they must not cross the boundary. All edges and relays of the triangulation must belong to triangles. The most common form of triangulation places relays of T on all vertices of P only. A *unit triangulation* is a triangulation in which all edges have at most length one. In order to achieve a unit triangulation, it is usually necessary to place relays in the interior and on boundary edges of P in addition to the vertices placed on P's vertices. These extra relays are called *Steiner points*.

For the *Minimum Relay Triangulation Problem* (MRTP), we are given a polygon P with vertex set V and a point $z \in P$. We want to compute a set, R (with $z \in R$ and $V \subset R$), of relays within P such that there exists a (Steiner) triangulation of P whose vertex set is exactly the set R and whose edges have length at most 1, i.e., a unit triangulation, that covers P. For the *Maximum Area Triangulation Problem* (MATP), we are given a polygon P, a point $z \in P$, and a number n of available relays. We want to compute a set, R (with $z \in R$), of n relays within P such that there exists a (Steiner) unit triangulation of P whose vertex set is exactly the set R, such that the total area of all triangles is maximized.

For the online version of the MRTP, the polygon P is unknown in advance. We want to compute a set R as for the MRTP. Here, the relays move into the polygon, starting from z. A relay extending the yet established subset $R' \subset R$ must stay within a distance of 1 to at least one relay $r \in R'$. Once it fixed its position it will not move again. The OMATP is defined analogously. For this paper P is an unknown polyomino, so we consider the online versions of MRTP and MATP. A strategy needs to place relays one by one, such that at any time there are at most two edges and two relays that are not part of a triangle.

We now state several useful lemmas that are related to triangulations.

Lemma 1. *The number of relays n in any triangulation of a polygon P with h holes is equal to $\frac{1}{2}(t + b) + 1 - h$, where t is the number of triangles and b the number of relays on the boundary. (The boundary of the holes is part of the polygon's boundary.)*

Proof. For any triangulation of P, let t denote the number of triangles, e the number of edges and n the number of relays, of which b are on the boundary of the polygon P. Counting the triangle-edge incidences we find that every triangle has three edges and every edge belongs to two triangles if it is inside the polygon P, or one triangle if it is on the boundary of P. The number of boundary edges is equal to b, the number of boundary relays. This yields $3t = 2e - b$. Not counting the one face outside the Polygon, Euler's formula gives $n + t + h = e + 1$ or $e = n + t - 1$. Substituting this into the previous equation yields $3t = 2n + 2t - 2 + 2h - b$, which can now be solved for $n = \frac{1}{2}(t + b) + 1 - h$. □

In this work, we consider unit triangulations. We therefore state a simple lower bound for the number of relays in such a triangulation.

Lemma 2. *Let A be the area, h the number of holes and B be the length of the boundary of the polygon P. Then the minimum number of relays in a unit triangulation of P is $n_{OPT} \geq \frac{1}{2}\left(\frac{4A}{\sqrt{3}} + B\right) + 1 - h$.*

Proof. We use the formula $n = \frac{1}{2}(t + b) + 1 - h$ from Lemma 1. $b \geq B$ and $t \geq \frac{4A}{\sqrt{3}}$ must hold for any triangulation, because the maximum side length in each triangle of the triangulation is 1, so the maximum area of a triangle is $\frac{\sqrt{3}}{4}$, and we have to cover the whole area and boundary of the Polygon P. We get $n_{OPT} \geq \frac{1}{2}\left(\frac{4A}{\sqrt{3}} + B\right) + 1 - h$ for the optimal solution. □

3   Solutions for Squares

We briefly sketch some results for the case in which P is a square of limited size. These results serve as stepping stones for the following online strategies.

Lemma 3. *The optimal solution of the MRTP is five for a unit square, twelve for a 2 × 2-square, and 21 for a 3 × 3-square.*

In addition, Fig. 1, left, shows the best solution we could find for a 4 × 4-square. It places the minimum number of 16 relays on the boundary. The relays of the second layer are placed greedily towards the middle. The four center relays are rotationally symmetric. Note that our solution for a 4×4-square can be enhanced into four optimal 2 × 2-solutions, or 16 optimal 1 × 1-solutions, by adding relays on the unit grid positions inside the square. This construction, which is shown in Fig. 1, plays a key role in our online algorithm.

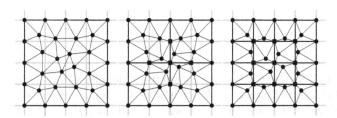

Fig. 1. Decomposition of our best solution for side length four into four optimal solutions of side length two and 16 optimal solutions for unit squares

4   Minimum Relay Triangulation in Polyominoes

4.1   A Strategy Using Optimal 1 × 1-Squares

Our first strategy is to apply the optimal solution for a unit square for each grid square of the polyomino, as shown in Fig. 2. Algorithm 4.1 picks a square, places the center relay, connects it to all existing grid point relays, places and connects the remaining grid point relays and moves on to the next square.

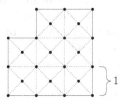

Fig. 2. Example of a triangulation as a result of Algorithm 4.1

Lemma 4. *Algorithm 4.1 achieves a competitive ratio of $\sqrt{3}$ for the OMRTP on any polyomino P.*

Algorithm 4.1. Deploying relays with the optimal unit square strategy

> **Input**   : Starting point on integer coordinates on the boundary of a yet
>   unknown polyomino P
> **Output**: Triangulation of P with at most OPT·$\sqrt{3}$ relays
> **while** *P is not completely triangulated* **do**
> > pick an empty grid square bordering the triangulated area;
> > place one relay in its center;
> > connect it to all existing relays on grid vertices in range;
> > **while** *the grid square is not completely triangulated* **do**
> > > place one relay on an empty vertex of the grid square;
> > > connect it with the existing relays in range;
> >
> > **end**
>
> **end**

Proof. We apply Lemma 1. Our algorithm uses triangles with an area of $\frac{1}{4}$, thus $t = 4A$, and places all boundary relays with the maximum distance of one to their neighbors, thus $b = B$. Overall the algorithm places $n_{ALG} = \frac{1}{2}(t+b) + 1 - h = \frac{1}{2}(4A+B) + 1 - h$ relays. According to Lemma 2, the optimal solution needs at least $n_{OPT} \geq \frac{1}{2}\left(\frac{4A}{\sqrt{3}} + B\right) + 1 - h$ relays.

The smallest hole in the polyomino P must itself be a polyomino and thus have a boundary length of at least 4. Therefore, $h < \frac{1}{4}B$ and, in particular, $B + 2 - 2h \geq 0$. Thus, the competitive ratio is

$$\frac{n_{ALG}}{n_{OPT}} \leq \frac{\frac{1}{2}(4A+B)+1-h}{\frac{1}{2}\left(\frac{4A}{\sqrt{3}}+B\right)+1-h} = \frac{4A+B+2-2h}{\frac{4A}{\sqrt{3}}+B+2-2h} \leq \sqrt{3}. \quad \square$$

4.2   A Strategy Using Optimal 2 × 2-Squares

Now we refine the previous strategy. The basic idea is to introduce an imaginary grid of width two and place the optimal solutions for squares of side length two according to this new grid, see Fig. 3, left. If such a 2 × 2 grid square turns out to be intersected by the boundary (because one of its unit squares is missing), an additional relay is placed in its middle and the strategy locally reverts to the 1 × 1-strategy. See Fig. 3 for an illustration; a detailed description is given by Algorithm 4.2.

We exploit the following properties of polyominoes.

Lemma 5. *In a polyomino without holes, $r = c - 4$, where r is the number of reflex vertices and c is the number of convex vertices.*

Lemma 6. *In a polyomino with $A > 1$ and $h = 0$, Algorithm 4.2 places at most $u \leq B - 4$ unit squares.*

Lemma 7. *In a polyomino with $A > 1$ and $h \geq 0$, Algorithm 4.2 places at most $u \leq B + 4h - 4$ unit squares.*

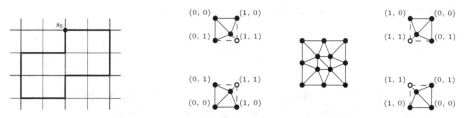

Fig. 3. A polyomino with its natural unit grid and the gray grid of side length two determined by the unit grid and the starting point. In this case, only one optimal 2×2-square will be placed, even though one could place two.

Algorithm 4.2. Deploying relays with the 2×2-square strategy

Input   : Starting point on integer coordinates on the boundary of a yet
             unknown polyomino P
Output: Triangulation of polyomino P with at most OPT·1.661 relays
while *P is not completely triangulated* **do**
> pick a grid square adjacent to the triangulated area;
> place one relay inside, according to the rules in Fig. 3;
> **if** *there is boundary at the square's (1,1) vertex* **then**
>> connect the relay to all grid relays in range;
>> **while** *the grid square is not completely triangulated* **do**
>>> place one relay on the vertex of the grid square closest to the
>>> existing triangulation;
>>> connect it with the existing relays in range;
>>
>> **end**
>
> **else**
>> connect the relay to all relays in range, without crossing existing
>> connections;
>> **while** *(0,0) or (0,1) or (1,0) does not have a relay* **do**
>>> place one relay on the vertex of the grid square closest to the
>>> existing triangulation except (1,1);
>>> connect it with the existing relays in range;
>>
>> **end**
>
> **end**

end

From these lemmas we gain $u \leq B$ and a new bound on the competitive ratio.

Theorem 8. *Algorithm 4.2 achieves a competitive ratio of $\frac{7}{8}\sqrt{3} \approx 1.516$ for the OMRTP in polyominoes with $h = 0$, $A > 1$.*

Proof. Lemma 1 yields $n_{ALG} = \frac{1}{2}(\frac{1}{2}u + \frac{7}{2}A + B) + 1$. In order to compare n_{ALG} with the lower bound for the optimal solution n_{OPT}, we have to eliminate u: $u \leq A$, because we cannot have more unit squares than area. Lemma 6 yields $u \leq B$. We proceed by another case-by-case analysis.

Case A ≤ B: In this case, we use $u \leq A$ and obtain

$$n_{ALG} = \frac{1}{2}(\frac{1}{2}u + \frac{7}{2}A + B) + 1 \leq \frac{1}{2}(4A + B) + 1 \Rightarrow \frac{n_{ALG}}{n_{OPT}} \leq \frac{\frac{1}{2}(4A + B) + 1}{\frac{1}{2}\left(\frac{4A}{\sqrt{3}} + B\right) + 1}.$$

We observe that the ratio increases if the area is large in comparison to the boundary length. Because we have $A \leq B$, we can obtain

$$\frac{n_{ALG}}{n_{OPT}} \leq \frac{\frac{1}{2}(4A + B) + 1}{\frac{1}{2}\left(\frac{4A}{\sqrt{3}} + B\right) + 1} \leq \frac{\frac{1}{2}(5B) + 1}{\frac{1}{2}\left(\frac{4B}{\sqrt{3}} + B\right) + 1} \leq \frac{5}{\frac{4}{\sqrt{3}} + 1} < \frac{7}{8}\sqrt{3} \approx 1.516$$

Case A ≥ B: In this case, we use $u \leq B$ and obtain

$$n_{ALG} = \frac{1}{2}(\frac{1}{2}u + \frac{7}{2}A + B) + 1 \leq \frac{1}{2}(\frac{7}{2}A + \frac{3}{2}B) + 1.$$

Thus,

$$\frac{n_{ALG}}{n_{OPT}} \leq \frac{\frac{7}{4}A + \frac{3}{4}B}{\frac{2}{\sqrt{3}}A + \frac{1}{2}B} \leq \frac{\frac{7}{4}A + \frac{7}{8}B}{\frac{2}{\sqrt{3}}A + \frac{1}{2}B} \leq \frac{\frac{7}{4}A + \frac{7}{8}B}{\frac{2}{\sqrt{3}}A + \frac{1}{\sqrt{3}}B} = \frac{7}{8}\sqrt{3} \approx 1.516.$$

A similar analysis provides the following result for polyominoes with holes. A detailed proof is omitted due to space constraints.

Theorem 9. *Algorithm 4.2 achieves a competitive ratio of $\frac{17\sqrt{3}}{16+\sqrt{3}} \approx 1.661$ for the OMRTP in polyominoes with holes.*

4.3   A Strategy Using Good Solutions for 4 × 4-Squares

We now proceed to give a slight improvement for the case of polyominoes without holes. As described in Section 3 and illustrated by Fig. 1, our best solution for the 4 × 4-square can be divided into smaller optimal solutions by placing relays on the grid points inside the 4 × 4-square. see Fig. 1. As in the 2 × 2-approach, our 4 × 4-solution can be adjusted when boundary pixels are encountered before a 4 × 4-solution is encountered. An example is shown in Fig. 4.

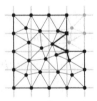

Fig. 4. A 4×4-square is cut by a polyomino's boundary. Adjustments are only necessary on the new boundary: adjusted edges are shown in bold, while the edges and relays of the original 4 × 4-solution are indicated in gray.

Our algorithm places the 4×4-solution on a grid of width four, starting at the entry point s_0, and adds boundary relays if necessary. That is, only if we cannot place the complete 4×4-square, as the boundary interesects this module, we switch to 2×2-squares and 1×1-squares. The layout of the 4×4-solution allows to do so with the limit on the vision range. The algorithm triangulates as much of each 4×4-grid square as possible before moving on to the next 4×4-square. Inside the 4×4-square, the inner unit squares are treated only after all boundary unit squares around them have been processed. This specified order of triangulation ensures that at most two relays are not part of the triangulation. Correctness is provided by Lemma 10.

Lemma 10. *At all times during the execution of the algorithm using 4×4-squares, at most two relays and edges are not part of the triangulation.*

For the competitive ratio estimate of this algorithm, we assume that a 4×4-square intersected by boundary is divided into four 2×2-squares and each 2×2-square intersected by boundary is divided into four unit squares, as in Section 4.2. The result is that if we place any additional relay at all, we will automatically assume that the additional relays at $(1, 2), (2, 1), (2, 2), (2, 3)$, and $(3, 2)$ have also been placed. Having assumed these extra relays, we can reuse some of the analysis techniques from Section 4.2.

For the case of polyominoes without holes, we can improve upon the result of Section 4.2. Proof details are omitted due to space constraints.

Theorem 11. *The algorithm using 4×4-squares achieves a competitive ratio of $\frac{19\sqrt{3}}{20+\sqrt{3}} \approx 1.514$ for the OMRTP in polyominoes with $h = 0$ and $A > 1$.*

For polyominoes with holes, we achieve the same result as in Section 4.2.

4.4 Lower Bound

Theorem 12. *For deterministic online algorithms, a lower bound on the competitive ratio for the OMRTP on polyominoes is $\frac{38}{37} \approx 1.027$.*

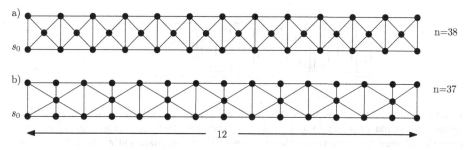

Fig. 5. The adversary's relay placement for a strip of length 12

Proof. We consider a unit strip with starting point at the lower left corner; details are due to space constraints. As shown in Fig. 5, it is possible to save a relay compared to a unit-square solution for a strip of length 12; however, if an algorithm tries this approach, an adversary can choose a shorter strip, for which the unit-square solution is still better.

5   Online Maximum Area Triangulation in Polyominoes

In the Online Maximum Area Triangulation Problem (OMATP) we are given a fixed number of relays, n, and our goal is to triangulate as much area as possible. As mentioned in the introduction, no competitive algorithm can exist for arbitrary polygons. However, for polyominoes we can achieve a factor of $\frac{2}{3\sqrt{3}}$. We use Algorithm 4.1, because it guarantees triangles of size $\frac{1}{4}$.

Theorem 13. *With $n \geq 5$ relays, Algorithm 4.1 achieves a competitive factor of $\frac{2}{3\sqrt{3}}$, which is asymptotically best possible for any deterministic online algorithm.*

Proof. We assess the number of triangles using Lemma 1: $n = \frac{1}{2}(t+b)+1-h \Rightarrow t = 2n - 2 + 2h - b$. Note that the boundary edges and holes are those of the triangulation and not of the polyomino. With $n \geq 4$, the optimal triangulation must have at least four outside boundary edges, because placing the fourth relay inside an already formed triangle cannot be optimal. Moreover, each hole must have more than two boundary edges, so $b \geq 2h + 4$ and $2h + 4 - b \leq 0$ so $t = 2n - 2 + 2h - b \leq 2n - 6 = 2(n-3)$. Since the maximal size of each triangle is $\frac{\sqrt{3}}{4}$ we obtain $A_{OPT} = \frac{\sqrt{3}}{2}(n-3)$ for the area covered by an optimal strategy.

For $n \geq 5$, Algorithm 4.1 fills at least one unit square, plus one unit square for at least every three additional relays, plus at least $\frac{1}{4}$ for each relay in the last, not completely triangulated unit square. Therefore $A_{ALG} \geq 1 + 1 \cdot \lfloor \frac{n-5}{3} \rfloor + \frac{1}{4}\left(\frac{n-5}{3} - \lfloor \frac{n-5}{3} \rfloor\right) \geq \frac{n-4}{3} + \frac{1}{2} \geq \frac{n-3}{3}$ (the penultimate inequality is derived by a case distinction for different remainders by the division by 3 and the resulting relay placements).

For the lower bound, see Figure 6.   □

Fig. 6. An online algorithm will fail to find the large polygonal region, where the offline optimum can places a large number of unit triangles. Thus, an online algorithm uses triangles of area $\frac{1}{4}$, while the optimum uses only an asymptotically small fraction of them, with the bulk being unit triangles.

6   Conclusion

We have given a number of competitive strategies for the OMRTP in orthogonal integral polygons. Our refined algorithms rely on optimal solutions for subsquares of limited size. A possible improvement could arise from a refined analysis of the algorithm using 4×4-square, where we do overestimate the number of placed relays. A further interesting extension could be an optimality proof for the 4×4-square; studying triangulations of general $k \times k$-squares is interesting in itself, but can be expected to serious difficulties even for moderate values of k, as it is similar in nature to the notoriously difficult problem of packing and covering with unit disks. For the OMRTP, it may be possible to raise the general lower bound; for the offline problem, the complexity is open, but we believe it to be NP-hard. Other possible extensions ask for a biconnected network when deploying the relays for the OMRTP in addition to the property that every relay and edge is part of a triangle.

There are also some open problems for the case of general polygons, where it may not only be possible to improve on the competitive factor of 3 for the OMRTP, but also achieve finite bounds for the OMATP, assuming bounded feature size.

References

1. Bern, M., Eppstein, D.: Mesh Generation and Optimal Triangulation. Computing in Euclidean Geometry 1, 23–90 (1992)
2. Bredin, J., Demaine, E., Hajiaghayi, M., Rus, D.: Deploying Sensor Networks with Guaranteed Fault Tolerance. IEEE/ACM Transactions on Networking (TON) 18(1), 216–228 (2010)
3. Brunner, J., Mihalák, M., Suri, S., Vicari, E., Widmayer, P.: Simple Robots in Polygonal Environments: A Hierarchy. In: Fekete, S.P. (ed.) ALGOSENSORS 2008. LNCS, vol. 5389, pp. 111–124. Springer, Heidelberg (2008)
4. Degener, B., Fekete, S., Kempkes, B., Meyer auf der Heide, F.: A Survey on Relay Placement with Runtime and Approximation Guarantees. Computer Science Review 5(1), 57–68 (2011)
5. Efrat, A., Fekete, S.P., Gaddehosur, P.R., Mitchell, J.S.B., Polishchuk, V., Suomela, J.: Improved Approximation Algorithms for Relay Placement. In: Halperin, D., Mehlhorn, K. (eds.) ESA 2008. LNCS, vol. 5193, pp. 356–367. Springer, Heidelberg (2008)
6. Fekete, S.P., Kamphans, T., Kröller, A., Mitchell, J.S.B., Schmidt, C.: Exploring and Triangulating a Region by a Swarm of Robots. In: Goldberg, L.A., Jansen, K., Ravi, R., Rolim, J.D.P. (eds.) APPROX/RANDOM 2011. LNCS, vol. 6845, pp. 206–217. Springer, Heidelberg (2011)
7. Fekete, S.P., Kamphans, T., Kröller, A., Schmidt, C.: Robot Swarms for Exploration and Triangulation of Unknown Environments. In: Proceedings of the 25th European Workshop on Computational Geometry, pp. 153–156 (2010)
8. Fekete, S.P., Mitchell, J., Schmidt, C.: Minimum Covering with Travel Cost. Journal of Combinatorial Optimization, 393–402 (2010)
9. Fekete, S.P., Schmidt, C.: Polygon Exploration with Time-Discrete Vision. Computational Geometry 43(2), 148–168 (2010)

10. Friedman, E.: Packing Unit Squares in Squares: A Survey and New Results. The Electronic Journal of Combinatorics (2009)
11. Hoffmann, F., Icking, C., Klein, R., Kriegel, K.: The Polygon Exploration Problem. SIAM Journal on Computing 31(2), 577–600 (2002)
12. Hsiang, T., Arkin, E., Bender, M., Fekete, S., Mitchell, J.: Algorithms for Rapidly Dispersing Robot Swarms in Unknown Environments. In: Algorithmic Foundations of Robotics V, pp. 77–94 (2004)
13. Icking, C., Kamphans, T., Klein, R., Langetepe, E.: Exploring Simple Grid Polygons. In: Wang, L. (ed.) COCOON 2005. LNCS, vol. 3595, pp. 524–533. Springer, Heidelberg (2005)
14. Kashyap, A., Khuller, S., Shayman, M.: Relay Placement for Fault Tolerance in Wireless Networks in Higher Dimensions. Comp. Geom. 44(4), 206–215 (2011)
15. McLurkin, J., Smith, J.: Distributed Algorithms for Dispersion in Indoor Environments using a Swarm of Autonomous Mobile Robots. In: Distributed Autonomous Robotic Systems 6, pp. 399–408 (2007)
16. Schmidt, C.: Algorithms for Mobile Agents with Limited Capabilities. Ph.d. thesis, Braunschweig Institute of Technology (2011)
17. Suri, S., Vicari, E., Widmayer, P.: Simple Robots with Minimal Sensing: From Local Visibility to Global Geometry. The International Journal of Robotics Research 27(9), 1055–1067 (2008)

A Competitive Strategy
for Distance-Aware Online Shape Allocation

Sándor P. Fekete, Nils Schweer, and Jan-Marc Reinhardt

Department of Computer Science, TU Braunschweig, Germany
{s.fekete,n.schweer,j-m.reinhardt}@tu-bs.de

Abstract. We consider the following online allocation problem: Given a unit square S, and a sequence of numbers $n_i \in \{0,1\}$ with $\sum_{j=0}^{i} n_j \leq 1$; at each step i, select a region C_i of previously unassigned area n_i in S. The objective is to make these regions compact in a distance-aware sense: minimize the maximum (normalized) average Manhattan distance between points from the same set C_i. Related location problems have received a considerable amount of attention; in particular, the problem of determining the "optimal shape of a city", i.e., allocating a *single* n_i has been studied, both in a continuous and a discrete setting. We present an online strategy, based on an analysis of space-filling curves; for continuous shapes, we prove a factor of 1.8092, and 1.7848 for discrete point sets.

Keywords: Clustering, average distance, online problems, optimal shape of a city, space-filling curves, competitive analysis.

1 Introduction

Many optimization problems deal with allocating point sets to a given environment. Frequently, the problem is to find compact allocations, placing points from the same set closely together. One well-established measure is the average L_1 distance between points. A practical example occurs in the context of grid computing, where one needs to assign a sequence of jobs i, each requiring an (appropriately normalized) number n_i of processors, to a subset C_i of nodes of a large square grid, such that the average communication delay between nodes of the same job is minimized; this delay corresponds to the number of grid hops [10], so the task amounts to finding subsets with a small average L_1, i.e., *Manhattan* distance. Karp et al. [7] studied the same problem in the context of memory allocation.

Even in an offline setting without occupied nodes, finding an optimal allocation for one set of size n_i is not an easy task; as shown in Fig. 1, the results are typically "round" shapes. If a whole sequence of sets has to be allocated, packing such shapes onto the grid will produce gaps, causing later sets to become disconnected, and thus leads to extremely bad average distances. Even restricting the shapes to be rectangular is not a remedy, as the resulting problem of deciding whether a set of squares (which are minimal with respect to L_1 average distance

S.K. Ghosh and T. Tokuyama (Eds.): WALCOM 2013, LNCS 7748, pp. 41–52, 2013.

among all rectangles) can be packed into a given square container is NP-hard [9]; moreover, disconnected allocations may still occur.

In this paper, we give a first algorithmic analysis for the *online* problem. Using an allocation scheme based on a space-filling curve, we establish competitive factors of 1.8092 and 1.7848 for minimizing the maximum average Manhattan distance within an allocated set, and non-trivial lower bounds for these factors.

Related Work

Compact location problems have received a considerable amount of attention. Krumke et al. [8] have considered the *offline* problem of choosing a set of n vertices in a weighted graph, such that the average distance is minimized. They showed that the problem is NP-hard (even to approximate); for the scenario in which distances satisfy the triangle inequality, they gave algorithms that achieve asymptotic approximation factors of 2. For points in two-dimensional space and Manhattan distances, Bender et al. [2] gave a simple 1.75-approximation algorithm, and a polynomial-time approximation scheme for any fixed dimension.

The problem of finding the "optimal shape of a city", i.e., a shape of given area that minimizes the average Manhattan distance, was first considered by Karp, McKellar, and Wong [7]; independently, Bender, Bender, Demaine, and Fekete [1] showed that this shape can be characterized by a differential equation for which no closed form is known. For the case of a finite set of n points that needs to be allocated to a grid, Demaine et al. [5] showed that there is an $O(n^{7.5})$ dynamic-programming algorithm, which allowed them to compute all optimal shapes up to $n = 80$. Note that all these results are strictly offline, even though the original motivation (register or processor allocation) is online.

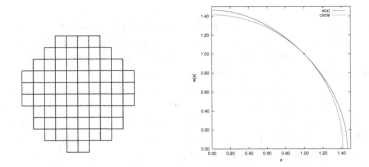

Fig. 1. Finding optimal individual shapes. (Left) An optimal shape composed of n=72 grid cells, according to [5]. (Right) The optimal limit curve $w(x)$, according to [2].

Space-filling curves for processor allocation with our objective function have been used before, see Leung et al. [10]; however, no algorithmic results and no competitive factor was proven. Wattenberg [15] proposed an allocation scheme for purposes of minimizing the maximum *Euclidean diameter* of an allocated shape; this is a different measure than the one established by [10]. Like other authors before (in particular, Niedermeier et al. [11] and Gotsman and Lindenbaum [6]), he considered *c-locality*: for a sequence $1, \ldots, i, \ldots, j, \ldots$ of points on a line, a space-filling mapping $h(.)$ will guarantee $L_2(h(i), h(j)) < c\sqrt{|j - i|}$, for a constant c that is $\sqrt{6} \approx 2.449$ for the Hilbert curve, and 2 for the so-called H-curve. One can use c-locality for establishing a constant competitive factor for our problems; however, given that their focus is on bounding the worst-case distance ratio for an embedding instead of the average distance, it should come as no surprise that the resulting values are significantly worse than ours. On a different note, de Berg, Speckmann, and van der Weele [4] consider treemaps with bounded aspect ratio. Other related work includes Dai and Su [3].

Our Results
We give a first competitive analysis for the online shape allocation problem within a given bounding box, with the objective of minimizing the maximum average Manhattan distance. In particular, we give the following results.

– We show that for the case of continuous shapes (in which numbers n_i correspond to area), a strategy based on a space-filling Hilbert curve achieves a competitive ratio of 1.8092.
– For the case of discrete point sets (in which numbers indicate the number of points that have to be chosen from an appropriate $N \times N$ orthogonal grid), we prove a competitive factor of 1.7848.
– We sketch how these factors may be further improved, but point out that a Hilbert-based strategy is no better than a competitive factor of 1.3504, even with an improved analysis.
– We establish a lower bound of 1.144866 for *any* online strategy in the case of discrete point sets, and argue the existence of a lower bound for the continuous case.

The rest of this paper is organized as follows. In Section 2, we give some basic definitions and fundamental facts. In Section 3, we provide a brief description of an allocation scheme based on a space-filling curve. Section 4 gives a mathematical study for the case of continuous allocations, proving that the analysis can be reduced to a limited number of shapes, and establishes a competitive factor of 1.8092. Section 5 sketches a similar analysis for the case of discrete allocations; as a result, we prove a competitive factor of 1.7848. Section 6 discusses lower bounds for online strategies. Final conclusions are presented in Section 7.

2   Preliminaries

We examine the problem of selecting shapes from a square, such that the maximum average L_1-distance of the shapes is minimized. We first formulate the problem more precisely. This covers both the continuous and the discrete case; the former arises as the limiting case of the latter, while the latter needs to be considered for allocations within a grid of limited size.

Definition 1. *A* city *is a (continuous) shape in the plane with fixed area. For a city C of area n, we call*

$$c(C) = \frac{1}{2} \iiiint_{(x,y),(u,v) \in C} (|x - u| + |y - v|) \, dv \, du \, dy \, dx \qquad (1)$$

the total Manhattan distance *between all pairs of points in C and*

$$\phi(C) = \frac{2\,c(C)}{n^{5/2}} \qquad (2)$$

the ϕ*-value* or *average distance of C. An n-town T is a subset of n points in the integer grid. Its normalized average Manhattan distance is*

$$\phi(T) = \frac{2c(T)}{n^{5/2}} = \frac{\sum_{s \in T} \sum_{t \in T} \|s - t\|_1}{n^{5/2}} \qquad (3)$$

The normalization with $n^{2.5}$ yields a dimensionless measure that remains unchanged under scaling (so it depends only on the shape, not on the size), and makes the continuous and the discrete case comparable; see [1].

Problem 2. *In the continuous setting, we are given a sequence $n_1, n_2, \ldots, n_k \in \mathbb{R}^+$ with $\sum_{i=1}^{k} n_i \leq 1$. Cities $C_1, C_2, \ldots, C_k$ of size $n_1, n_2, \ldots, n_k$ are to be chosen from the unit square, such that $\max_{1 \leq i \leq k} \phi(C_i)$ is minimized.*

In the discrete setting, we are given a sequence $n_1, n_2, \ldots, n_k \in \mathbb{N}^+$ with $\sum_{i=1}^{k} n_i \leq N^2$. Towns $C_1, C_2, \ldots, C_k$ of size $n_1, n_2, \ldots, n_k$ are to be chosen from the $N \times N$ grid, such that $\max_{1 \leq i \leq k} \phi(C_i)$ is minimized.

Although it has not been formally proven, the offline problem is conjectured to be NP-hard, see [13]; if we restrict city shapes to be rectangles, there is an immediate reduction from deciding whether a set of squares can be packed into a larger square [9]. (A special case arises from considering integers, which corresponds to choosing grid locations.) Our approximation works online, i.e., we choose the cities in a specified order, and no changes can be made to previously allocated cities; clearly, this implies approximation factors for the corresponding offline problems.

There are lower bounds for $\max_{1 \leq i \leq k} \phi(C_i)$ that generally cannot be achieved by any algorithm. One important result is the following theorem.

Theorem 3. *Let C be any city. Then $\phi(C) \geq 0.650245$.*

A proof can be found in [1]. For $n_1 = 1$ any algorithm must select the whole unit square, thus $2/3$, the ϕ-value of a square, is a lower bound for the achievable ϕ-value. We will discuss better lower bounds in the conclusions.

3   An Allocation Strategy

While long and narrow shapes tend to have large ϕ-values, shapes that fill large parts of an enclosing rectangle with similar width and height usually have better average distances; however, one has to make sure that early choices with small average distance do not leave narrow pieces with high average distance, or even disconnected pieces, making the normalized ϕ-values potentially unbounded.

Our approach uses the recursive Hilbert family of curves in order to yield a provably constant competitive factor. That family is based on a recursive construction scheme and becomes space filling for infinite repetition of said scheme [12]. For a finite number r of repetitions, the curve traverses all points of the used grid. For $1 \leq r \leq 3$, the curve is shown in Fig. 2. Thus, the Hilbert curve provides an order for the cells of the grid, which is then used for allocation, as illustrated in Fig. 3. More formal details of the recursive definition of the Hilbert family (e.g. with text-rewriting rules, such as the ones in [14]) go beyond the scope of this extended abstract.

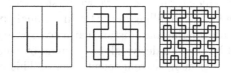

Fig. 2. Hilbert curve with $1 \leq r \leq 3$

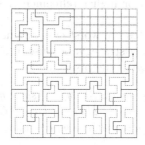

Fig. 3. A sample allocation according to our strategy

More technically, the unit square is recursively subdivided into a grid consisting of $2^r \times 2^r$ grid cells, for an appropriate refinement level $r > 0$, as shown in Fig. 2. For the sake of concise presentation within this short abstract, we assume that every input n_i is an integral multiple of $c = 4^{-R}$, for an appropriately large $R > 0$. (We will mention in the Conclusions how this assumption can be removed, based on Lemma 6.) Similar to the recursive structure of quad-trees, the actual subdivision can be performed in a self-refining manner, whenever a grid cell is not completely filled. This means that during the course of the online allocation, we may use different refinement levels in different parts of the layout; however, this will not affect the overall analysis, as further refinement of the grid does not change the quality of existing shapes.

Definition 4. *For a given refinement level r, an r-pixel P is a grid square of size $2^{-r} \times 2^{-r}$. For a given allocated shape C_i, a pixel is full if $P \subseteq C_i$; it is fractional, if $P \cap C_i \neq \emptyset$ and $P \not\subseteq C_i$.*

Now the description of the algorithm is simple: for every input n_i, choose the next set of $n_i/2^R$ R-pixels traversed by the Hilbert curve as the city C_i, starting in the upper left corner of the grid. For an illustration, see Fig. 3.

The following lemma is a consequence of the recursive structure of the Hilbert family. We use it in the following section for deriving upper bounds.

Lemma 5. *Let C be a city generated by our strategy with area at most $n \leq l\,4^j\,4^{-R}$ for $j \in \{0, 1, \ldots, R\}, l \in \mathbb{N}$. Then at any refinement level r, C contains at most two fractional r-pixels.*

4 Analysis

For the analysis of our allocation scheme we will first make use of Lemma 5. As noted in the following Lemma 6, filling in the two fractional pixels of an allocated shape yields an estimate for the total distance at a coarser refinement level. In a second step, this will be used to derive global bounds by computing the worst-case bounds for shapes of at most refinement level 3. This reduces the task of providing a general upper bound on the competitive factor to considering a finite number of shapes of limited size. (As discussed in the Conclusions, carrying out the computations on a lower or higher refinement level gives looser or tighter results.)

In the following, W_n denotes the worst case among all cities of n pixels that can be produced by our Hilbert strategy; because of the normalized nature of ϕ, this is independent on the size of the pixels, and only the shape matters.

Lemma 6. *Let C be a city generated by our strategy with area at most $n \leq l\,4^r\,4^{-R}$ for $r \in \{0, 1, \ldots, R\}, l \in \mathbb{N}$. Then we have $c(C) \leq c(W_{l+1})$, where W_{l+1} is a worst case among all cities produced by our allocation scheme that consists of $(l+1)$ r-pixels.*

Proof. By Lemma 5, we know that only the first and the last pixel of C may be fractional. Therefore C cannot intersect more than $l+1$ r-pixels. By replacing the two fractional pixels by full pixels, we get a city W that consists of $l+1$ full r-pixels, and $c(C) \leq c(W)$. By definition, $c(W) \leq c(W_{l+1})$, and the claim holds. $\square$

Therefore, we can give upper bounds for the worst case by considering the values of W_n at some moderate refinement level. The W_n can be found by enumeration; as described in the full version of the paper, a speed-up can be achieved by making use of the recursive construction of the W_n. We determined the shapes and ϕ-values of the W_n for $n \leq 65$; by Lemma 6, this suffices to provide upper bounds for all cities with area up to $64 * 2^{-r}$, i.e., these computational results give an estimate for the round-up error using refinement level 3. The full table of average distances can be found in the full version of the paper; the worst cases among the examined ones are W_{56} and W_{14}, which have the same shape, shown in Fig. 4.

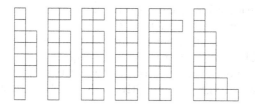

Fig. 4. Worst cases W_n for $12 \leq n \leq 17$

Theorem 7. *A Hilbert strategy guarantees* $\max_{1 \leq n \leq k} \phi(C_n) \leq 1.1764$.

Proof. Consider a city C of size n generated by our strategy. If n is sufficiently small, i.e., smaller than an $R - r$-pixel, $r \geq 0$, C consists of at most 4^r cells and its average distance can be bounded by the worst case for that particular number of cells. In the case that C has a larger, more refined shape, an analysis of a finite number of shapes is still sufficient:

We know that $n > 4^r c$ and we can assume that $n \leq 4^{r+1} c$ (or else we use the analysis on the less refined $(R - (r + 1))$-pixels). Thus, there must be an l such that $l4^r c < n \leq (l + 1)4^r c$ with $l = 1, \ldots, 3$. Yet, we can get closer to n, as we know that an $(R - r)$-pixel consists of 4^r cells. We get the inequality $l4^{r-k} < n \leq (l + 1)4^{r-k}c$, $k \leq r$, $l = 4^k, \ldots, 4^{k+1} - 1$.

Hence, a city of arbitrary size n corresponds to at most $(l + 1)$ sub-squares of a certain size (depending on the precision of the analysis), i.e., a city of size at most $(l + 1)4^{r-k}c$. Now we can use Lemma 5 to bound the average distance of the city, yielding

$$\phi(C) \leq \frac{2\,c(W)}{(l\,4^{r-k}c)^{5/2}} = \frac{\phi(W_{l+2})((l + 2)\,4^{r-k}c)^{5/2}}{(l\,4^{r-k}c)^{5/2}} \tag{4}$$

$$= \phi(W_{l+2})\left(1 + \frac{2}{l}\right)^{5/2} =: \Phi(W_l). \tag{5}$$

The resulting bound is $\max(\{\phi(W_i) : 1 \leq i \leq 4^r\} \cup \{\Phi(W_l) : 4^k \leq l \leq 4^{k+1} - 1\})$. Note that the number of shapes considered is at most 4^{k+1}.

We conducted the calculations for $k = 2$; as it turns out, the maximum is attained for $\Phi(W_{16}) = 1.1764$. See the full version of the paper for details.   □

Corollary 8.- *Our strategy achieves a competitive factor of 1.8092.*

Proof. According to Theorem 3, no algorithm can guarantee a better ϕ-value than 0.650245. Our strategy yields an upper bound of 1.1764. This results in a factor of $1.1764/0.650245 \approx 1.8092$.   □

5   Discrete Point Sets

Our above analysis relies on continuous weight distributions, which imply the lower bound on ϕ-values stated in Theorem 1. This does not include the discrete scenario, in which each value n_i indicates the number of integer grid points that have to be chosen from an appropriate $N \times N$-grid. As discussed in the paper [5], considering discrete weight distributions may allow lower average distances; e.g., a single point yields a ϕ-value of 0. As a consequence, *towns* (subsets of the integer grid) have lower average distances than cities of the equivalent total weight. However, we still get a competitive ratio for the case of online towns.

Theorem 9. *For n-towns, a Hilbert-curve strategy guarantees a competitive factor of at most 1.7848 for the ϕ-value.*

Proof. Lemma 5 still holds, so analogously to Theorem 7, we consider the values up to $n = 64$, and show that the worst case is attained for $n = 16$, which yields an upper bound of 1.123. See the full version for detailed numbers.

For a lower bound, the general value of 0.650245 for ϕ-values cannot be applied, as discrete point sets may have lower average distance. Instead, we verify that the ratio $\rho(n)$ of achieved ϕ to optimal ϕ, is less than 1.7848. This is the same as $c(T_n)/c_{town}(n)$ for $n \leq 64$; see the full version of the paper. For $65 \leq n \leq 80$, the optimal values in [5] allow us to verify that $\phi \geq 0.6292$; see the full version of the paper.

Thus, we have to establish a lower bound for ϕ for $n \geq 81$. We make use of equation (5), p. 89 of [5]; see Fig. 5: for a given number n of grid points, the difference between the optimal total Manhattan distance $c_{city}(n)$ for a city consisting of n unit squares and the optimal total distance $c_{town}(n)$ for a town consisting of n grid points is equal to $\Lambda(n) := \frac{1}{6} \left(\sum_i c_i^2 + \sum_j r_j^2 \right)$, where c_i is the number of grid points in column i, and r_j is the number of grid points in row j. Because $\frac{2c_{city}(n)}{n^{2.5}}$ is bounded from below by $\psi = 0.650245$, we get a lower bound of $\psi - \frac{2\Lambda(n)}{n^{2.5}} \leq \frac{2c_{town}(n)}{n^{2.5}}$ for the ϕ-value of an n-town.

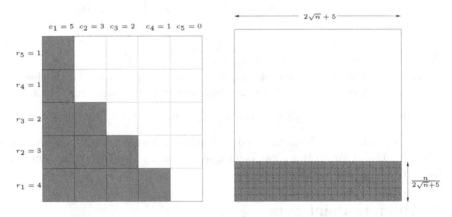

Fig. 5. Establishing a lower bound for ϕ: Defining $\Lambda(n)$; an arrangement that maximizes $\Lambda(n)$

This leaves the task of providing an upper bound for $2\Lambda(n)/n^{2.5}$. According to Lemma 5 of [5], the bounding box of an optimal n-town does not exceed $2\sqrt{n}+5$. Therefore, we have $c_i \leq 2\sqrt{n}+5$; as $\sum_i c_i = n$ and the function $\sum_i c_i^2$ is superlinear in the c_i, we conclude that $\sum_i c_i^2$ is maximized by subdividing n into $\frac{n}{2\sqrt{n}+5}$ columns of $2\sqrt{n}+5$ points each, so $\sum_i c_i^2 \leq n(2\sqrt{n}+5)$. Analogously, we have $\sum_j r_j^2 \leq n(2\sqrt{n}+5)$, so $2\Lambda(n)/n^{2.5} \leq \frac{2}{3}(\frac{2}{n} + \frac{5}{n^{1.5}})$. For $n \geq 81$, this implies $2\Lambda(n)/n^{2.5} \leq \frac{4}{243} + \frac{10}{2187} = 0.0210333...$ or $\phi(n) \geq 0.6292$. This yields an overall competitive ratio of not more than $1.123/0.6292$, i.e., 1.7848.   □

A more refined analysis of $\Lambda(n)$ considers maximizing $\sum_i c_i^2 + \sum_j r_j^2$ all at once, instead of $\sum_i c_i^2$ and $\sum_j r_j^2$ separately, for a maximum value of $n(2\sqrt{n}+5) + \frac{n^2}{2\sqrt{n}+5}$. For $n \geq 81$, this yields $2\Lambda(n)/n^{2.5} \leq \frac{2}{243} + \frac{5}{2187} + \frac{2}{621} = 0.0137373...$ As the resulting competitive ratio of 1.7643 is only very slightly better, we omit further details from this extended abstract. If instead we rely on the unproven conjecture in [5] that $\frac{2c_{town}}{n^{2.5}} \approx \psi - \frac{0.410}{n}$, we get $\phi \geq 0.6451$, which corresponds to experimental evidence; the resulting competitive factor is 1.7406.

6   Lower Bounds

We demonstrate that there are non-trivial lower bounds for a competitive factor. We start by considering the discrete online scenario for towns.

Theorem 10. *No online strategy can guarantee a competitive factor below* $\frac{64}{\sqrt{5}^5} = 1.144866...$

Proof. Consider a 3×3 square, and let $n_1 = 4$; see Fig. 6. If (a) the strategy allocates a 2×2 square (for a total distance of 8), then $n_2 = 5$, and the resulting L-shape has a total distance of 20 and a ϕ-value of $40/5^{2.5} = 0.715541...$ Allocating (b) the first town with an L-shape of total distance 10 results in $\phi = 20/32 = 0.625$, and the second with a total distance of 16, or $\phi = 32/5^{2.5} = 0.572433...$

If instead, (c) the first town is allocated different from a square, the total distance is at least 10, and $\phi \geq 20/32$; then (d) $n_2 = n_3 = n_4 = n_5 = n_6 = 1$, and an optimal strategy can allocate the first town as a 2x2 square, with $\phi = 0.5$. This bounds the competitive ratio, as claimed.   □

For the case of continuous allocations, we claim the following.

Theorem 11. *There is* $\delta > 0$, *such that no online strategy can guarantee a competitive factor* $1 + \delta$.

Proof. Consider $n_1 = 1/2$, in combination with the two possible scenarios

(a) $n_2 = 1/2$;
(b) $n_2 = n_3 = ... = \varepsilon$.

In scenario (a), an adversary can assign two $(1 \times 1/2)$-rectangles, for a ϕ-value of $0.707...$; in scenario (b), an adversary can assign all shapes as squares, for a ϕ-value of $0.666...$ If the player chooses a square of size $\sqrt{2}/2$ first, the adversary

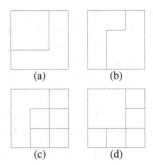

Fig. 6. The four cases considered in Theorem 10; the left column shows the choices by an algorithm, the right the corresponding optimal choices for the ensung sequence

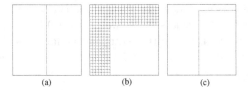

Fig. 7. The scenarios considered in Theorem 11, and a possible choice for the player

can choose scenario (a), causing the second allocation to be in L-shape with ϕ-value $\frac{2}{3}(7 - 4\sqrt{2}) = 0.895431...$, as opposed to the optimal value of 0.707... If the player chooses a $(1 \times 1/2)$-rectangle first, the adversary chooses scenario (b), for a ratio of 1.06066... The existence of the claimed lower bound follows from continuity, as the ϕ-value changes continuously with continuous deformation of the involved shapes.   □

The precise value arising from the scenarios in Theorem 11 is complicated. It can be obtained by computing the optimal intermediate value for the player that allows him to protect against both scenarios at once. For example, optimizing over the family of allocations shown in Figure 7 (c) yields a competitive ratio that is better than 1.06; however, the player may do even better by using curved boundaries. The involved computational effort for the resulting optimization problem promises to be at least as complicated as computing the "optimal shapes of a city", for which no closed-form solution is known, see [7,1].

7   Conclusions

We have established a number of results for the online shape allocation problem. In principle, further improvement could be achieved by replacing the computational results for level 3 (i.e., $n = 16, \ldots, 64$) by level 4 (i.e., $n = 65, \ldots, 256$). (Conversely, a simplified analysis with level 2, i.e., $n = 4, \ldots, 16$; yields a worse factor of 3.6525.) However, the highest known optimal ϕ-values are for $n = 80$, obtained by using the $O(n^{7.5})$ algorithm of [5]. In any case, there is a threshold of 1.3504 for Hilbert-based strategies, which we believe to be tight: this is the ratio between the upper bound of 0.8768 for $n = 14$ (and for $n = 56, 224, \ldots$) and the asymptotic lower bound of 0.650245; because asymptotically, continuous and discrete case converge, this also applies to the discrete case. Other open problems are to raise the lower bound of 1.144866 for the discrete case, and establish definitive values for the continuous case.

As noted in Section 3, we can eliminate the assumption of all n_i being multiples of some 2^{-R}, by making use of Lemma 6, and allocating a small round-off fraction to a fractional pixel maintains the same bounds. However, the formal aspects of describing the resulting allocation scheme become somewhat tedious and would require more space than provided for this short abstract.

The offline problem is interesting in itself: for given n_i, $i = 0, \ldots, m$, allocate disjoint regions of area n_i in a square, such that the maximum average Manhattan distance for each shape is minimized. As mentioned, there is some indication that this is an NP-hard problem; however, even relatively simple instances are prohibitively tricky to solve to optimality, making it hard to give a formal proof. Clearly, our online strategy provides a simple approximation algorithm; however, better factors should be possible by exploiting the a-priori information of knowing all n_i, e.g., by sorting them appropriately.

Fig. 8. A possible worst-case scenario for the offline problem

Another interesting open question for the offline scenario is the maximum optimal ϕ-value for any set $n_1, \ldots, n_m$. A simple lower bound is $2/3 = 0.666\ldots$, as that is the average distance of the whole square. A better lower bound is provided by dividing the square into two or three equal-sized parts. For the case $n_1 = n_2 = 1/2$, we can use symmetry and convexity to argue that an optimum can be obtained by a vertical split, yielding $\phi = \sqrt{2}/2 = 0.707$. We believe the global worst case is attained for $n_1 = n_2 = n_3 = 1/3$. Unfortunately, it is no longer possible to exploit only symmetry for arguing global optimality. Figure 8 shows an allocation with $\phi = 0.718736\ldots$ for all three regions[1]. We conjecture that this is the best solution for $n_1 = n_2 = n_3 = 1/3$, as well as the worst case for any optimal partition of the unit square.

---

[1] More precisely, the involved values can be expressed as $a = \frac{1}{108}\left(55 - \frac{791}{\theta} + \theta\right)$ and $\phi = \dfrac{\left(9602477 - 13416\sqrt{585705}\right)\theta + \left(202679 + 204\sqrt{585705}\right)\psi^2 + 82133\theta^3}{77760\sqrt{3}\theta}$ with $\theta :=$ $\left(-16253 + 36\sqrt{585705}\right)^{1/3}$.

Acknowledgments. A short abstract based on preliminary results of this paper appears in the informal, non-competitive Workshop EuroCG. (Standard disclaimer of that workshop: "This is an extended abstract of a presentation given at EuroCG 2011. It has been made public for the benefit of the community only and should be considered a preprint rather than a formally reviewed paper. Thus, this work is expected to appear in a conference with formal proceedings and/or in a journal.")

We thank Bettina Speckmann for pointing out references [15] and [4], and other colleagues for helpful hints to improve the presentation of this paper.

References

1. Bender, C.M., Bender, M.A., Demaine, E.D., Fekete, S.P.: What is the optimal shape of a city? J. Physics A: Mathematical and General 37(1), 147–159 (2004)
2. Bender, M.A., Bunde, D.P., Demaine, E.D., Fekete, S.P., Leung, V.J., Meijer, H., Phillips, C.A.: Communication-Aware Processor Allocation for Supercomputers: Finding Point Sets of Small Average Distance. Algorithmica 50(2), 279–298 (2008)
3. Dai, H.K., Su, H.C.: On the Locality Properties of Space-Filling Curves. In: Ibaraki, T., Katoh, N., Ono, H. (eds.) ISAAC 2003. LNCS, vol. 2906, pp. 385–394. Springer, Heidelberg (2003)
4. de Berg, M., Speckmann, B., van der Weele, V.: Treemaps with bounded aspect ratio. CoRR, abs/1012.1749 (2010)
5. Demaine, E.D., Fekete, S.P., Rote, G., Schweer, N., Schymura, D., Zelke, M.: Integer point sets minimizing average pairwise L1 distance: What is the optimal shape of a town? Comp. Geom. 40, 82–94 (2011)
6. Gotsman, C., Lindenbaum, M.: On the metric properties of discrete space-filling curves. IEEE Transactions on Image Processing 5(5), 794–797 (1996)
7. Karp, R.M., McKellar, A.C., Wong, C.K.: Near-Optimal Solutions to a 2-Dimensional Placement Problem. SIAM J. Computing 4(3), 271–286 (1975)
8. Krumke, S., Marathe, M., Noltemeier, H., Radhakrishnan, V., Ravi, S., Rosenkrantz, D.: Compact location problems. Theor. Comput. Sci. 181(2), 379–404 (1997)
9. Leung, J.Y.-T., Tam, T.W., Wing, C.S., Young, G.H., Chin, F.Y.: Packing squares into a square. J. Parallel Distrib. Comput. 10(3), 271–275 (1990)
10. Leung, V.J., Arkin, E.M., Bender, M.A., Bunde, D.P., Johnston, J., Lal, A., Mitchell, J.S.B., Phillips, C.A., Seiden, S.S.: Processor Allocation on Cplant: Achieving General Processor Locality Using One-Dimensional Allocation Strategies. In: Proc. IEEE CLUSTER 2002, pp. 296–304 (2002)
11. Niedermeier, R., Reinhardt, K., Sanders, P.: Towards optimal locality in mesh-indexings. Discrete Applied Mathematics 117(1-3), 211–237 (2002)
12. Sagan, H.: Space-Filling Curves. Springer, New York (1994)
13. Schweer, N.: Algorithms for Packing Problems. PhD thesis, Braunschweig (2010)
14. Siromoney, R., Subramanian, K.: Space-filling Curves and Infinite Graphs. In: Ehrig, H., Nagl, M., Rozenberg, G. (eds.) Graph Grammars 1982. LNCS, vol. 153, pp. 380–391. Springer, Heidelberg (1983)
15. Wattenberg, M.: A note on space-filling visualizations and space-filling curves. In: Proceedings of the IEEE Symposium on Information Visualization, INFOVIS, pp. 181–186 (2005)

Base Location Problems
for Base-Monotone Regions

Jinhee Chun[1], Takashi Horiyama[2], Takehiro Ito[1],
Natsuda Kaothanthong[1], Hirotaka Ono[3], Yota Otachi[4],
Takeshi Tokuyama[1], Ryuhei Uehara[4], and Takeaki Uno[5]

[1] Graduate School of Information Sciences, Tohoku University
Sendai 980-8579, Japan
{jinhee,natsuda,tokuyama}@dais.is.tohoku.ac.jp, takehiro@ecei.tohoku.ac.jp
[2] Graduate School of Science and Engineering, Saitama University
Saitama 338-8570, Japan
horiyama@al.ics.saitama-u.ac.jp
[3] Department of Economic Engineering, Kyushu University. 6-19-1 Hakozaki
Higashi-ku, Fukuoka 812-8581, Japan
hirotaka@en.kyushu-u.ac.jp
[4] School of Information Science, Japan Advanced Institute of Science and
Technology. Asahidai 1-1, Nomi, Ishikawa 923-1292, Japan
{otachi,uehara}@jaist.ac.jp
[5] National Institute of Informatics, 2-1-2 Hitotsubashi, Chiyoda-ku,
Tokyo, 101-8430, Japan
uno@nii.ac.jp

Abstract. The problem of decomposing a pixel grid into base-monotone regions was first studied in the context of image segmentation. It is known that for a given pixel grid and baselines, one can compute in polynomial time a maximum-weight region that can be decomposed into disjoint base-monotone regions [Chun et al. ISAAC 2009]. We continue this line of research and show the NP-hardness of the problem of optimally locating k baselines in a given $n \times n$ pixel grid. We also present an $O(n^3)$-time 2-approximation algorithm for this problem. We then study two related problems, the k base-segment problem and the quad-decomposition problem, and present some complexity results for them.

1 Introduction

Let P be an $n \times n$ pixel grid. A *pixel* (i, j) of P is the unit square whose top-right corner is the grid point $(i, j) \in \mathbb{Z}^2$. For example the bottom-left cell of P is $(1, 1)$ and the top-right cell is (n, n). Each pixel $p = (i, j)$, where $1 \le i, j \le n$, has its *weight* $w(p) \in \mathbb{Z}$. Now we define the following general problem.

Problem: MAXIMUM WEIGHT REGION PROBLEM (MWRP)
Instance: An $n \times n$ pixel grid P.
Objective: Find a region $R \in \mathcal{F}$ maximizing the weight $w(R) = \sum_{p \in R} w(p)$,
where $\mathcal{F} \subseteq 2^P$ be a fixed family of pixel regions.

S.K. Ghosh and T. Tokuyama (Eds.): WALCOM 2013, LNCS 7748, pp. 53–64, 2013.
© Springer-Verlag Berlin Heidelberg 2013

Fig. 1. Image segmentation via k baseline MWRP. We first convert a picture to a gray scale image. Next, with some suitable function, we construct a pixel grid in which each dark pixel has positive weight and each light pixel has negative weight. Finally we solve the k baseline MWRP to segment the back ground from the objects. In this example, the edges of the picture is used as baselines ($k = 4$). For example, the red region in the third figure (from left) uses the top edge as its baseline.

The general problem MWRP has been studied for several families $\mathcal{F}$ that are related to practical problems. Observe that if $\mathcal{F} = 2^P$, then R can be arbitrarily chosen, and thus the answer is the set of all positive cells. On the other hand, if $\mathcal{F}$ is the family of connected regions (in the usual 4-neighbor topology), then the problem becomes NP-hard [2]. For the complexity of MWRP for other families, see the paper by Chun et al. [4] and the references therein.

Motivated by the image segmentation problem, Chun et al. [5] studied more complicated family of pixel regions for MWRP (see Fig. 1). A *baseline* of an $n \times n$ pixel grid P is a vertical line $x = b$ or horizontal line $y = b$, where $0 \le b \le n$. A pixel region R is a *based x-monotone region* if there is a horizontal baseline $y = b$ such that $(i, j) \in R$ implies $(i, j-1) \in R$ for $j > b+1$, and $(i, j) \in R$ implies $(i, j+1) \in R$ for $j < b$ (see Fig. 2). *Based y-monotone regions* are analogously defined. Based x- and y-monotone regions are *base-monotone regions*. Given a set of k baselines, a region R is *base-monotone feasible* if it can be decomposed into pairwise disjoint base-monotone regions with respect to the baselines. The *k baseline MWRP* is MWRP in which we are given k (vertical or horizontal) baselines, and we find a maximum-weight base-monotone feasible region with respect to the baselines.

Chun et al. [5] observed that the complement of a maximum-weight base-monotone feasible region represents an object in a picture nicely if the baselines are located reasonably (see Figs. 1 and 3). They showed that the k baseline MWRP can be solved in polynomial time. They also studied the *k base-segment MWRP*, in which we are given k segments and find a region decomposable into base-monotone regions with respect to the given base-segments. (We define this problem more precisely in the next section.) They showed some partial results on the complexity of this problem. For other approaches, see e.g. [6,9].

In the setting of the k baseline MWRP, we are given k baselines. Thus a natural question would be *"What if baselines are not given?"* In other words, *"How can we divide the pixel grid into subgrids with vertical and horizontal lines?"* We study this problem and show that the problem of optimally locating k baselines is NP-hard but can be approximated within factor 2. Next we study the k base-segment MWRP and present sharp contrasts of its computational complexity. Finally, we propose another way for dividing the pixel grid into subgrids, and show that this variant can be solved in polynomial time.

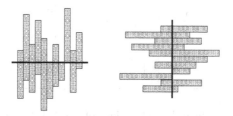

Fig. 2. A based x-monotone region (left) and a based y-monotone region (right)

Fig. 3. The complement of a base-monotone feasible region may represent an object in a picture nicely. By additional baselines, the result may be improved.

2   Definitions of the Three Problems

2.1   Baseline Location

The first and main problem is defined as follows.

Problem: BASELINE LOCATION
Instance: An $n \times n$ pixel grid P and positive integers k and w.
Question: Are there k baselines in P such that a maximum-weight base-monotone feasible region has weight at least w?

There are only $\binom{2n+2}{k}$ possible allocations of k baselines. Thus BASELINE LO-CATION can be solved in $O(2^k n^{k+3})$ time. However, this is impractical if k is a part of the input. We want to solve this problem in $O(\text{poly}(k + n))$ time or in $O(f(k) \cdot \text{poly}(n))$ time, where $f(k)$ is a computable function that depends only on k. Unfortunately, the former, $O(\text{poly}(k + n))$ time, is very unlikely as we prove the problem to be NP-hard if k is a part of the input. The latter, $O(f(k) \cdot \text{poly}(n))$ time, remains unsettled in this paper.

2.2   The k Base-Segment MWRP

Consider a segment s contained in a baseline ℓ. If a base-monotone region R with baseline ℓ intersects ℓ only in s, then R has s as its *base-segment*. Chun et al. [5] also studied k *base-segment MWRP*, in which k base-segments are given, and one wants to find a region that can be decomposed into disjoint monotone regions with respect to the given base-segments. They also studied two-directional version of this problem in which the region can be built only on the right side of each vertical base-segment and on the upper side of each horizontal base-segment. They showed the following results.

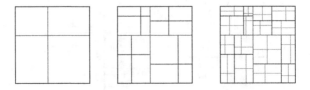

Fig. 4. Quad decompositions of depth 1, 2, and 3

Theorem 2.1 ([5]). *The k base-segment MWRP can be solved in $O(n^{O(k)})$ time. The two-directional version can be solved in $O(k^{O(k)}n^4)$ time.*

It was not known whether the problem is NP-hard when k is a part of the input and whether the two-directional version can be solved in polynomial time with both n and k. We will present affirmative answers to these questions.

2.3   Quad Decomposition

Chun et al. [4] showed that solving the k baseline MWRP is equivalent to solving the following problem for each subgrid obtained by the given baselines.

Problem: ROOM-EDGE PROBLEM
Instance: An $m \times n$ pixel grid P.
Objective: Find a maximum-weight base-monotone feasible region with the four baselines $x = 0$, $x = m$, $y = 0$, and $y = n$.

They presented an $O(mn^2)$-time algorithm for the problem above [4]. We solve the ROOM-EDGE PROBLEM for each subgrid, and then answer their total weight as one for the baseline MWRP. From this point of view, we propose another problem QUAD DECOMPOSITION. For an $n \times m$ pixel grid P and a point $p = (i, j)$, we can divide P naturally into four subgrids: the bottom-left, bottom-right, upper-left, and upper-right parts with respect to the point p. We call the resultant set of subgrids the *quad decomposition* of P at p. If we recursively apply this decomposition d times (at arbitrarily chosen points), then we will have 4^d subgrids of P (see Fig. 4) We call the resultant set of subgrids a *depth d quad decomposition* of P. Now our problem is defined as follows.

Problem: QUAD DECOMPOSITION
Instance: An $n \times n$ pixel grid P and positive integers d and w.
Objective: Find a depth d quad decomposition of P that maximizes the total sum of the weight of the optimum solution of ROOM-EDGE PROBLEM for the subgrids in the decomposition.

Note that we can assume $d \leq \log_2 n$ since otherwise the problem becomes trivial (we can take all the positive cells). We will show that this problem can be solved in polynomial time. In the context of image segmentation, we may expect that quad decompositions work well compared to k baseline decompositions. This is because, by using quad decompositions, we can place many bases in complicated parts of the image.

Fig. 5. A baseline forcer: forcing one baseline

3 NP-Hardness of Baseline Location

Here we prove the following theorem.

Theorem 3.1. BASELINE LOCATION *is NP-complete in the strong sense.*

The problem is clearly in NP. We prove its NP-hardness by reducing INDEPENDENT SET to this problem. An *independent set* of a graph is a set of pairwise non-adjacent vertices. The following problem is known to be NP-complete [7].

Problem: INDEPENDENT SET
Instance: A graph G and a positive integer s.
Question: Does G have an independent set of size at least s?

3.1 Gadgets

We first define two small gadgets for forcing baselines into restricted zones. Throughout this paper, each red $\times$ in a pixel grid represents a huge (but polynomially bounded) negative weight whose absolute value is equal to the sum of all the positive weights in the grid. Also, each blue $\bullet$ represents a (not necessarily large) positive weight. All the other cells have weight 0.

Our first gadget is the 3×3 grid depicted in Fig. 5. If we want to take the positive cell at the center, we need one baseline as in the figure. Since we cannot take any huge negative cell, the possible locations of the baselines are restricted to the four positions in the figure. We call this gadget a *baseline forcer*. The *weight* of a baseline forcer is the weight of the positive cell, and the *position* of a baseline forcer is the position of its bottom-left cell.

Next we consider a similar gadget depicted in Fig. 6. To take all the positive cells and not to take any negative cell, we need either one vertical baseline or two horizontal baselines. Therefore, if we need to minimize the number of baselines, then we have to use one vertical baseline. We call this gadget a *vertical baseline forcer*. By rotating this gadget, we can also obtain a gadget for forcing two vertical baselines or one horizontal baseline. We call it a *horizontal baseline forcer*. Two positive cells in this gadget have the same weight, and their weight is the *weight* of the vertical or horizontal baseline forcer. The *position* of a vertical or horizontal baseline forcer is the position of its bottom-left cell.

Vertical and horizontal baseline forcers work even if we insert some space between columns or rows as in Fig. 6. The location of the baseline is restricted to the area depicted in the figure. We say that a vertical (horizontal) baseline forcer *intersects* a vertical (horizontal resp.) baseline if the baseline is in the

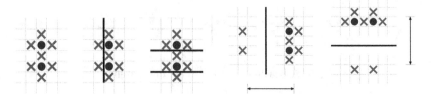

Fig. 6. (Left) A vertical baseline forcer: forcing one vertical baseline. (Right) Forced baselines are restricted to the area indicated by double headed arrows.

restricted area; that is, a base monotone shape with the vertical or horizontal baseline can contain the positive cells in the vertical or horizontal baseline forcer. The number of the columns used by a vertical baseline forcer is its *width*, and the number of rows used by a horizontal baseline forcer is its *height*. For example, the original vertical baseline forcer in Fig. 6 is of width 3.

3.2   Reduction

Given an instance (G, s) of INDEPENDENT SET, we construct an instance (P, k, w) of BASELINE LOCATION as follows. It is easy to see that the reduction below can be done in polynomial time, and the absolute values of the weights are bounded by a polynomial of the input size.

In the following, we assume $|V(G)| = |E(G)|$ for notational convenience. (It is easy to see that INDEPENDENT SET is NP-hard even if $|V(G)| = |E(G)|$.) Let $V(G) = \{v_1, \ldots, v_m\}$ and $E(G) = \{e_1, \ldots, e_m\}$. We set the number of baselines $k = 2m$ and the required weight $w = 8m^3 + 8m^2 + s$. The grid P is the $(20m + 20) \times (20m + 20)$ pixel grid with the following entries (see Fig. 7).

Vertex gadgets. For each vertex v_i, we put a vertical baseline forcer of width 5 and weight $2m^2 + m$, denoted VF_i, at the position $(10i, 5i)$. We also put a baseline forcer of weight 1, denoted BF_i, at the position $(10i - 1, 20m + 15)$.

Edge gadgets. Let $e_h = \{v_i, v_j\} \in E(G)$ be an edge with $i < j$. We put a horizontal baseline forcer of height 10 and weight $2m^2 + m$, denoted HF_h, at the position $(10m + 5h, 5m + 15h)$. Next we put two horizontal baseline forcers $HF_{h,i}$ and $HF_{h,j}$ of height 3 and weight m at the positions $(10i - 3, 5m + 15h - 1)$ and $(10j - 3, 5m + 15h + 8)$, respectively. Also, we put two baseline forcers $BF_{h,i}$ and $BF_{h,j}$ of weight m at the positions $(10i + 3, 5m + 15h + 2)$ and $(10j + 3, 5m + 15h + 5)$, respectively.

The weight of negative cells. We have the following positive cells in the grid:

- $4m$ cells of weight $2m^2 + m$,
- $6m$ cells of weight m, and
- m cells of weight 1.

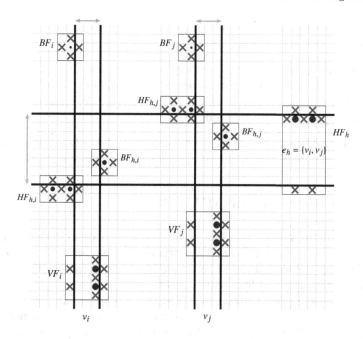

Fig. 7. Gadgets for an edge $\{v_i, v_j\}$: black thick lines are the candidates of required baselines, two vertical and one horizontal

The total weight of the positive cells is $W = 4m(2m^2 + n) + 6m^2 + n = 8m^3 + 10m^2 + m$. We set the weight of the negative cells to $-W$ so that these cells cannot be taken in any solution with a positive total weight.

Lemma 3.2. (G, s) *is a yes-instance of* INDEPENDENT SET *if and only if* (P, k, w) *is a yes-instance of* BASELINE LOCATION.

Since the weight of each cell is polynomially bounded, the problem is NP-hard in the strong sense.

4   A 2-Approximation Algorithm for Baseline Location

Our approximability result is based on the polynomial-time solvability of the following problem.

Problem: VERTICAL BASELINE LOCATION
Instance: An $n \times n$ pixel grid P and a positive integer k.
Objective: Find k vertical baselines in P that maximize the weight of an optimal base-monotone feasible region with respect to these baselines.

The problem HORIZONTAL BASELINE LOCATION is defined analogously. We show that these problems can be solved in $O(n^3)$ time.

Theorem 4.1. VERTICAL BASELINE LOCATION *and* HORIZONTAL BASELINE LOCATION *can be solved in* $O(n^3)$ *time.*

We solve VERTICAL BASELINE LOCATION with k vertical baselines and HORIZONTAL BASELINE LOCATION with k horizontal baselines in $O(n^3)$ time, independently. We output the better one of these solutions. We can show that the output is a 2-approximation solution.

Theorem 4.2. *There is an $O(n^3)$-time 2-approximation algorithm for locating k baselines to maximize the weight of optimum base-monotone feasible region.*

5   The k Base-Segment MWRP

Here we extend the results of Chun et al. [5] (see Theorem 2.1). We first reduce the two-directional version to WEIGHTED INDEPENDENT SET in bipartite graphs. We next reduce INDEPENDENT SET in planar graphs to the original k base-segment MWRP. The first reduction implies that the two-directional version can be solved in polynomial time, and the second implies that the original problem is NP-hard, since INDEPENDENT SET can be solve in polynomial time [11] for bipartite graphs, and is NP-hard for planar graphs [8].

5.1   Two-Directional Version

To prove the following theorem, we reduce the problem to WEIGHTED INDEPENDENT SET for bipartite graphs.

Theorem 5.1. *The two-directional k base-segment MWRP can be solved in $O(k^3 n^6 \log kn)$ time.*

We first divide each base-segment of length ℓ into ℓ unit base-segments. This refinement does not change the optimum value. Now we have $O(kn)$ base-segments of length 1. We identify a base-segment s with (i,j) if s is the left or bottom edge of a pixel (i,j).

For each vertical base-segment $s = (i,j)$, we define its *range* as follows: if there is no vertical base-segment $s' = (i',j)$ with $i' > i$, then the range of s is $[i,n]$; otherwise the range of s is $[i, i'-1]$, where i' is the smallest index for which such a segment exists (see Fig. 8). We define the *range* of a horizontal base-segment analogously.

Let $s = (i,j)$ be a vertical base-segment with range $[i,i']$. Let $a_s(0) = i-1$, and for $p \geq 1$, let $a_s(p)$ be the minimum index h such that $a_s(p-1) < h \leq i'$ and $\sum_{a_s(p-1)<q\leq h} w(q,j)$ is positive. If there is no such index, then $a_s(p)$ is undefined. If $a_s(p)$ is defined for some $p \geq 1$, then let $w_s(p) = \sum_{a_s(p-1)<q\leq a_s(p)} w(q,j)$. See Fig. 8. For each horizontal base-segment s', we also define the sequence $a_{s'}(\cdot)$ analogously.

Now we construct a bipartite graph $G = (U, V; E)$. Let $s = (i,j)$ be a vertical base-segment. Assume that r is the largest index such that $a_s(r)$ is defined.

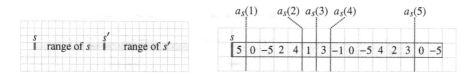

Fig. 8. (Left) The ranges of vertical base-segments s and s'. (Right) Example of $a_s(p)$. The corresponding weights $w_s(1), \ldots, w_s(5) = 5, 1, 1, 3, 3$.

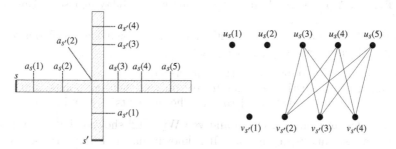

Fig. 9. The bipartite graph construction. The vertices corresponding to the crossing ranges of two base-segments induce the disjoint union of an independent set and a complete bipartite graph.

Now all $a_s(0), \ldots, a_s(r)$ are defined by the definition. If $r = 0$, then this segment s is useless and ignored. Otherwise, we put vertices $u_s(p)$, $1 \leq p \leq r$, with weight $w_s(p)$ into U. For each horizontal base-segment $s' = (i', j')$, we put vertices $v_{s'}(p')$ into V in the same way. Next we define the edge set E. Two vertices $u_s(p) \in U$ and $v_{s'}(p') \in V$ are adjacent if and only if two base-monotone regions with base-segments s and s' have nonzero area intersection if they contain $(a_s(p), j)$ and $(i', a_{s'}(p'))$, respectively. More precisely, this can be stated as: $i \leq i' \leq a_s(p)$ and $j' \leq j \leq a_{s'}(p')$. See Fig. 9 for an example.

Lemma 5.2. *An optimum solution of an instance of the two-directional k base-segment MWRP has weight at least W if and only if the corresponding bipartite graph G has an independent set of weight at least W.*

5.2 NP-Hardness of the k Base-Segment MWRP

We now show the following theorem.

Theorem 5.3. *The k base-segment MWRP is NP-complete in the strong sense.*

The problem is clearly in NP, and thus it suffices to show the NP-hardness. We reduce INDEPENDENT SET for planar graphs to the k base-segment MWRP. A graph is *planar* if it can be drawn in the plane without edge crossings. It is known that INDEPENDENT SET is NP-hard even for planar graphs [8].

Nice visibility representations. A $w \times h$ *grid* is the subset $\{1, 2, \ldots, w\} \times \{1, 2, \ldots, h\}$ of the plane. A *visibility representation* of a planar graph G maps each vertex of G

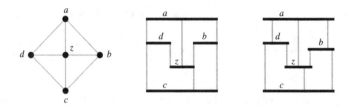

Fig. 10. A planar graph. Its visibility and nice visibility representations

to a horizontal segment with endpoints in a grid and each edge of G to a vertical segment with endpoints in a grid such that

1. no segments of two distinct vertices intersect,
2. segments of two distinct edges intersect only at their endpoints, and
3. the segment of an edge $\{u, v\}$ touches the segments of u and v.

See Fig. 10 for an example. Otten and van Wijk [12] showed that every planar graph has a visibility representation. It is known that a visibility representation of a planar graph in an $O(n) \times O(n)$ grid can be found in linear time (see [13]). Additionally, we need the following conditions for representations:

4. no two vertical segments have the same x-coordinate,
5. no two horizontal segments have the same y-coordinate, and
6. no two endpoints of segments have the same position.

We call a visibility representation satisfying the three additional conditions a *nice visibility representation*. Given a visibility representation of a planar graph, we can obtain a nice visibility representation of the graph in polynomial time by refining each cell of the grid to an $O(n) \times O(n)$ subgrid, slightly extending each horizontal segment, and slightly shifting each vertical segment. Note that each segment in this representation has length at least $2n$.

Reduction. Let (G, s) be an instance of INDEPENDENT SET, where G is a planar graph with n vertices and m edges. Note that we do not assume $n = m$ here. We first construct a nice visibility representation $R = (A, B)$ of G in polynomial time, where A is the set of horizontal segments and B is the set of vertical segments. We construct a pixel grid P from R as follows (see Fig. 11).

For each vertex $u \in V$ with the corresponding horizontal segment $a_u = [x_1, x_2] \times \{y\} \in A$, we put a vertical base-segment (x_1, y) and set the weight 1 to the cell (x_2, y). For each edge $e = \{v, w\} \in E$ with the corresponding vertical segment $b_u = \{x\} \times [y_1, y_2] \in B$, we put horizontal base-segments (x, y_1) and $(x, y_2 + 1)$ and set the weight n to the cell (x, y_e), where the y-coordinate y_e is not used by any vertical base-segment and $y_1 < y_e < y_2$. Such a coordinate can be chosen since each segment has length at least $2n$. Note that the weight of a cell is at most n and there is no negative-weight cell.

Equivalence. We now show that (G, s) is a yes-instance if and only if the optimum value of k base-segment MWRP on P is at least $mn + s$. Since the weight of each cell is polynomially bounded, the problem is NP-hard in the strong sense.

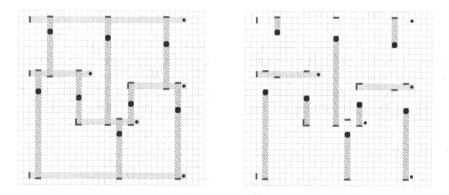

Fig. 11. Each green thick segment is a base-segment. In the right figure $S = \{b, d\}$.

The three-directional version. In the reduction above, we may assume without loss of generality that the region can be built only on the right side of each vertical base-segment, on the upper sides of some horizontal base-segments, and on the lower sides of the remaining horizontal base-segments. We call this version the three-directional k base-segment MWRP.

Corollary 5.4. *The three-directional k base-segment MWRP is NP-complete in the strong sense.*

6 Polynomial-Time Algorithm for Quad Decomposition

Recall that QUAD DECOMPOSITION is the problem of finding a depth d quad decomposition of P that maximizes the total sum of the weight of the optimum solution of ROOM-EDGE PROBLEM for the subgrids in the decomposition.

A dynamic programming approach allows us to have the following result.

Theorem 6.1. QUAD DECOMPOSITION *can be solved in $O(n^7)$ time.*

The bottleneck of the running time above is the first phase of solving ROOM-EDGE PROBLEM for all the possible $O(n^4)$ subgrids. Using techniques developed in the study of the all-pairs shortest path problem, we can slightly improve the running time of the first phase.

Given $s \times t$ and $t \times r$ real matrices $A = (a_{i,j})$ and $B = (b_{i,j})$, the *funny* matrix product $A \odot B$ is the $s \times r$ matrix $C = (c_{i,j})$ with $c_{i,j} = \max_{1 \leq k \leq n}(a_{i,k} + b_{k,j})$. It is known that the computational complexity of funny matrix multiplication is equivalent to that of all-pairs shortest path problem in weighted directed graphs (see [1, Section 5.9]). We can show that the first phase involves funny matrix multiplication. Using the current best algorithm for funny matrix multiplication by Han and Takaoka [10], we have the following result.

Theorem 6.2. QUAD DECOMPOSITION *can be solved in $O(n^7 \log \log n / \log^2 n)$ time.*

7   Concluding Remarks

BASELINE LOCATION and related problems are studied as formulations of image segmentation problems. However, in this paper, we focused on their theoretical aspects and studied their computational complexity. We believe that these problems can arise in practical settings. Experimental results of k-baseline MWRP and QUAD DECOMPOSITION for image segmentation can be found in [3,4].

References

1. Aho, A.V., Hopcroft, J.E., Ullman, J.D.: The Design and Analysis of Computer Algorithms. Addison-Wesley (1974)
2. Asano, T., Chen, D.Z., Katoh, N., Tokuyama, T.: Efficient algorithms for optimization-based image segmentation. Internat. J. Comput. Geom. Appl. 11, 145–166 (2001)
3. Chun, J., Horiyama, T., Ito, T., Kaothanthong, N., Ono, H., Otachi, Y., Tokuyama, T., Uehara, R., Uno, T.: Algorithms for computing optimal image segmentation using quadtree decomposition. To appear in TJJCCGG 2012 (2012)
4. Chun, J., Kaothanthong, N., Kasai, R., Korman, M., Nöllenburg, M., Tokuyama, T.: Algorithms for computing the maximum weight region decomposable into elementary shapes. Comput. Vis. Image Und. 116, 803–814 (2012)
5. Chun, J., Kasai, R., Korman, M., Tokuyama, T.: Algorithms for Computing the Maximum Weight Region Decomposable into Elementary Shapes. In: Dong, Y., Du, D.-Z., Ibarra, O. (eds.) ISAAC 2009. LNCS, vol. 5878, pp. 1166–1174. Springer, Heidelberg (2009)
6. Felzenszwalb, P.F., Huttenlocher, D.P.: Efficient graph-based image segmentation. Int. J. Comput. Vis. 59, 167–181 (2004)
7. Garey, M.R., Johnson, D.S.: Computers and Intractability: A Guide to the Theory of NP-Completeness. Freeman (1979)
8. Garey, M.R., Johnson, D.S., Stockmeyer, L.: Some simplified NP-complete graph problems. Theoret. Comput. Sci. 1, 237–267 (1976)
9. Gibson, M., Han, D., Sonka, M., Wu, X.: Maximum Weight Digital Regions Decomposable into Digital Star-Shaped Regions. In: Asano, T., Nakano, S.-I., Okamoto, Y., Watanabe, O. (eds.) ISAAC 2011. LNCS, vol. 7074, pp. 724–733. Springer, Heidelberg (2011)
10. Han, Y., Takaoka, T.: An $O(n^3 \log \log n / \log^2 n)$ Time Algorithm for All Pairs Shortest Paths. In: Fomin, F.V., Kaski, P. (eds.) SWAT 2012. LNCS, vol. 7357, pp. 131–141. Springer, Heidelberg (2012)
11. Hochbaum, D.S.: Efficient bounds for the stable set, vertex cover and set packing problems. Discrete Appl. Math. 6, 243–254 (1983)
12. Otten, R.H.J.M., van Wijk, J.G.: Graph representations in interactive layout design. In: IEEE Internat. Symp. on Circuits and Systems, pp. 914–918 (1978)
13. Wang, J.J., He, X.: Compact visibility representation of plane graphs. In: 28th International Symposium on Theoretical Aspects of Computer Science, STACS 2011. LIPIcs, vol. 9, pp. 141–152 (2011)

Counting Maximal Points in a Query Orthogonal Rectangle[*]

Ananda Swarup Das[1], Prosenjit Gupta[2], and Kannan Srinathan[1]

[1] International Institute of Information Technology, Hyderabad, India
[2] Heritage Institute of Technology, Kolkata, India
anandaswarup@gmail.com, prosenjit_gupta@acm.org,
srinathan@iiit.ac.in

Abstract. In this work, we propose a solution with sub-logarithmic query time for counting the number of maximal points in an axis parallel query rectangle. The problem has been previously studied in [3] and [5]. To the best of our knowledge, this is the first sub-logarithmic query time solution for the problem. Our model of computation is the word RAM with word size of $\Theta(\log n)$ bits.

1 Introduction

Range Searching is one of the most widely studied topics in computational geometry. An orthogonal range query is defined as: Given a set P of geometrical objects (lines, points etc), efficiently report all the objects that are intersecting the query rectangle q. However, in today's world of information explosion, the definition may itself be not sufficient. An end-user in today's world is often more willing to see the interesting (meaningful) results instead of the entire result set. In literature, aggregate functions like *sum of weights, count, maximum weight, minimum weight, average weight, top k weighted points* have been used to find interesting results in a query rectangle. In this work, we use the notion of counting *skyline points* as our aggregate function to find the interesting points. In computational geometry, the skyline points are nothing but the set of maximal points for a data set. A point p is said to be dominating point p' if $p_x \geq p'_x$ and $p_y \geq p'_y$. Given a point set R, the set of maximal points R' is the set of points which are not dominated by any other points in the set R.

In this work, we study the problem of counting the number of maximal points in a query rectangle. Our model of computation is the word-RAM (see [7,9]) with word size of $\Theta(\log n)$ bits.

2 Problem Definition

In this work, we provide an efficient solution for the following problem.

---

[*] A preliminary version of the paper was accepted in Canadian Conference on Computational Geometry 2012 but was not presented in the venue. Consequently, the paper has been removed from the conference program and the official proceedings.

S.K. Ghosh and T. Tokuyama (Eds.): WALCOM 2013, LNCS 7748, pp. 65–76, 2013.

Problem 1. *Given a set R of n points on an $n \times n$ grid, that is $R = \{(x_i, y_i) | x_i \in \{1, n\}, \ y_i \in \{1, n\}\}$, preprocess them into a data structure such that given an orthogonal query rectangle q, the count of maximal points in $q \cap R$ can be reported efficiently.*

We assume all the points to have distinct x and y-coordinates. In the word RAM model, Das et al. [3] presented a data structure for the above problem that requires a storage space of $O(n \frac{\log^2 n}{\log \log n})$ and can be queried in time $O(\frac{\log^{\frac{3}{2}} n}{\log \log n})$. In the comparison model, Kalavagattu et al. [5] gave a data structure for the same problem that requires a storage space of $O(n \log n)$ and can be queried in time $O(\log n)$. Efficient solutions for the reporting version of the problem have been proposed in [1,3]. In this paper we manage to obtain the first solution with sub-logarithmic query time counting algorithm for the problem.

Hereafter we denote by $A[i, j]$, the elements of the array A from position i to position j.

2.1   The Solution Sketch

Consider an instance of the two-level binary search tree of [6]. We denote the tree as T_x. Given a query rectangle $q = [a, b] \times [c, d]$, the segment $[a, b]$ can be allocated to at most $O(\log n)$ canonical nodes of the tree T_x. These nodes can be ordered in order of their position from right to left in the tree T_x. We visit these canonical nodes from the right to the left. In the course of our visit, if we can count in $O(1)$ time, the number of maximal points in q coming from the subtree rooted at the currently visited canonical node, then in $O(\log n)$ time, we can count all the maximal points in q.

To achieve a sub-logarithmic query time counting algorithm, we increase the internal degree of each internal node of T_x from two to $\sqrt{\log n}$, thereby decreasing the height of the tree to $O(\frac{\log n}{\log \log n})$. Given a query rectangle $[a, b] \times [c, d]$, allocate the segment $[a, b]$ to a node $\mu \in T_x$ if $int(\mu) \subseteq [a, b]$ but $int(parent(\mu)) \nsubseteq [a, b]$. As stated in [7], the set of all such canonical nodes can be grouped into $O(\frac{\log n}{\log \log n})$ groups with each group containing some children $v_l \ldots v_k$ for some node v. We call the node v, a group leader for the group $g(i)$ and denote it as $GL(i)$. Let $\{GL(1) \ldots, GL(O(\frac{\log n}{\log \log n}))\}$ be the set of group leaders arranged in order of their positions from right to left in the modified tree T_x. If we visit each group leader node $GL(i)$ from right to left and in the process count in $O(1)$ time the number of maximal points in q coming from the nodes of the group $g(i)$ for which $GL(i)$ is the group leader, we can count the maximal points in q in $O(\frac{\log n}{\log \log n})$. What follows next in the rest of the paper are the details of how to materialize the idea.

3   Subproblems

In order to solve Problem 1, we will need solutions for the following subproblems.

Problem 2. *Given a set S of n points from a universe of $[1, \log^\rho n] \times [1, n]$ for $0 < \rho \le \frac{1}{2}$, preprocess the points into a data structure such that given an axis parallel rectangle $q = [a, b] \times [c, d]$ for $(a, b) \in [1, \log^\rho n] \times [1, \log^\rho n]$ and $(c, d) \in [1, n] \times [1, n]$, we can efficiently count the number of maximal points in $S \cap q$.*

In the complete solution, an efficient technique for the above problem will help us to count in $O(1)$ time the number of maximal points coming from the nodes $\{v_i \ldots, v_j\}$ of the group $g(i)$ in a given query rectangle.

Problem 3. *Given a sorted array A which is a union of $\sqrt{\log n}$ distinct sorted sub-arrays, each of size $\frac{n}{\log n}$ (that is $A = A_0 \cup A_1 \cup A_2 \cup \ldots \cup A_{\sqrt{\log n} - 1}$), preprocess A into a data structure such that given two indices $[i, j]$ for the array A and two integer values $a, b \in [0, \sqrt{\log n} - 1]$, we can efficiently find the the smallest value t such that $a \le t \le b$ and $A_t[1, |A_t|] \cap A[i, j] \ne \emptyset$.*

The solution for the above problem will help us to effectively emulate fractional cascading in the modified tree T_x with each internal node having a degree of $O(\sqrt{\log n})$.

3.1   Solution for Problem 2

Preprocessing

Let all n points have distinct y-coordinates. We sort the points in decreasing order of their y-coordinates in an array A.

1. Construct a height-balanced binary search tree T_y whose leaf nodes store the values in the array A and are at the same level. At each internal node $m \in T_y$, store a key value y_m which is the median of the points stored in the subtree rooted at m.
2. For each pair of possible points $(i_1, i_2) \in [1, \log^\rho n] \times [1, \log^\rho n]$, form an interval $[i_1, i_2]$ (including $i_1 = i_2$).
3. For each value $y_2 \in A$ and for each possible interval $[i_1, i_2]$, form 3-sided anchored rectangles $[i_1, i_2] \times [y_2, \infty)$ and $[i_1, i_2] \times (-\infty, y_2]$. Next, for the interval $[i_1, i_2]$ and the value y_2, do the following:
   (a) visit the ancestors of y_2 in the tree T_y. At each ancestor m, find the value y_m stored as the key value.
   (b) if $y_m < y_2$, form an axis-parallel rectangle $R_1 = [i_1, i_2] \times [y_m, y_2]$.
      i. For the points in $S \cap R_1$, compute the subset of points that are not dominated by any other point. We call such a subset a maximal chain. Next, we denote by p_{ymax} and p_{xmax} respectively, the topmost point and the bottommost point of the maximal chain.
      ii. Count the number of maximal points in the chain p_{xmax} to p_{ymax}. We denote the value as $|p_{xmax}, p_{ymax}|$. Store the value in a variable denoted by $count_m(y_2)$.
      iii. Next, for the point p_{xmax}, find the topmost point p_{nodom} in $[i_1, i_2] \times (-\infty, y_m]$ such that p_{nodom} is not dominated by p_{xmax}.

  iv. Create a tuple $\langle count_m(y_2), p_{nodom} \rangle$ and store it with reference to the rectangle R_1 in a hash table. Here the suffix m denotes the index for the node $m \in T_y$ that is an ancestor of the leaf node storing the value y_2.

  v. **Special Cases**

   A. If no points are present in the rectangle R_1, store $\langle 0, NULL \rangle$.

   B. If the point p_{nodom} does not exist, store $\langle count_m(y_2), NULL \rangle$.

 (c) On the other hand, if $y_m > y_2$, form a rectangle $R_2 = [i_1, i_2] \times [y_2, y_m]$ and then find the topmost point p'_{ymax} and the bottommost point p'_{xmax} in the maximal chain for the points in $S \cap R_2$. Count the number of maximal points in the chain from p'_{xmax} to p'_{ymax} that is count $|p'_{xmax}, p'_{ymax}|$. Store it in a tuple $\langle count'_m(y_1), p'_{xmax} \rangle$.

 (d) **Special Case**

  i. If there are no points in R_2, store $\langle 0, NULL \rangle$.

4. Maintain the data structure of [4] to find the least common ancestor for two given leaf nodes of the tree T_y in $O(1)$ time.

Lemma 1. *The total storage space needed by the data structure is $O(n \log^{1+2\rho} n)$ words.*

Query Algorithm

1. Given a query rectangle $[a, b] \times [c, d]$ such that $(a, b) \in [1, \log^\rho n] \times [1, \log^\rho n]$ for $0 < \rho \leq \frac{1}{2}$ and $(c, d) \in [1, n] \times [1, n]$, find the least common ancestor m for the values d, c in the tree T_y. Let the key value stored at m be y_m.

2. Consider the rectangle $[a, b] \times [y_m, d]$ and the corresponding tuple $\langle count_m(d), p_{nodom} \rangle$. The suffix m has the same meaning as specified in step 3(b)iv of the preprocessing algorithm.

3. Let $p_{nodom}(y)$ be the y-coordinate of the point p_{nodom}. If $p_{nodom} \neq NULL$ and $c \leq p_{nodom}(y)$, we do the following:

 (a) Consider the rectangle $[a, b] \times [p_{nodom}(y), y_m]$. Choose the value $count'_m(p_{nodom}(y))$ from the corresponding tuple for the rectangle $[a, b] \times [p_{nodom}(y), y_m]$.

 (b) For the rectangle $[a, b] \times [c, y_m]$ choose the value $count'_m(c)$.

 (c) Return $count'_m(c) - count'_m(p_{nodom}(y)) + 1 + count_m(d)$.

4. If $p_{nodom} \neq NULL$ but $p_{nodom}(y) < c$, then

 (a) Return the value $count_m(d)$.

5. **Special Cases**

 (a) If $count'_m(c) = 0$, return $count_m(d)$.

 (b) If $count_m(d) = 0$, return $count'_m(c)$.

 (c) If $count'_m(c) \neq 0$ but $p_{nodom} = NULL$, return $count_m(d)$.

3.2   Analysis of the Query Algorithm

We divide the analysis of the query algorithm into three cases:

Case 1: $p_{nodom} \neq NULL$ and $c \leq p_{nodom}(y) < d$.

Lemma 2. *If $count'_m(c) \neq 0$, then the point p_{nodom} belongs to the subtree rooted at $m \in T_y$.*

Proof. Since $c \leq p_{nodom}(y) \leq d$ and m is the least common ancestor of c and d in T_y, the point p_{nodom} belongs to the subtree rooted at $m \in T_y$.   □

Lemma 3. *If $count'_m(c) \neq 0$, then the maximal chain in the rectangle $[a, b] \times [c, y_m]$ will pass through the point p_{nodom}.*

Proof. As the point p_{nodom} is the topmost point below y_m and above c and is not dominated by the point p_{xmax} in $[a, b] \times [y_m, d]$, the point cannot be dominated by any other point in the rectangle in $[a, b] \times [c, y_m]$. Else, the point dominating p_{nodom} would have been the topmost point not dominated by p_{xmax}.   □

Lemma 4. *The number of maximal points inside the rectangle $[a, b] \times [c, d]$ is equal to the number of maximal points in the rectangle $[a, b] \times [c, p_{nodom}(y)]$ plus the number of maximal points in $[a, b] \times [p_{xmax}(y), d]$.*

Proof. As the point p_{nodom} is not dominated by p_{xmax}, $p_{nodom}(x) > p_{xmax}(x)$. Here $p_{nodom}(x)$ and $p_{xmax}(x)$ are the x-coordinates of the points p_{nodom} and p_{xmax} respectively. Any point in the maximal chain inside the rectangle $[a, b] \times [c, p_{nodom}(y)]$ must have an x- coordinate greater than $p_{nodom}(x)$, otherwise the point will be dominated by p_{nodom}. Also, it should be noted that any point in the maximal chain in the rectangle $[a, b] \times [p_{xmax}(y), d]$ is not dominated by any other point in the rectangle $[a, b] \times [p_{xmax}(y), d]$ or $[a, b] \times [c, p_{xmax}(y)]$.   □

Lemma 5. *See Figure 1. Let p'_{xmax} and p'_{ymax} be respectively the two points with maximum x and y-coordinates in the rectangle $[a, b] \times [c, y_m]$. Let $|p'_{xmax}, p_{nodom}|$ denote the number of maximal points between p'_{xmax} and p_{nodom} (including p'_{xmax} and p_{nodom}) inside the the rectangle $[a, b] \times [c, y_m]$. Then $|p'_{xmax}, p_{nodom}| = |p'_{xmax}, p'_{ymax}| - |p'_{ymax}, p_{nodom}| + 1$.*

Proof. By Lemma 3, the point p_{nodom} is a point in the maximal chain from p'_{xmax} to p'_{ymax}. As the rectangle $[a, b] \times [p_{nodom}(y), y_m]$ is contained in $[a, b] \times [c, y_m]$, any maximal point in $[a, b] \times [p_{nodom}(y), y_m]$ is also a maximal point in $[a, b] \times [c, y_m]$. Therefore, $|p'_{xmax}, p_{nodom}| = |p'_{xmax}, p'_{ymax}| - |p'_{ymax}, p_{nodom}| + 1$.   □

Case 2: $p_{nodom} \neq NULL$ but $p_{nodom}(y) < c$.

Lemma 6. *The number of maximal points inside the rectangle $[a, b] \times [c, d]$ is equal to the number of maximal points in the rectangle $[a, b] \times [y_m, d]$.*

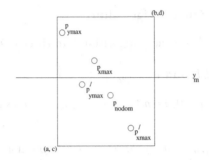

Fig. 1. The rectangle $[a, b] \times [c, d]$ is split into two parts (a) $[a, b] \times [y_m, d]$ and (b) $[a, b] \times [c, y_m]$. The points p_{xmax}, p_{ymay} (respectively p'_{xmax}, p'_{ymay}) are the bottommost and the topmost points of the maximal chain in $[a, b] \times [y_m, d]$ (respectively $[a, b] \times [c, y_m]$). The point p_{nodom} is the topmost point in the anchored rectangle $[a, b] \times (-\infty, y_m)$ which is not dominated by p_{xmax}.

Proof. As p_{nodom} is the topmost point below y_m and not dominated by p_{xmax}, the case of $p_{nodom}(y) < c$ is possible only if (a) there are no points in the rectangle $[a, b] \times [c, y_m]$ or (b) any point in the rectangle $[a, b] \times [c, y_m]$ is dominated by the point p_{xmax}. In any case, no maximal point for the rectangle $[a, b] \times [c, d]$ is contained in $[a, b] \times [c, y_m]$.   □

Case 3: Special Cases.

Lemma 7. *If $count'_m(c) = 0$, then there are no points in $S \cap [a, b] \times [c, y_m]$.*

Lemma 8. *If $count_m(d) = 0$, then there are no points in $S \cap [a, b] \times [y_m, d]$.*

Lemma 9. *If $count'_m(c) \neq 0$ but $p_{nodom} = NULL$, then any point below y_m in $[a, b] \times (-\infty, y_m]$ is dominated by p_{xmax}.*

Query Time

Lemma 10. *The query algorithm takes $O(1)$ time to count the number of maximal points in a query rectangle.*

Proof. As all n points have distinct y-coordinates and the y-coordinates of the points are in the range $[1, n]$, the values c, d is present in array A. Thus, locating the indices (as well as the leaf nodes) storing these two values can be done in $O(1)$ time. Finding the least common ancestor m can also be done in $O(1)$ time. Thus, we are left with is to find the respective tuples for the rectangles $[i_1, i_2] \times [c, y_m]$ and $[i_1, i_2] \times [y_m, d]$ and then counting the maximal points in the rectangle $[i_1, i_2] \times [c, d]$. All these operations can be done in $O(1)$ time.   □

By Lemmas 1 and 10, we conclude the following.

Theorem 1. *Given a set S of n points from a universe of $[1, \log^\rho n] \times [1, n]$ for $0 < \rho \leq \frac{1}{2}$, we can preprocess the points into a data structure of size $O(n \log^{1+2\rho} n)$ words, such that given an axis parallel rectangle $q = [a, b] \times [c, d]$ for $(a, b) \in [1, \log^\rho n] \times [1, \log^\rho n]$ and $(c, d) \in [1, n] \times [1, n]$, we can count the number of maximal points in $S \cap q$ in $O(1)$ time.*

3.3 Solution for Problem 3

The problem was also studied in [3]. However, the solution we present here is much simpler compared to the previous solution.

Preprocessing

1. Given the array A, for an element y stored in the index i, $i = 1, \ldots, |A|$, create a 2-d point (i, k) if and only if $y \in A \cap A_k$ (note that all the arrays are distinct).
2. Store these points in a linear-space data structure RS supporting range successor queries (see [9]) of the form $[x_1, x_2] \times [y_1, \infty)$. Given a set S of 2-d points on a grid of size $n \times n$ and a query $q = [x_1, x_2] \times [y_1, \infty)$, a range successor query returns the point with the smallest y-coordinate in $S \cap q$. It is known from [9] that a range successor query can be answered in $O(1)$ time if the y-coordinates of the points are in the range $[1, O(\frac{\log n}{(\log \log n)^2})]$. In our case, the points (i, k) belong to a grid of size $n \times \sqrt{\log n}$.

Query Algorithm

1. On getting the two values a, b where $(a, b) \in \sqrt{\log n} \times \sqrt{\log n}$ and the two indices i, j of the array A where $(i, j) \in n \times n$, we search the data structure RS with the query rectangle $[i, j] \times [a, \infty)$ and in return, we get a value t in $O(1)$ time.
2. If $t \leq b$, we return t, otherwise we return $NULL$.

Thus, we conclude the following.

Theorem 2. *Given a sorted array A which is a union of $\sqrt{\log n}$ sorted subarrays (that is $A = A_0 \cup A_1 \ldots \cup A_{\sqrt{\log n}-1}$) , we can preprocess A into a data structure of size $O(|A|)$ words such that given two indices i, j and two integer values $a, b \in [0, \sqrt{\log n} - 1]$, we can find the smallest tag t such that $a \leq t \leq b$ and $A_t[1, |A_t|] \cap A[i, j] \neq \emptyset$ in $O(1)$ time.*

4 Solution for Problem 1

Preprocessing

1. Construct a tree T_x, the leaf nodes of which are at the same level (height). The leaf nodes of the tree store the x-coordinates of the points in the set R in non-decreasing order of their values.
2. Each internal node $\mu \in T_x$ has $O(\sqrt{\log n})$ children, the left most child being numbered as $\sqrt{\log n} - 1$ while the right most child being numbered as 0. Each internal node $\mu \in T_x$ is assigned an interval $int(\mu)$ which is equal to the union of the discrete intervals induced on the x-axis by the values stored at the leaf nodes of the subtree rooted at μ.
3. Next, the following arrangement has to be done for all the internal nodes of the tree T_x except the root.

(a) Each internal node μ has an auxiliary array A_μ which stores the y-coordinates of the points, the x-coordinates of which are present in the leaf nodes of the subtree rooted at μ. Thus, $A_\mu = \bigcup A_i : i = 0 \ldots \sqrt{\log n} - 1$. Here, A_i is the auxiliary array for the node v_i which is a child μ. A_μ is sorted in non-increasing order of its values.

(b) Each element of $A_i, i = 0, \ldots, \sqrt{\log n} - 1$ will point to its corresponding position in the array A_μ.

(c) However, there will be no pointers from the elements of the array A_μ to the elements in the arrays of its children. Rather, the array A_μ will be preprocessed into a data structure $D_{\mu(1)}$ which is an instance of the data structure of Theorem 2. While constructing the data structure $D_{\mu(1)}$, we have to perform the following step:

  i. For the value y_j stored in $A_\mu[j]$, create a 2-d point (v_i, y_j) provided the value y_j belongs to the array A_i, the auxiliary array associated with the child v_i of the node μ. It should be noted that v_i is a value in the range $[0, \sqrt{\log n} - 1]$. Thus, we have a set of points from a grid of $[0, \sqrt{\log n} - 1] \times [1, n]$.

(d) Each child v_i of μ maintains a binary string denoted by *lookup* of size $|A_\mu|$. The zth most significant bit of the string *lookup* is set to one if the element stored in $A_\mu[z]$ belongs to the array A_i, the auxiliary array associated with the node v_i.

(e) The string *lookup* should support *rank*() and *select*() queries of [2]. A similar binary string is also maintained in the data structure of [3].

(f) Maintain RM_{A_μ}, a range maxima data structure of [10] such that given two indices i, j of the array A_μ, we can return the maximum x-coordinate for the points whose y-coordinates are stored between $A_\mu[i]$ to $A_\mu[j]$.

(g) For the values of the array A_μ, construct the following two auxiliary trees at the node μ.

  i. VT_μ which is an instance of the van Emde Boas tree [8].

  ii. A height balanced binary search tree $T_{\mu,y}$. Any node $\phi \in T_{\mu,y}$ stores the median of the values stored in the leaf nodes of its subtree.

4. For the root of T_x, we do the following

(a) Each index i of the auxiliary array A_{root} will have $2\sqrt{\log n}$ pointers of which $\sqrt{\log n}$ pointers will be pointing to the smallest elements greater than $A_{root}[i]$ in each of the arrays A_j, the auxiliary arrays associated with the nodes v_j which are the children of the root node.

(b) Similarly the other $\sqrt{\log n}$ pointers will be pointing to the largest elements greater than $A_{root}[i]$ in the arrays A_j.

(c) Construct a range maxima data structure $RM_{A_{root}}$ such that given two indices i, j of A_{root}, we can return the maximum x-coordinate among the points whose y-coordinates are stored between $A_{root}[i]$ to $A_{root}[j]$.

5. Construct an instance of the data structure of Theorem 1 at each internal node $\mu \in T_x$ for the points, the x-coordinates of which are in the subtree of μ. This is done as follows:

(a) While considering a horizontal interval $[i, j]$ (including $i = j$), we consider all the points present in the subtrees rooted at the children $v_i, \ldots, v_j$ of the node μ. Denote such a set as S'.

(b) Next, for each element $y \in A_\mu$ and each value y_m stored in the ancestors for the value y in the tree $T_{\mu,y}$ at node μ (see step 3(g)ii), we form a rectangle $R_1 = [i,j] \times [y, y_m]$, assuming $y_m > y$ (respectively $R_1 = [i,j] \times [y_m, y]$, if $y_m < y$).

(c) We then find

    i. The topmost and the bottommost point of the maximal chain in $R_1 \cap S'$ Denote these points as (p_{top}, p_{bottom})

    ii. The count of the maximal points in $R_1 \cap S'$.

(d) This information is stored in a tuple $Tuple_{R_1} = \langle p_{top}, p_{bottom}, count \rangle$ for the rectangle R_1.

Lemma 11. *The total storage space needed by the data structure for the counting problem is* $O(n \frac{\log^3 n}{\log \log n})$.

Query Algorithm

Before considering our counting algorithm, let us take a look of the decomposition of the query rectangle.

Decomposition of the Rectangle

1. Given a query rectangle $[a,b] \times [c,d]$, the segment $[a,b]$ is allocated to a node $\mu \in T_x$ if $int(\mu) \subseteq [a,b]$ but $int(parent(\mu)) \not\subseteq [a,b]$. There will be $O(\frac{\log^{\frac{3}{2}} n}{\log \log n})$ such nodes. Denote the set of such nodes as V.

2. As stated in [7], the set of all such canonical nodes can be grouped into $O(\frac{\log n}{\log \log n})$ sets with each set $g(i)$ containing some children $v_l \ldots v_k$ for some node v. We will refer the node v as *group leader* $GL(i)$. Let $G = \{GL(1), \ldots, GL(O(\frac{\log n}{\log \log n}))\}$ be the set of group leaders stored in order of their positions from right to left in the tree T_x.

3. For counting the maximal points in the rectangle $[a,b] \times [c,d]$, we decompose the rectangle into $O(\frac{\log n}{\log \log n})$ smaller rectangles. The rightmost rectangle is denoted as R_1 while the leftmost rectangle is denoted by R_z for $z = O(\frac{\log n}{\log \log n})$. All these rectangles have the same height. However, the width of the rectangles are defined as follows:

(a) Consider the set V of nodes to which the segment $[a,b]$ is allocated. Let $v_i \in V$ be the rightmost node in the tree T_x among all the nodes in V. Then:

    i. Consider the group leader $GL(1)$ for which the node v_i is a child. Remember that the children of $GL(1)$ to which $[a,b]$ is allocated are grouped in a set $g(1)$.

    ii. The horizontal interval for the rectangle R_1 is equal to $\bigcup int(v_j)$ provided

        A. v_j is a child of $GL(1)$ and

        B. $int(v_j) \subset [a,b]$. In other words, $v_j \in g(1)$.

(b) Next, consider the reduced set $V' = V - g(1)$ for which let v_k be the rightmost node. Consider the group leader $GL(2)$ and the corresponding set $g(2)$ for the node v_k. The width of the rectangle R_2 is the union of the intervals of the nodes in $g(2)$.

(c) Similarly, we can define the widths of all the rectangles.

Counting

1. Find the least common ancestor (lca) for the leaf nodes storing the values a, b in the tree T_x. Visit the node lca and search the auxiliary array A_{lca} to find the indices i, j such that $A_{lca}[i]$ has the smallest value $\geq c$ and $A_{lca}[j]$ has the largest value $\leq d$. This can be done by searching the van Emde Boas tree maintained at the node lca.

2. Let the leader node $GL(1)$ be in the subtree rooted at the child node v_m for the node lca.

3. Consider the binary string $lookup$ for the node v_m. Then, compute the number of $ones$ present in the string $lookup$ till $(i-1)^{th}$ most significant bit (msb). This can be done by using the $rank$ operation of [2]. Let there be t $ones$ till $(i-1)^{th}$ msb of the string $lookup$. Then, find the position z of the $(t+1)^{th}$ one in the string $lookup$ by using $select$ operation of [2]. The largest value smaller than d in A_{v_m} is stored at the index z. In the similar fashion, we can find the largest element $y_2' \leq d$ and the smallest element $y_1' \geq c$ in $A_{GL(1)}$.

4. For the node $GL(1)$, let v_i be its rightmost child to which the segment $[a, b]$ is allocated. Then, all the children of the node $GL(1)$ starting from v_i to $v_{\sqrt{\log n}-1}$ are in the set $g(1)$. This is because,
   (a) the children of $GL(1)$ are numbered from 0 to $\sqrt{\log n}-1$ in order of their positions from right to left and
   (b) the least common ancestor lca is an ancestor for $GL(1)$.

5. Next, we search the data structure $D_{GL(1)}$ which is an instance of the data structure of Theorem 2 to find the smallest index $m \in [i, \sqrt{\log n}-1]$ such that the node v_m has the rightmost maximal point in the rectangle $[a, b] \times [y_1', y_2']$. To find the tag m, we run the range successor query for $[i, \infty) \times [y_1', y_2']$. As $i \in [0, \sqrt{\log n}-1]$, the range successor query can be performed in $O(1)$ time (See [9]).

6. We then, count the number of maximal points in the rectangle $[m, \sqrt{\log n}-1] \times [y_1', y_2']$ by searching the instance of the data structure of Theorem 1 (see step 5 of preprocessing) at node $GL(1)$. Also, we find $p_{ymax} = (p_{ymax}(x), p_{ymax}(y))$, the point with the maximum y-coordinate in $[m, \sqrt{\log n}-1] \times [y_1', y_2']$ (see step 5d of preprocessing). Store this count in a variable $Total_1$.

7. Next, we move to the leader node $GL(2)$. Let the group $g(2)$ be $\{v_w, v_{w+1}, \ldots, v_{q-1}, v_q\}$. Then, at the node $GL(2)$, we will count the number of maximal points in the rectangle $[w, q] \times [p_{ymax}(y), d]$. Store the count in a variable $Total_2$.

8. It should be noticed that the node $GL(2)$ will satisfy one of the following three conditions.
   (a) $GL(2)$ is in the path from $GL(1)$ to lca.
   (b) $GL(2)$ is in the path from $GL(z)$ to lca.
   (c) $GL(2)$ is the lca.

9. Let $GL(2)$ be present in the path from $GL(1)$ to lca. Notice that any element in $A_{GL(1)}$ has a pointer to its corresponding position in the auxiliary array attached to the parent of $GL(1)$. Thus, finding $y_2' \leq d$ in $A_{GL(2)}$ can be done easily by following pointers. Similarly, the first element greater than $p_{ymax}(y)$ can also be found in $p_{ymax}(y)$.

10. If $GL(2)$ is in the path from lca to $GL(z)$, then, we first move to the node lca by following pointers from $GL(1)$ and then by using the techniques as described in step 3, we descend to $GL(2)$.

11. Finally, if $GL(2)$ is the lca, we move to the node lca by following pointers from $GL(1)$.

12. We repeat similar steps until we have visited all the *group leader* nodes.

13. At the end of visiting all the *group leader* nodes, return $Total = \sum Total_i$: $i = 1, \ldots, O(\frac{\log n}{\log \log n})$.

5   Query Time Analysis and Correctness Proof

Lemma 12. *Our query algorithm correctly counts the number of maximal points inside the query rectangle.*

Proof. We decompose the query rectangle into $O(\frac{\log n}{\log \log n})$ smaller rectangles. The rightmost rectangle is R_1 whereas the leftmost rectangle is R_z for $z = O(\frac{\log n}{\log \log n})$. We start our search from the rightmost rectangle R_1 and then, sequentially visit the rectangles from right to left. Let $p_{ymax} = (p_{ymax}(x), p_{ymax}(y))$ be the topmost point of the rectangle R_1. The point is sure to be in the maximal chain as all points in the rectangles $R_2 \ldots R_z$ have x-coordinates less than $p_{ymax}(x)$. Thus, for the rectangle R_2, while counting the number of maximal points, we consider points with y-coordinates in the range $[p_{ymax}(y), d]$. The process is continued till we visited all the rectangles. In each individual rectangle R_i, the counting is equivalent to counting maximal points in a narrow query rectangle $[x_1, x_2] \times [y_1, y_2]$, where $(x_1, x_2) \in [\sqrt{\log n} \times \sqrt{\log n}]$ and $(y_1, y_2) \in n \times n$. By Theorem 1, we can correctly count the number of maximal points in a query rectangle $[x_1, x_2] \times [y_1, y_2]$ in $O(1)$ time, if $(x_1, x_2) \in [\sqrt{\log n} \times \sqrt{\log n}]$ and $(y_1, y_2) \in n \times n$. We repeat similar steps until all the decomposed rectangles are visited. The final answer that is returned is the sum of the counts of the maximal points in each of the visited rectangles.   □

Lemma 13. *The query algorithm takes $O(\frac{\log n}{\log \log n})$ time to count the number of maximal points in the query rectangle.*

Proof. By Theorem 1, we know that counting the number of maximal points in a query rectangle $[i, j] \times [c, d]$ for $(i, j) \in [1, \log^\rho n] \times [1, \log^\rho n]$ for $\rho \leq \frac{1}{2}$ can be done in $O(1)$ time. Thus, if we can show that all the group leaders can be visited in $O(\frac{\log n}{\log \log n})$ time, we are done. There are two possible scenarios. (a) The first scenario is where the lca is not a *group leader*. We start our counting from the node $GL(1)$. The next node $GL(2)$ will be either in the path from $GL(1)$ to lca or in the path from lca to $GL(z)$ for $z = O(\frac{\log n}{\log \log n})$. Let $GL(2)$ be in the path from $GL(1)$ to lca. Once we have discovered the point p_{ymax} with maximum

y-coordinate for the rectangle R_1 at node $G(1)$, finding the position of the element $p_{ymax}(y)$ in the array $A_{GL(2)}$ is easy as there are pointers from the elements of the array $A_{GL(1)}$ to their corresponding positions at $A_{parent(GL(1))}$(step 9). On the other hand, if $GL(2)$ is in the path from lca to $G(z)$ for $z = O(\frac{\log n}{\log \log n})$, finding the smallest value greater than or equal to $p_{ymax}(y)$ could be done easily by using step 10 at the nodes in the path. Thus, all the group nodes in the path from $GL(1)$ to lca as well as from lca to $GL(z)$ for $z = O(\frac{\log n}{\log \log n})$ can be visited in $O(\frac{\log n}{\log \log n})$ time as the length of any path is $O(\frac{\log n}{\log \log n})$; (b) The second scenario is where lca is the *group leader* $GL(2)$. In that case, we can move from $GL(1)$ to $GL(2)$ by following the pointers. It should be noticed that no path is visited more than once. If the lca is not a *group leader*, we carry out search operations on the *group leaders* on the two paths. On the other hand, if the lca is a *group leader*, we need to search one additional *group leader* along with the ones on the two paths. The length of any path is $O(\frac{\log n}{\log \log n})$.   □

By Lemmas 11 and 13, we conclude the following.

Theorem 3. *Given a set S of n points from a universe of $n \times n$ where all the points have distinct x and y-coordinates, we can preprocess the points into a data structure of size $O(n \frac{\log^3 n}{\log \log n})$ words, such that given an axis parallel rectangle $q = [a, b] \times [c, d]$, we can count the number of maximal points in $S \cap q$ in time $O(\frac{\log n}{\log \log n})$.*

References

1. Brodal, G.S., Tsakalidis, K.: Dynamic Planar Range Maxima Queries. In: Aceto, L., Henzinger, M., Sgall, J. (eds.) ICALP 2011. LNCS, vol. 6755, pp. 256–267. Springer, Heidelberg (2011)
2. Clark, D.R., Munro, J.I.: Efficient suffix trees on secondary storage (extended abstract). In: SODA, pp. 383–391 (1996)
3. Das, A.S., Gupta, P., Kalavagattu, A.K., Agarwal, J., Srinathan, K., Kothapalli, K.: Range Aggregate Maximal Points in the Plane. In: Rahman, M. S., Nakano, S.-I. (eds.) WALCOM 2012. LNCS, vol. 7157, pp. 52–63. Springer, Heidelberg (2012)
4. Harel, D., Tarjan, R.E.: Fast algorithms for finding nearest common ancestors. SIAM J. Comput. 13(2), 338–355 (1984)
5. Kalavagattu, A.K., Agarwal, J., Das, A.S., Kothapalli, K.: On Counting Range Maxima Points in Plane. In: Arumugam, S., Smyth, B. (eds.) IWOCA 2012. LNCS, vol. 7643, pp. 263–273. Springer, Heidelberg (2012)
6. Kalavagattu, A.K., Das, A.S., Kothapalli, K., Srinathan, K.: On finding skyline points for range queries in plane. In: CCCG (2011)
7. Nekrich, Y.: A linear space data structure for orthogonal range reporting and emptiness queries. Int. J. Comput. Geometry Appl. 19(1), 1–15 (2009)
8. van Emde Boas, P.: Preserving order in a forest in less than logarithmic time. In: FOCS, pp. 75–84 (1975)
9. Yu, C.C., Hon, W.K., Wang, B.F.: Improved data structures for the orthogonal range successor problem. Comput. Geom. 44(3), 148–159 (2011)
10. Yuan, H., Atallah, M.J.: Data structures for range minimum queries in multidimensional arrays. In: SODA, pp. 150–160 (2010)

Voronoi Game on Graphs

Sayan Bandyapadhyay, Aritra Banik, Sandip Das, and Hirak Sarkar

Indian Statistical Institute
Kolkata, India
{sayan.bandyapadhyay,aritrabanik,hiraksarkar.cs}@gmail.com,
sandipdas@isical.ac.in

Abstract. Voronoi game is a geometric model of competitive facility location problem, where each market player comes up with a set of possible locations for placing their facilities. The objective of each player is to maximize the region occupied on the underlying space. In this paper we consider one round *Voronoi game* with two players. Here the underlying space is a road network, which is modeled by a graph embedded on $\mathbb{R}^2$. In this game each of the players places a set of facilities and the underlying graph is subdivided according to the *nearest neighbor rule*. The player which dominates the maximum region of the graph wins. Given a placement of facilities by Player 1, we have characterized the optimal placement by Player 2. At first we dealt with the case when Player 2 places a constant number of facilities and provided an algorithm for the same. Next we have proved that finding the optimal placement of k facilities by Player 2 is $\mathcal{NP}$-hard where k is given. Lastly we presented a 1.58 factor approximation algorithm for the above mentioned problem.

1 Introduction

A situation often arises in market where the competitive service providers (Hotel Chains, Supermarkets etc.) want to occupy a big area in a locality so that they could attract as much customers as possible. The game-theoretic analogue of competitive facility location problem is *Voronoi Game* which was proposed by Ahn et al. [1]. In this game the main objective of a player is to cover maximum area by placing its facilities on the underlying space. A point on the underlying space is always served by its nearest facility. Different versions of this game can be modeled by changing the underlying space like line segment, circular arc, graph and 2D-plane.

Ahn et al. [1] have discussed the case where the game is restricted to 1-dimensional continuous domain. Cheong et al. [3] and Fekete et al. [4] have dealt with 2-dimensional case but for one round. Banik et al. [2] have discussed the one round discrete version of this game on lines. Demaine et al. [8] have dealt with the discrete version of the game on graphs where the users and facilities are constrained to be located on vertices. A special case of this game when the underlying space is a path have considered by Kiyomi et al. [7].

S.K. Ghosh and T. Tokuyama (Eds.): WALCOM 2013, LNCS 7748, pp. 77–88, 2013.

In this paper we consider a game where the underlying space is a road network, described by a graph $G(V, E)$. With each edge $(u, v) \in E$, a positive weight $w(u, v)$ is associated which can be considered as the length of the edge (u, v). Throughout the paper we will assume that an embedding of G on $\mathbb{R}^2$ is given. As any edge (u, v) is having a positive weight $w(u, v)$, we can map it to a closed interval $[0, w(u, v)]$ of length $w(u, v)$. Thus for any point p on this interval we can define its distance to u as $|p|$ and to v as $w(u, v) - |p|$. One thing to note here is that, throughout the paper, by a *point* p *on* G we mean either $p \in V$ or p belongs to any edge (u, v). For any two points p and q in G, the *distance* between p and q is considered as the shortest path distance between them and is denoted by $d(p, q)$. A weight w_v is associated with each vertex v in G and the total weight of G is defined as,

$$W = \sum_{(u,v) \in E} w(u, v) + \sum_{v \in V} w_v$$

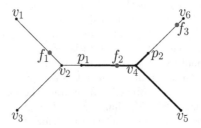

Fig. 1. Service zone of f_2

We refer to a portion of an edge as *sub-edge*. As we have defined earlier that an edge e can be modeled as an interval of length $w(e)$, a sub-edge may be defined as a sub-interval of the interval with length defined accordingly. We define a *sub-graph* $G' = (V', E')$ of a Graph G, such that E' is a finite subset of edges and sub-edges of G. Hence V' can contain some vertices of V or some points belong to edges of G. Weight of a sub-edge is same as the length of the interval correspond to it. For any sub-graph G' of G we define the weight of G', $W(G')$ as the sum of the weights of the edges, sub-edges and vertices present in G'.

Like any other versions of Voronoi game, here facilities are modeled as points in the underlying space. Given any set of facilities F on G, service zone $G_F(f)$ of any facility $f \in F$ is defined as the set of points in G that are closer to f than any other facility $f' \in F$. In case if a point is equidistant to its nearest facilities of P1 and P2, then it is included in the service zone of P2. Observe that for any facility f, $G_F(f)$ is a connected sub-graph of G. It would be more appropriate to refer $G_F(f)$ as a subset of G, because $G_F(f)$ may contain portions of some edges. For example in Figure 1, service zone of f_2 (shown in bold) contains the portions of the edges (v_2, v_4) and (v_4, v_6) where p_1 be the point such that $d(f_1, p_1) = d(p_1, f_2)$ and for p_2, $d(f_2, p_2) = d(p_2, f_3)$. For a set of facilities $F' \subseteq F$ define the *service zone* of F', $G_F(F') = \cup_{f \in F'} G_F(f)$.

In this paper we will consider the *One-Round (m, k) Voronoi Game on Graphs*. The game consists of a weighted graph $G(V, E)$ and two players P1 and P2 respectively. Initially P1 places m facilities, followed by which P2 places k facilities in G. For any set of facilities F and S by P1 and P2 respectively, the payoff of P1, $Q_1(F, S)$ is defined as $W(G_{F \cup S}(F))$ and the payoff of P2, $Q_2(F, S)$ is defined as $W - Q_1(F, S)$. Let $\nu(F) = \max_S Q_2(F, S)$ where maximum is taken over any placement of k facilities S by P2. The *One-Round (m, k) Voronoi Game on Graphs* can be formally stated as follows.

One-Round (m, k) Voronoi Game on Graphs: Given a graph $G = (V, E)$ and two players P1 and P2 having m and k facilities respectively, P1 chooses a set F^* of m facility locations following which P2 chooses a set S^* of k facility locations disjoint from F^* in G, such that:

   (i) $\max_S Q_2(F^*, S)$ is attained at $S = S^*$;

   (ii) $\min_F \nu(F)$ is attained at $F = F^*$, where the minimum is taken over all possible set of facility locations F of P1.

The paper is organized as follows. In the next section, we give a lower bound on optimal payoff of first player (P1) on trees. In section 3, we will characterize the optimal strategy of P2 for *One-Round Voronoi Game on Graphs* and propose an algorithm for finding an optimal placement of facilities by P2, where the number of facilities placed by P2 is constant. In section 4, we will prove that the problem of finding an optimal strategy of P2 on general graphs is $\mathcal{NP}$-hard. Finally in section 5, we will propose a 1.58 factor approximation algorithm for finding an optimal strategy of P2 for *One-Round Voronoi Game on Graphs*.

2   Voronoi Game on Trees

In this section we will consider the game where the underlying space is a tree, say $T = (V, E)$. Denote the total weight of T by W. Let $P = \{p_1, p_2, \ldots, p_\tau\}$ be any set of points on T. Observe that $T \setminus P$ is a set of sub-trees of T. We refer to those sub-trees as partitions of T. For example in Figure 2, four partitions of an example tree with respect to the set of points $\{p_1, p_2, p_3\}$ has shown. Let us denote $T \setminus P$ by $T(P)$. Observe that for any set of m facilities in T placed by P1 partitions T into at least $m + 1$ partitions. By placing one or more facility in a partition, P2 can get only a portion of that partition. Now we have the following lemma.

Lemma 2.1. *For any tree T, there exists a set of points $P = \{p_1, p_2, \ldots, p_\tau\}$ which partitions T into at least $\tau + 1$ sub-trees, such that weight of each sub-tree $T_i \in T(P)$ is at most $\frac{W}{\tau + 1}$, where τ is any positive integer.*

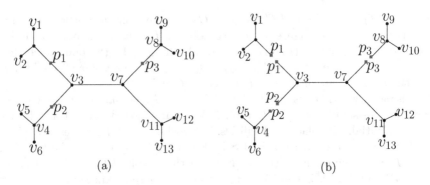

Fig. 2. Example of partition of a tree: (a) Original Tree T and (b) Partitions of T

Proof. Observe that it is enough to prove that given any weighted tree T and a positive integer τ there exist a point $\mathring{p}$, which partitions the tree into two or more parts such that weight of one part is less than or equal to $\frac{\tau * \mathcal{W}}{\tau+1}$ and weight of all other parts are less than or equal to $\frac{\mathcal{W}}{\tau+1}$. Choose an arbitrary vertex of tree as the root of T. Define an extended weight function w_T from the vertices of T to $\mathbb{R}$, such that, weight of any leaf node v is its original weight w_v and weight of any internal node v_i is equal to $\sum_{v_j}(w_{v_i} + w_T(v_j) + w(v_i, v_j))$, where v_j is child of v_i and $w(v_i, v_j)$ be the weight of the edge (v_i, v_j). Now observe that there will always be a node with weight greater than or equal to $\frac{\mathcal{W}}{\tau+1}$ and all of its children are having weight less than $\frac{\mathcal{W}}{\tau+1}$. Denote that vertex by $\check{v}$. Let the children of $\check{v}$ be $\{v_1, v_2, \ldots, v_k\}$. Now if for all $1 \le i \le k$, $w_T(v_i) + w(\check{v}, v_i)$ is less than $\frac{\mathcal{W}}{\tau+1}$, then $\mathring{p} = \check{v}$. Otherwise there exist a child v_j of $\check{v}$, such that, $w_T(v_j) + w(\check{v}, v_j) > \frac{\mathcal{W}}{\tau+1}$, but $w_T(v_j) < \frac{\mathcal{W}}{\tau+1}$. But in that case observe that there exists a point p on the edge $(\check{v}, v_j)$ which partitions the tree into two parts, one having weight $\frac{\mathcal{W}}{\tau+1}$ and other having weight $\frac{\tau * \mathcal{W}}{\tau+1}$. Thus $\mathring{p} = p$ and the result follows. □

Corollary 1. *There exists a placement strategy of P1 such that it always achieves at least $\frac{m-k+1}{m+1}\mathcal{W}$ as its payoff for One-Round (m,k) Voronoi Game on Trees.*

Proof. We prove this corollary by proposing an placement strategy of P1. By Lemma 2.1 we know that there exists a set F' such that F' partition the tree T in a manner such that each of the partition is having a weight at most $\frac{\mathcal{W}}{m+1}$, where $|F'| = m$. Suppose P1 places its facilities on the points of F'. As each partition is bounded by some facility of P1, placing k facilities on such partitions limit the optimal payoff of P2 to $\frac{\mathcal{W}}{m+1}k$. Hence the payoff of P1 is at least $\frac{m-k+1}{m+1}\mathcal{W}$, which completes the proof of this corollary. □

Now consider a restricted version of this game where $k = 1$, i.e P2 places only one facility. Also consider the class of *Star* trees with $m + 1$ edges of equal weight. For this case, an optimal strategy of P1 is to place a facility at the *central vertex* and the remaining $m - 1$ to anywhere on the *Star*. On the other hand P2 chooses a point as close as possible to the *central vertex*, on some edge, which doesn't contain any facility of P1, as its optimal strategy. Thus service zone of P2 is limited within an edge and payoff of P1 is $\frac{m}{m+1}\mathcal{W}$. Hence the bound of Observation 1 is tight.

3   Optimal Facility Locations of P2 on Graphs

Before we move on to the general problem on graph let us first consider a restricted game. Let $G(V, E)$ be any graph and $F = \{f_1, f_2, \ldots f_m\}$ be a set of facilities placed by P1. Now suppose P2 wants to place only one facility. Goal is to find the optimal placement by P2 in G. Let $V = \{v_1, v_2, \ldots, v_n\}$ and denote the edge joining any two vertices v_i and v_j by e_{ij}. Define *arc* to be an edge or a portion of an edge. An arc between two points u and v are denoted by $\langle u, v \rangle$. For any vertex $v_i \in V$ denote the facility closest from v_i among the facilities in F by $f(v_i)$ and the distance between v_i and $f(v_i)$ by d_i. Let $\Gamma(v_i)$ be the set of points in G which are at a distance d_i from v_i. Now observe that any edge can contain at most two points from $\Gamma(v_i)$. Hence for any vertex v_i, $|\Gamma(v_i)|$ contains $O(|E|)$ many points. Let $\Gamma = \cup_{1 \leq i \leq n} \Gamma(v_i)$. Thus Γ contains $O(|V||E|)$ many points.

Let us assume for each arc $\langle f_i, v_j \rangle$, where $f_i \in F$ and $v_j \in V$, there exists a point p very close to f_i such that distance between p and f_i is small enough to be considered as zero. We include all those points into Γ and we have the following observation.

Observation 3.1. *Number of points in Γ is bounded by $O(|V||E|)$.*

Let s be any placement of a facility by P2 located on an arbitrary edge e_{ij}. Consider any path λ between s and any facility $f_l \in F$, such that half of the points of λ are closer to f_l than any other facility in $F \cup \{s\}$ and the rest of the points are closer to s than any other facility in F. Denote all such paths by $\pi(s)$. For example in Figure 3 the path between s and f_2 is in $\pi(s)$, but the path between s and f_1 is not in $\pi(s)$.

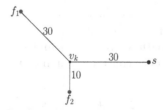

Fig. 3. Example of facilities placed by P1 and P2

Observe that for any path $\lambda \in \pi(s)$, λ contains at least one of v_i or v_j. Let $\pi_1(s)$ be the set containing all those paths of $\pi(s)$ which contain v_i, but not v_j and $\pi_2(s)$ be the set containing all those paths of $\pi(s)$ which contain v_j, but not v_i. For the paths $\lambda \in \pi(s)$, such that λ contain both of v_i and v_j, include λ in $\pi_1(s)$ if v_i is preceded by v_j in λ, otherwise include it in $\pi_2(s)$. Observe that $\pi_1(s) \cap \pi_2(s) = \emptyset$. Define the set $B_1(s)$ and $B_2(s)$ such that they contain the midpoints of the paths in $\pi_1(s)$ and $\pi_2(s)$ respectively. We refer to those midpoints as *bisectors*.

Observation 3.2. *Each edge contains at most one point of $B_1(s) \cup B_2(s)$.*

Proof. Suppose there exists two paths λ_1 and λ_2 in $\pi(s)$ such that the bisectors of λ_1 and λ_2 belong to the same edge e_{ab}. Let b_1 and b_2 be the bisectors of λ_1 and λ_2 respectively. Without loss of generality assume the paths λ_1 and λ_2 start at s and end at f_k and f_l respectively. Note that a path is a sequence of vertices. Suppose along the path λ_1, the vertex v_a precedes the vertex v_b. Now there will be two cases.

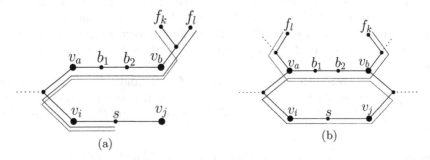

Fig. 4. Figure showing two cases for the proof of Observation 3.2

Case 1: Along the path λ_2, v_a precedes v_b(see Figure 4(a)). Suppose distance of b_1 and f_k is δ_1 along λ_1 and distance of b_2 and f_l is δ_2 along λ_2. Now there are two possibilities, $\delta_2 < \delta_1$ or $\delta_2 \geq \delta_1$. The first possibility contradicts the fact that the arc $\langle b_1, b_2 \rangle$ is served by P1 considering the path λ_1. The second possibility contradicts the fact that the arc $\langle b_1, b_2 \rangle$ is served by P2 considering the path λ_1.

Case 2: Along the path λ_2, v_b precedes v_a(see Figure 4(b)). If we consider the path λ_1, then by definition of bisector the arc $\langle b_2, v_b \rangle$ will be served by $P1$. But if we consider the path λ_2, then by definition of bisector the arc $\langle b_2, v_b \rangle$ will be served by $P2$. Hence contradiction and the result follows. □

Fig. 5. Positions of s, p_s and f_l

Let $p_s \in \langle s, v_j \rangle$ be the point closest to s, such that $p_s \in \Gamma \cup V$(see Figure 5). Let $\lambda \in \pi_2(s)$ be a path between s and f_l, where $f_l \in F$ and m_s be the midpoint of λ. Let p be any point on the arc $\langle s, p_s \rangle$. Suppose the distance between s and p_s along e_{ij} is equals to δ. Observe the length of the path is now reduced and hence the mid point is shifted from m_s to a new point m_p. Now we have the following observation.

Observation 3.3. *Distance between m_s and m_p is equal to $\delta/2$ along $e_{\alpha\beta}$.*

Observation 3.3 holds for any path $\lambda \in \pi_2(s)$. Similarly for any path $\lambda' \in \pi_1(s)$ consider the point p'_s closest to s, such that $p'_s \in \Gamma \cup V$. Observe that if the facility of P2 is shifted from s to any point $p \in \langle p'_s, s \rangle$ the midpoint of the path λ' is moved to a distance $\delta'/2$, where δ' is the distance between p and s along e_{ij}.

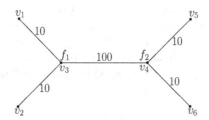

Fig. 6. Example of facilities placed by P1 and P2

One thing to note here is that there might be more than one optimal placement by P2, even there might be infinitely many optimal placements by P2. In Figure 6 where P1 has placed two facilities at v_3 and v_4, any point on the edge joining v_3 and v_4 is an optimal placement by P2. Now we have the following theorem.

Theorem 3.1. *There exists an optimal strategy of P2 which belongs to $\Gamma \cup V$.*

Proof. Let $\mathring{s}$ be any optimal placement by P2 such that $\mathring{s} \notin \Gamma \cup V$. Suppose $\mathring{s}$ belongs to the edge e_{ij}. Let $p_l \in \langle v_i, \mathring{s} \rangle$ be the point closest to $\mathring{s}$, such that $p_l \in \Gamma \cup V$. Similarly define $p_r \in \langle \mathring{s}, v_j \rangle$ be the point closest to $\mathring{s}$, such that $p_r \in \Gamma \cup V$ (see Figure 7). Now observe that it is enough to show that either $\mathcal{P}(F, \mathring{s}) \leq \mathcal{P}(F, p_l)$ or $\mathcal{P}(F, \mathring{s}) \leq \mathcal{P}(F, p_r)$.

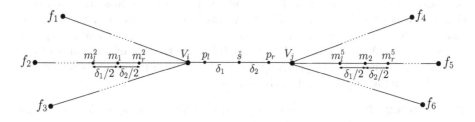

Fig. 7. Positions of $\mathring{s}$, p_l and p_r

Now suppose $\mathcal{P}(F, \mathring{s}) > \mathcal{P}(F, p_l)$ and $\mathcal{P}(F, \mathring{s}) > \mathcal{P}(F, p_r)$. Recall that for any placement of facility s by P2, $B_1(s)$ and $B_2(s)$ are the sets of bisectors correspond to the set of paths $\pi_1(s)$ and $\pi_2(s)$. Now based on the emptiness of $B_1(\mathring{s})$ and $B_2(\mathring{s})$ two cases can arise.

Case 1: $B_1(\mathring{s}) = \emptyset$ or $B_2(\mathring{s}) = \emptyset$. Without loss of generality assume $B_2(\mathring{s}) = \emptyset$. Observe that as $F \neq \emptyset$, $B_1(\mathring{s}) \neq \emptyset$. Now there is no path between $\mathring{s}$ and any facility of P1 via v_j. Thus $\mathcal{P}(F, \mathring{s}) \leq \mathcal{P}(F, p_l)$, which contradicts our basic assumption and hence the result follows.

Case 2: $B_1(\mathring{s}) \neq \emptyset$ and $B_2(\mathring{s}) \neq \emptyset$. Suppose distance between $\mathring{s}$ and p_l be δ_1 and distance between $\mathring{s}$ and p_r be δ_2(see Figure 7). Further let $|B_1(\mathring{s})| = k_1$ and $|B_2(\mathring{s})| = k_2$. Consider any path λ in $\pi_1(\mathring{s})$. Let $m_1 \in B_1(\mathring{s})$ be the midpoint of λ. Observe that instead of placing the facility at $\mathring{s}$, if P2 would have placed it at p_l, the length of the path λ between two facilities is reduced by δ_1. From Observation 3.3 we know the new midpoint is at a distance $\delta_1/2$ from m_1 along λ(see Figure 7). Now as $|B_1(\mathring{s})| = k_1$, k_1 many such paths are there. Hence along all such paths the payoff of $P2$ will be decreased by $\frac{k_1 * \delta_1}{2}$. Similarly along the paths in $\pi_2(s)$ the payoff will be increased by $\frac{k_2 * \delta_1}{2}$. Hence,

$$\mathcal{P}(F, p_l) = \mathcal{P}(F, \mathring{s}) + (k_2 - k_1) * \delta_1/2 \tag{1}$$

Similarly if P2 would have placed the facility at p_r, the payoff of P2,

$$\mathcal{P}(F, p_r) = \mathcal{P}(F, \mathring{s}) + (k_1 - k_2) * \delta_2/2 \tag{2}$$

Now as, $\mathcal{P}(F, \mathring{s}) > \mathcal{P}(F, p_l)$ and $\mathcal{P}(F, \mathring{s}) > \mathcal{P}(F, p_r)$, from Equation 1 and 2 we get, $(k_2 - k_1) * \delta_1/2 < 0$ and $(k_1 - k_2) * \delta_2/2 < 0$. As $\delta_1, \delta_2 > 0$, we get $(k_1 - k_2) < 0$ and $(k_2 - k_1) < 0$, hence contradiction and the result follows. □

Now consider the general problem where P2 is interested in placing $k(> 1)$ facilities. Again the goal is to find the optimal placement by P2 on G. Consider any set of placements S by P2. Let $s \in S$ be any arbitrary facility location. Without loss of generality we assume s is on the edge e_{ij}. We refine the definition of $\pi(s)$ by saying that $\pi(s)$ is the set of paths between s and any facility of P1 such that for each path $\lambda \in \pi(s)$, half of the points of λ are closer to some $f_i \in F$ than any other facility point in $F \cup S$ and the rest of the points are closer to s than any other facility point in $F \cup S$. Similarly define $\pi_1(s)$ and $\pi_2(s)$ as the disjoint subset of $\pi(s)$, such that the paths in $\pi_1(s)$ and $\pi_2(s)$ contains v_i and v_j respectively. Accordingly let $B1(s)$ and $B2(s)$ are the sets of midpoints of the paths in $\pi_1(s)$ and $\pi_2(s)$ respectively. Next we present a theorem whose proof is somewhat similar to the proof of theorem 4.1.

Theorem 3.2. *For One-Round (m, k) Voronoi Game on Graphs there exists an optimal strategy of P2 which belongs to $\Gamma \cup V$.*

Proof. Let $\mathring{S}$ be any optimal placement by P2. We show that there exists a placement $S' \subseteq \Gamma \cup V$ by P2 such that $\mathcal{P}(F, \mathring{S}) \leq \mathcal{P}(F, S')$. Let there exists a placement point $\mathring{s} \in \mathring{S}$ such that $\mathring{s} \notin \Gamma \cup V$. Without loss of generality assume

$\mathring{s}$ belongs to the edge e_{ij}. Let $p_l \in \langle v_i, \mathring{s} \rangle$ be the point closest to $\mathring{s}$, such that $p_l \in \Gamma \cup V$. Similarly define $p_r \in \langle \mathring{s}, v_j \rangle$ be the point closest to $\mathring{s}$, such that $p_r \in \Gamma \cup V$.

Instead of placing a facility at $\mathring{s}$ if P2 would have placed it at p_l or p_r, then using a similar argument like in proof of theorem 4.1, we can prove that either $\mathcal{P}(F, \mathring{S}) \leq \mathcal{P}(F, \mathring{S} \setminus \{\mathring{s}\} \cup p_l)$ or $\mathcal{P}(F, \mathring{S}) \leq \mathcal{P}(F, \mathring{S} \setminus \{\mathring{s}\} \cup p_r)$.

By using this construction repeatedly we substitute each of such $\mathring{s} \in \mathring{S}$, such that $\mathring{s} \notin \Gamma \cup V$, by a point in $\Gamma \cup V$. We end up getting a set $S' \subseteq \Gamma \cup V$, such that $\mathcal{P}(F, \mathring{S}) \leq \mathcal{P}(F, S')$, which completes the proof of this theorem. □

One thing here to note, that it is possible to design a simple algorithm, which by checking all subsets of size k of the set $\Gamma \cup V$, finds out the optimal strategy of P2. But when k is not constant and considered as part of input, the running time of this algorithm is exponential. In the next section we prove that when k is considered as part of input, the problem is $\mathcal{NP}$-hard.

4   Computational Complexity for Graphs

In this section we will prove that given a placement of m facilities by P1 determining the optimal placement by P2 in *One-Round* (m, k) *Voronoi Game on Graphs* is $\mathcal{NP} - hard$. Let us call the problem of finding the optimal placement of P2 in *One-Round* (m, k) *Voronoi Game on Graphs* as *Maximum payoff problem*. Now consider the decision version of *Maximum payoff problem*. Given a graph $G = (V, E)$, a set of m facilities F by P1 in G, and a real number δ, we have to decide whether there exists a set of k points S disjoint from F in G such that the payoff of P2, $\mathcal{Q}_2(F, S) \geq \delta$ or not. Clearly the problem is in $\mathcal{NP}$ as given any placement of facility by P1 and P2 time it is possible to find out $\mathcal{Q}_2(F, S)$ in polynomial time. To prove that the problem is $\mathcal{NP}$-hard we will reduce *Maximum payoff problem* from *Minimum Dominating Set* problem which is known to be $\mathcal{NP}$-hard[5]. But before that let us first define the *Minimum Dominating Set* problem.

Minimum Dominating Set Problem: Given a graph $G = (V, E)$ a *dominating set* is a set of vertices $S \subseteq V$ such that each vertex in graph G is either in S or is a neighbor of at least one element of S. The problem asks to find such S with minimum cardinality.

Given a graph G and an integer k, the decision version of Minimum Dominating Set Problem asks whether there exist a *dominating set* of size k or not. Now we have the following theorem.

Theorem 4.1. *Decision version of Maximum Payoff Problem is $\mathcal{NP}$-complete.*

Proof. Let $\mathcal{I} = (G, k)$ be any valid instance of the *minimum dominating set* where G is an un-weighted graph and k is an integer. We will construct a new weighted graph $G' = (V', E')$ from G by adding a pendant vertex to each of the vertices. Figure 8 is showing the construction for an example graph. Let $\tilde{F}$ be the set

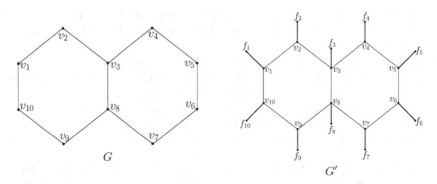

Fig. 8. Construction of G' from an example graph G

of $|V|$ new vertices. Now $V' = V \cup \tilde{F}$ and $E' = E \cup (v_i, f_i) \, \forall v_i \in V$. We will assign weight $w_e < \frac{1}{|V|+|E|+k}$ to each edge and weight $w_v = 1$ for each vertex $v \in V'$. Now consider the *Maximum Payoff Problem* on G' where P1 has placed its facilities at $\tilde{F}$. Now we claim that there exists a dominating set of size k in G if and only if there exists a set of k points S in G' such that $\mathcal{Q}_2(\tilde{F}, S) \geq |V|$.

Let S be a set of k points in G' such that $\mathcal{Q}_2(\tilde{F}, S) \geq |V|$. Now from Theorem 3.2 we know that there exists an optimal placement by P2 which belongs to $\Gamma \cup V$. Without loss of generality assume $S \subseteq \Gamma \cup V$. Recall that we have assumed for each edge (f_i, v_i) there exists a point p_i very close to f_i such that distance between p_i and f_i is small enough to be considered as zero. Denote the set of all such points as P. Now observe that as weight of each edge is same $\Gamma \subseteq V \cup P$. Hence $S \subseteq P \cup V$. Now we will construct a new set of placements of facilities S' from S as follows. For all points $s_i \in S$ such that $s_i \in V$ add s_i to S'. For all points $s_i \in S$ such that $s_i \in P$ if $v_j \notin S$ where v_j is adjacent to s_i add v_j to S' else add any vertex $v \in V$ such that $v \notin S$ (see Figure 9). Observe $S' \subset V$ and $\mathcal{Q}_2(\tilde{F}, S') > \mathcal{Q}_2(\tilde{F}, S) - k * w_e$. Now the payoff $\mathcal{Q}_2(\tilde{F}, S)$ can be written as $\mathcal{Q}_{E'} + \mathcal{Q}_V$ where $\mathcal{Q}_{E'}$ sum of the length of all the arcs those are served by P2 and $\mathcal{Q}_V$ is the number of vertices in the service zone of P2. Observe now $\mathcal{Q}_{E'} \leq (|V| + |E|) * w_e$. Hence $\mathcal{Q}_V \geq \mathcal{Q}_2(\tilde{F}, S) - k * w_e - (|V| + |E|) * w_e$. But $w_e < \frac{1}{|V|+|E|+k}$, that is $w_e * (|V| + |E| + k) < 1$, $|V|$ and $\mathcal{Q}_V$ are integers. Further $\mathcal{Q}_2(\tilde{F}, S) \geq |V|$. Therefore $\mathcal{Q}_V \geq |V|$. Now any vertex $v_i \in V$ will be served by a facility $s_j \in S'$ if and only if s_j is neighbor of v_i. Hence S' is a dominating set for G of size k.

Fig. 9. Formation of S' from S in proof of Theorem 4.1

Now consider the case where the graph G has a dominating set D of size k. In graph G', D can be used for placements by $P2$. Every vertex in V is adjacent to one of the vertices of D. So the payoff by $P2$ is at least $|V|$. Hence the result follows. $\square$

5   Approximation Bound for Optimal Payoff of P2 on Graphs

In this section we will provide an 1.58 factor approximation algorithm for the problem of finding the optimal strategy of P2 on graphs. We reduce our problem to *Weighted Maximum Coverage Problem* and will use the existing approximation algorithm of *Weighted Maximum Coverage Problem* to derive an approximation algorithm for our problem. But before that let us define the *Weighted Maximum Coverage Problem*.

Weighted Maximum Coverage Problem: Given an universe $X = \{x_1, x_2 \ldots x_n\}$, a family S of subsets of X, an integer τ and a weight function w_i associated with each $x_i \in X$ the *Weighted Maximum Coverage Problem* is to find τ sets such that total weight of the covered elements is maximized.

The *Weighted Maximum Coverage Problem* is $\mathcal{NP}$-hard, and cannot be approximated within $\frac{e-1}{e} - o(1) \approx 1.58$ under standard assumptions [6]. There is a greedy approximation algorithm for the *Weighted Maximum Coverage Problem*, which at each stage chooses a set, which contains the maximum weighted uncovered elements. Now we are having the following theorem.

Theorem 5.1. *[6] The greedy algorithm for Weighted Maximum Coverage Problem achieves an approximation ratio of $\frac{e-1}{e}$.*

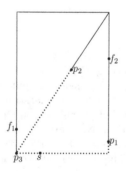

Fig. 10. Service zone of s

Now consider any instance of our problem. Let $G = (V, E)$ be any graph and F be any set of facilities placed by P1 in G. P2 wants to place k new facilities. For the sake of simplicity we assume that the weight of each vertex is zero. But by a simple modification our algorithm can be extended to handle the case when the vertices are having non-zero weights. Now from Theorem 3.2 we know that there exists an optimal placement by *P2* which belong to $\Gamma \cup V$. Now consider any placement of facility $s \in \Gamma \cup V$ by P2. Let Ω_s be the set of bisectors correspond to s. For example in Figure 10, P1 has placed two facilities f_1 and f_2 and P2 has placed the facility s. The service zone of P2 is shown with dotted lines. Here the set Ω_s will be equal to $\{p_1, p_2, p_3\}$. Define

$$\Omega = \{\cup_{s \in \Gamma \cup V} \Omega_s\} \cup \Gamma$$

From Observation 3.2 it is implied that the cardinality of Ω is bounded by $O((\Gamma \cup V)E)$, that is $O((V + E)^2)$. Now from $G = (V, E)$ construct another graph $G' = (V', E')$ as follows. $V' = V \cup \Omega \cup F$. For each edge $e_{ij} \in E$ which does not contains any point of Ω include that edge to E'. Any edge e_{ij}, which contains one or more points of Ω, $\{\omega_1, \omega_2, \ldots \omega_l\}$ sorted along v_i to v_j, add the edges $(v_i, \omega_1), (\omega_1, \omega_2) \ldots (\omega_{l-1}, \omega_l)$ to E'. Now observe that for any placement s by P2, payoff of P2 will be equals to a set of edges $\tilde{E} \subset E'$ of G'. Now consider the set system where X is equal to E' and for each point $p_i \in \Gamma \cup V$ define the set $S_i \subseteq E'$, such that, S_i is the set of edges that is in service zone of the facility of P2 at p_i. For each edge $e_i \in E'$ the weight of e_i is equal to the length of e_i. Now run the greedy algorithm for the *Weighted Maximum Coverage Problem* on this set system for $\tau = k$. Now we have the following lemma which follows from the construction.

Lemma 5.1. *Any α factor approximation algorithm for the Weighted Maximum Coverage Problem will produce an α factor approximation for our problem.*

Now from Theorem 5.1 and Lemma 5.1 we have the following theorem.

Theorem 5.2. *There exist an 1.58 factor approximation algorithm for Maximum Payoff Problem.*

References

1. Ahn, H.-K., Cheng, S.-W., Cheong, O., Golin, M.J., van Oostrum, R.: Competitive facility location: the voronoi game. Theor. Comput. Sci. 310(1-3), 457–467 (2004)
2. Banik, A., Bhattacharya, B.B., Das, S.: Optimal strategies for the one-round discrete voronoi game on a line. Journal of Combinatorial Optimization, 1–15 (2012)
3. Cheong, O., Har-Peled, S., Linial, N., Matousek, J.: The one-round voronoi game. Discrete & Computational Geometry 31(1), 125–138 (2004)
4. Fekete, S.P., Meijer, H.: The one-round voronoi game replayed. Comput. Geom. 30(2), 81–94 (2005)
5. Garey, M.R., Johnson, D.S.: Computers and Intractability: A Guide to the Theory of NP-Completeness. W. H. Freeman (1979)
6. Hochbaum, D.S.: Approximation algorithms for $\mathcal{NP}$-Hard problems. PWS Publishing Company (1996)
7. Kiyomi, M., Saitoh, T., Uehara, R.: Voronoi game on a path. IEICE Transactions 94-D(6), 1185–1189 (2011)
8. Teramoto, S., Demaine, E.D., Uehara, R.: The voronoi game on graphs and its complexity. J. Graph Algorithms Appl. 15(4), 485–501 (2011)

Approximation Schemes for Covering and Packing*

Rom Aschner[1], Matthew J. Katz[1], Gila Morgenstern[2], and Yelena Yuditsky[1]

[1] Department of Computer Science, Ben-Gurion University
{romas,matya,yuditsky}@cs.bgu.ac.il
[2] Caesarea Rothschild Institute, University of Haifa
gilamor@cri.haifa.ac.il

Abstract. The local search framework for obtaining PTASs for NP-hard geometric optimization problems was introduced, independently, by Chan and Har-Peled [6] and Mustafa and Ray [17]. In this paper, we generalize the framework by extending its analysis to additional families of graphs, beyond the family of planar graphs. We then present several applications of the generalized framework, some of which are very different from those presented to date (using the original framework). These applications include PTASs for finding a *maximum l-shallow set* of a set of fat objects, for finding a *maximum triangle matching* in an l-shallow unit disk graph, and for *vertex-guarding* a (not-necessarily-simple) polygon under an appropriate shallowness assumption.

We also present a PTAS (using the original framework) for the important problem where one has to find a minimum-cardinality subset of a given set of disks (of varying radii) that covers a given set of points, and apply it to a class cover problem (studied in [3]) to obtain an improved solution.

1 Introduction

In their break-through papers, Chan and Har-Peled [6] and Mustafa and Ray [17] showed, independently, how a simple local-search-based algorithm can be employed to obtain a PTAS for an NP-hard geometric optimization problem. Chan and Har-Peled [6] used local search to obtain a PTAS for finding a maximum independent set of pseudo disks, and Mustafa and Ray [17] used it to obtain a PTAS for finding a minimum hitting set (from a given set of points) for half-spaces in $\mathbb{R}^3$ and for r-admissible regions in $\mathbb{R}^2$. This technique turned out to be very powerful, and since its publication, it was applied to a variety of additional problems. Gibson et al. [13] used it to obtain a PTAS for the 1.5D terrain

* Work by R. Aschner was partially supported by the Lynn and William Frankel Center for Computer Sciences. Work by R. Aschner and M.J. Katz was partially supported by the Israel Ministry of Industry, Trade and Labor (consortium COR-NET). Work by M.J. Katz was partially supported by grant 1045/10 from the Israel Science Foundation, and by grant 2010074 from the United States – Israel Binational Science Foundation. Work by G. Morgenstern was partially supported by the Caesarea Rothschild Institute (CRI).

S.K. Ghosh and T. Tokuyama (Eds.): WALCOM 2013, LNCS 7748, pp. 89–100, 2013.
© Springer-Verlag Berlin Heidelberg 2013

guarding problem, and Gibson and Pirwani used it to obtain a PTAS for finding a dominating set in disk graphs [14].

The local-search-based algorithm, as described in [6, 17], receives an integer parameter k and proceeds as follows. It starts with any feasible solution and performs a series of local improvements, where each such improvement involves only $O(k)$ objects. The analysis relies on the existence of a planar bipartite graph G, whose vertices on one side correspond to the objects found by the local search algorithm ("blue vertices") and on the other side to the objects in an optimal solution ("red vertices"), and whose ("blue-red") edges satisfy an appropriate *locality property* relating the two solutions. Chan & Har-Peled and Mustafa & Ray showed that for the problems mentioned above such a planar graph G exists, and applied the planar separator theorem to relate the size of the local search solution with that of the optimal solution.

In this paper we generalize the local search technique by extending its analysis to additional families of graphs. We show that one can achieve a PTAS even in cases where the graph whose vertices are the elements of the two solutions and which satisfies the locality property, is **not** planar, but belongs to a family of graphs that has a separator property. It is well known that there are many such families of graphs, e.g., graphs with forbidden minors [1] and various intersection graphs [10, 16, 18]. We also present several interesting applications of our extended analysis, some of which are very different from those presented in the past (using the original analysis).

These applications, which are presented in Section 3, include finding a *maximum l-shallow set* of a set of fat objects, finding a *maximum triangle matching* in a unit disk graph, and *guarding a polygon with limited range of sight*. We briefly discuss each of them.

Maximum l-shallow set. Let D be a set of n fat objects in $\mathbb{R}^2$ and let $l > 0$ be a constant. An *l-shallow* subset of D is a subset S of D, such that the depth of the arrangement of S is at most l, i.e. every point in the plane is covered by at most l objects of S. In the maximum l-shallow subset problem, one is asked to find a maximum-cardinality l-shallow subset S of D. Notice that for $l = 1$, S is a maximum independent subset of D. Chan [4] presented a PTAS for the problem of finding a maximum independent subset of D. Later, as mentioned above, Chan and Har-Peled [6] presented a local-search-based PTAS for finding a maximum independent set of pseudo-disks (and also of D). Notice that a maximum l-shallow set can be larger than the set that is obtained by repeatedly finding a maximum independent set of the set of remaining objects. We show that our generalized analysis enables us to obtain a PTAS for the maximum l-shallow subset problem, and emphasize that it is essential whenever $l > 1$.

Maximum triangle matching. Given a graph G, one has to find a maximum-cardinality collection of vertex-disjoint triangles in G. Baker [2] presented a PTAS for the important case where G is planar. We give a PTAS for the case where G is an l-shallow unit disk graph. This PTAS also holds in more general settings, see below.

Guarding. We apply our generalized analysis to several guarding problems. Many guarding problems are known to be APX-hard (e.g., guarding a simple n-gon P with as few vertex guards as possible [9]), and as such do not admit a PTAS. Ghosh [12] presented an $O(\log n)$-approximation algorithm for vertex-guarding a polygon. Subsequently, Efrat and Har-Peled [8] presented an $O(\log |OPT|)$-approximation algorithm for this problem. In contrast to these results, we obtain a PTAS for this problem, under the assumption that each vertex guard g has a limited range of sight r_g, such that every point in P is covered by at least one of the guards and the set of disks centered at the guards is l-shallow. We also discuss several other versions, including guarding through walls and guarding a 1.5D terrain with bounded range of sight.

Finally, in Section 4, we consider the important problem known as discrete coverage of points by disks, for which we obtain a PTAS within the original local search framework. Let P be a set of points in the plane and let D be a set of disks that covers P. One needs to find a minimum-cardinality subset $D' \subseteq D$ that covers P. Notice that the special case where D is a set of unit disks, is the dual of the discrete hitting set problem for unit disks, for which Mustafa and Ray [17] presented a PTAS (actually, for arbitrary disks). However, in the general case, where D is a set of disks of varying radii, coverage and hitting are not dual. We present a PTAS for the general case that is inspired by the work of Gibson and Pirwani [14]. We also apply this result to the a class cover problem, improving a result of Bereg et al. [3].

2   PTAS via Local Search

We begin by generalizing the local search technique to additional families of graphs, beyond the family of planar graphs, such as intersection graphs and graphs with forbidden minors. Actually, we describe the technique for any family of graphs that has a separator property, similar to the separator property of planar graphs.

Let $\mathcal{F}$ be a monotone family of graphs, i.e., all subgraphs of a graph $G \in \mathcal{F}$ are also in $\mathcal{F}$. Assume that $\mathcal{F}$ has a separator property, i.e., for any graph $G = (V, E)$ in $\mathcal{F}$, one can partition the vertex set V, where $|V| = n$, into three sets A, B and S, such that (i) $|S| \leq cn^{1-\delta}$, where $0 < \delta < 1$ and $c > 0$ is a constant, (ii) $|A|, |B| \leq \alpha n$, where $1/2 \leq \alpha < 1$, and (iii) the sets A and B are disconnected, i.e., there is no edge in E between a vertex of A and a vertex of B.

Frederickson [11] defined the notion of an r-division for planar graphs, and showed how to obtain an r-division for a given planar graph G, by repeatedly applying the planar separator algorithm. It is straight-forward to adapt Frederickson's construction and analysis to the family $\mathcal{F}$. Essentially, one needs only to replace the exponent $1/2$ in the size of the separator by $1 - \delta$. Thus, $\mathcal{F}$ has the following property. Let r be a parameter, $1 \leq r \leq n$, then for any connected graph $G = (V, E)$ in $\mathcal{F}$, one can find a collection of $\Theta(n/r)$ pairwise disjoint subsets $V_1, V_2, \ldots$ of V, such that (i) $|V_i| \leq c_2 r$, where $c_2 > 0$ is a constant, (ii) $|\Gamma(V_i)| \leq c_3 r^{1-\delta}$, where $\Gamma(V_i)$ is the set of vertices in $V \setminus V_i$ that are adjacent to a vertex in V_i and $c_3 > 0$ is a constant, and (iii) $\cup \Gamma(V_i) = S$, where $S = V \setminus \cup V_i$; in particular, $\Gamma(V_i) \subseteq S$, and, for any $j \neq i$, V_i and V_j are disconnected. It follows that $|S| \leq \Theta(n/r) \cdot c_3 r^{1-\delta} = c_1 \frac{n}{r^\delta}$, where $c_1 > 0$ is a constant. Also, it is easy to verify that

Claim 1. *If* $r = \frac{k}{c_2 + c_3}$, *where* k *is a sufficiently large constant, then, for any index* i, $|V_i| + |\Gamma(V_i)| < k$.

We distinguish between minimization problems and maximization problems.

2.1 Minimization Problems

Let us now recall the local search technique as it is used, e.g., in the context of geometric piercing or covering. Assume that we are considering a minimization problem $\mathcal{P}$, that is, one needs to find a minimum-cardinality subset of a given set X that is a solution for $\mathcal{P}$. Let $X_0 \subseteq X$ be an initial not-necessarily-optimal solution for $\mathcal{P}$, and let k be a sufficiently large constant. The local search technique checks whether there exists a subset $X' \subseteq X_0$ of size k and a subset $X'' \subset X$ of size $k - 1$, such that $(X_0 \setminus X') \cup X''$ is still a solution for $\mathcal{P}$. If yes, then it replaces X' by X'' (i.e., it performs this local improvement) and resumes the search. Otherwise, it halts.

Let B and R be two solutions for $\mathcal{P}$, where B was obtained by applying local search, and assume that $B \cap R = \emptyset$. (Otherwise, we can remove the elements that belong to both B and R.) Moreover, let $G = (V, E)$ be a graph, such that (i) $V = B \cup R$, and (ii) for each object o that needs to be "solved", there exists an edge $e \in E$ between a vertex $b \in B$ and a vertex $r \in R$, where both b and r "solve" o. This requirement is called the *locality condition*. We prove below that if G belongs to some family $\mathcal{F}$ that has a separator property, then $|B|$ is not much greater than $|R|$. Parts of the proof appear already in [6,17] and are included here for completeness. We first construct an r-division of G, for $r = \frac{k}{c_2 + c_3}$, and set $B_i = B \cap V_i$ and $R_i = R \cap V_i$.

Claim 2. *For any index* i, *the set* $(B \setminus B_i) \cup \Gamma(B_i)$ *is also a solution for* $\mathcal{P}$.

Proof. Fix i and let o be an object that needs to be "solved". If all vertices of B that "solve" o belong to B_i, then, by the locality condition, there exists $r \in \Gamma(B_i)$ that "solves" o. Otherwise, there is a vertex $b \in B \setminus B_i$ that "solves" o. We conclude that in both cases there is a vertex in $(B \setminus B_i) \cup \Gamma(B_i)$ that "solves" o.

Claim 3. *For any index* i, $|B_i| \leq |R_i| + |\Gamma(V_i)| < k$.

Proof. Observe first that by arguments similar to those in the proof of Claim 2, the set $(B \setminus B_i) \cup (R_i \cup \Gamma(V_i))$ is also a solution for $\mathcal{P}$. Moreover, by Claim 1, $|R_i| + |\Gamma(V_i)| \leq |V_i| + |\Gamma(V_i)| < k$. So, if $|B_i| > |R_i| + |\Gamma(V_i)|$, then the local search algorithm would have replaced B_i (or a subset of B_i of size k, if $|B_i| > k$) by $R_i \cup \Gamma(V_i)$ before halting. Since it has not done so, we conclude that $|B_i| \leq |R_i| + |\Gamma(V_i)|$.

Theorem 4. *For any* ε, $0 < \varepsilon < 1$, *one can choose a constant* k, *such that* $|B| \leq (1 + \varepsilon)|R|$.

Proof. Set $c' = 2c_1(c_2 + c_3)^\delta$, $\varepsilon' = \frac{\varepsilon}{5}$, $k = \frac{c'^{1/\delta}}{\varepsilon'^{1/\delta}}$, and recall that $r = \frac{k}{c_2 + c_3}$. Then, $|B| \leq |S| + \sum_i |B_i| \leq |S| + \sum_i |R_i| + \sum_i |\Gamma(V_i)| \leq \frac{c_1 n}{r^\delta} + |R| + \frac{c_1 n}{r^\delta} = |R| + \frac{2c_1 n}{r^\delta} = |R| + \frac{|R| + |B|}{k^\delta} 2c_1(c_2 + c_3)^\delta = |R| + \varepsilon'(|R| + |B|)$. We thus have, $|B| \leq |R| \cdot \frac{1 + \varepsilon'}{1 - \varepsilon'} = |R|(1 + \varepsilon')(1 + \varepsilon' + \varepsilon'^2 + \ldots) \leq |R|(1 + \varepsilon')(1 + \varepsilon' + \varepsilon') = |R|(1 + 3\varepsilon' + 2\varepsilon'^2) \leq |R|(1 + 5\varepsilon') = |R|(1 + \varepsilon)$.

2.2 Maximization Problems

Let us now recall the local search technique as it is used, e.g., in the context of geometric packing. Assume that we are considering a maximization problem $\mathcal{P}$, that is, one needs to find a maximum-cardinality subset of a given set X that is a solution for $\mathcal{P}$. Let $X_0 \subseteq X$ be an initial not-necessarily-optimal solution for $\mathcal{P}$, and let k be a sufficiently large constant. The local search technique checks whether there exists a subset $X' \subseteq X_0$ of size at most $k - 1$ and a subset $X'' \subset X$ of size k, such that $(X_0 \setminus X') \cup X''$ is still a solution for $\mathcal{P}$. If yes, then it replaces X' by X'' (i.e., it performs this local improvement) and resumes the search. Otherwise, it halts.

Let B and R be two solutions for $\mathcal{P}$, where B was obtained by applying local search, and assume that $B \cap R = \emptyset$. (Otherwise, we can remove the elements that belong to both B and R.) Moreover, let $G = (V, E)$ be a graph, such that (i) $V = B \cup R$, and (ii) there exists an edge $e \in E$ between a vertex $b \in B$ and a vertex $r \in R$ if and only if b and r intersect.

Theorem 7 below states that if G belongs to some family $\mathcal{F}$ that has a separator property, then $|B| \geq (1 - \varepsilon)|R|$. The proof is similar to the one in Section 2.1; we include it for completeness. We first construct an r-division of G, for $r = \frac{k}{c_2 + c_3}$, and set $B_i = B \cap V_i$ and $R_i = R \cap V_i$.

Claim 5. *For any index* i, *the set* $(B \setminus (B_i \cup \Gamma(V_i))) \cup R_i$ *is also a solution for* $\mathcal{P}$.

Proof. Fix i, and let $r \in R_i$. By definition, r intersects only its neighbors in G, i.e., only elements in $(B_i \cup \Gamma(V_i)) \cup R_i$. Thus, for each $b \in B \setminus (B_i \cup \Gamma(V_i)))$, $r \cap b = \emptyset$. Therefore, $(B \setminus (B_i \cup \Gamma(V_i))) \cup R_i$ is also a solution for $\mathcal{P}$.

Claim 6. *For any index* i, $|R_i| \leq |B_i| + |\Gamma(V_i)| < k$.

Proof. By Claim 5, the set $(B \setminus (B_i \cup \Gamma(V_i))) \cup R_i$ is also a solution for $\mathcal{P}$. Moreover, by Claim 1, $|B_i \cup \Gamma(V_i)| = |B_i| + |\Gamma(V_i)| \leq |V_i| + |\Gamma(V_i)| < k$. So, if $|R_i| > |B_i \cup \Gamma(V_i)|$, then the local search algorithm would have replaced $B_i \cup \Gamma(V_i)$ by R_i (or a subset of R_i of size k, if $|R_i| > k$) before halting. Since it has not done so, we conclude that $|R_i| \leq |B_i| + |\Gamma(V_i)|$.

Theorem 7. *For any* ε, $0 < \varepsilon < 1$, *one can choose a constant* k, *such that* $|B| \geq (1 - \varepsilon)|R|$.

Proof. Set $c' = 2c_1(c_2 + c_3)^\delta$, $k = (c'(\frac{2}{\varepsilon} - 1))^{1/\delta}$, and recall that $r = \frac{k}{c_2 + c_3}$. Then,

$$|R| \leq |S| + \sum_i |R_i| \leq |S| + \sum_i |B_i| + \sum_i |\Gamma(V_i)| \leq \frac{c_1 n}{r^\delta} + |B| + \frac{c_1 n}{r^\delta} = |B| + \frac{2c_1 n}{r^\delta}$$

$$= |B| + \frac{|R| + |B|}{k^\delta} 2c_1(c_2 + c_3)^\delta = |B| + \frac{1}{\frac{2}{\varepsilon} - 1}(|R| + |B|) = |B| + \frac{\varepsilon}{2 - \varepsilon}(|R| + |B|) \ .$$

Rearranging, we get that $|B| \geq (1 - \varepsilon)|R|$.

3   Applications

In this section we describe several original applications of our generalized local search technique. The applications in Sections 3.1 and 3.2 use the maximization version and the applications in Section 3.3 use the minimization version.

3.1   Maximum *l*-Shallow Set for Fat Objects

We consider the *maximum l-shallow set* problem for a set of fat objects. Let D be a set of n fat objects and let $l > 0$ be a constant. An *l-shallow set* of D is a subset S of D, such that the depth of the arrangement of S is at most l. In the maximum *l*-shallow set problem, one has to find a maximum-cardinality subset S of D, such that S is *l*-shallow. Notice that for $l = 1$, S is a maximum independent set of D.

We show that our generalization of the local search technique enables us to apply local search to find a $(1 - \varepsilon)$-approximation of a maximum *l*-shallow set of D. Indeed, let B be the set of fat objects obtained by applying local search and let R be an optimal set. (We may assume that $B \cap R = \emptyset$, since otherwise, we can remove the objects that appear in both sets.) Notice that the analysis of Chan and Har-Peled [6] does not immediately apply here (assuming $l > 1$), since the intersection graph of $B \cup R$ is not necessarily planar (even if one removes the "monochromatic" edges) However, this graph does have a separator of size $O(\sqrt{n})$. This follows from results of Miller et al. [16] and Smith and Wormald [18], who show that the intersection graph of a set of *l*-shallow fat objects has a separator of size $\sqrt{ln}$.

Recall that the local search algorithm begins with the empty solution, and in each iteration it improves the current solution by replacing a subset of size at most $k - 1$ of the current solution with a larger subset of size at most k of D, such that the resulting set is still a solution (i.e., it is still *l*-shallow). Since k is a constant, each iteration can be performed in polynomial time, and therefore the total running time of the local search algorithm is polynomial in n.

As for the size of B, since both B and R are *l*-shallow, the set $B \cup R$ is $2l$-shallow, and therefore the intersection graph of $B \cup R$ has a separator of size $\sqrt{2ln} = O(\sqrt{n})$. Hence, by Theorem 7, we conclude that $|B| \geq (1 - \varepsilon)|R|$.

Finally, we note that Chan's separator-based method [4] can be used to obtain a $(1 - \varepsilon)$-approximation of a maximum *l*-shallow set of D. However, as also mentioned in [6], the disadvantage of his method is that it explicitly applies an algorithm for finding a separator, while we apply it only for analysis purposes.

3.2 Maximum Triangle Matching in Unit Disk Graphs

Next, we consider the *maximum triangle matching* problem in unit disk graphs. Given a graph $G = (V, E)$, find a maximum-cardinality collection of pairwise-disjoint subsets $V_1, \ldots, V_m$ of V, each consisting of exactly 3 vertices, such that for each $V_i = \{u_i, v_i, w_i\}$, $1 \le i \le m$, the three edges $(u_i, v_i), (v_i, w_i), (w_i, u_i)$ belong to E. Baker [2] presented a PTAS for the important case where G is planar. We give a PTAS for the case where G is an l-shallow unit disk graph. This PTAS also holds in more general settings; see remark below.

Let S be a set of n points in the plane, and let $UDG(S)$ be the graph over S, in which there is an edge between u and v if and only if the Euclidean distance between u and v is at most 1. Let $\mathcal{D}$ be the set of disks of radius $1/2$ centered at the points of S, and assume that the depth of the arrangement of $\mathcal{D}$ is at most some constant l. Then, there is an edge between u and v if and only if $d_u \cap d_v \neq \emptyset$, where d_u, d_v are the disks of $\mathcal{D}$ centered at u, v, respectively.

Our generalization of the local search technique enables us to apply local search to find a $(1 - \varepsilon)$-approximation of a maximum triangle matching of $UDG(S)$. We begin with the empty set of triangles, and in each iteration we replace a subset of at most $k - 1$ triangles of the current solution with a larger subset of at most k triangles, such that the resulting set is still a solution (i.e., it is still a triangle matching).

Consider the set of triangles $\mathcal{B}$ obtained by applying local search, and the set of triangles $\mathcal{R}$ of a maximum matching. (Each $\triangle \in \mathcal{B} \cup \mathcal{R}$ is a triangle $\triangle = \{a, b, c\}$, where a, b, c are points in S and $(a, b), (b, c), (c, a)$ are edges of $UDG(S)$.) For each triangle $\triangle$ in $\mathcal{B}$ (resp. in $\mathcal{R}$), select an arbitrary unique representative point $p_\triangle$ that lies in the interior of $\triangle$, and denote by B (resp. R) the resulting set of representative points.

Now, consider the graph $G = (B \cup R, E)$, where $(p, q) \in E$ if and only if the Euclidean distance between p and q is at most 2. Notice that G satisfies the following locality condition. For any two points $b \in B$ and $r \in R$, whose corresponding triangles are $\triangle u_b v_b w_b$ and $\triangle u_r v_r w_r$, respectively, and such that $\{u_b, v_b, w_b\} \cap \{u_r, v_r, w_r\} \neq \emptyset$, the edge (b, r) belongs to E. This is true since the distance from b (alt., r) to any of the its triangle corners is at most 1.

It remains to show that G has a separator. Let $\mathcal{D}'$ be the set of unit disks centered at the points of $B \cup R$. Then, $(p, q) \in E$ if and only if the disks of $\mathcal{D}'$ centered at p and q, respectively, intersect. It is easy to see that the depth of the arrangement of $\mathcal{D}'$ is at most some constant $c = c(l)$. This follows from the assumption that the depth of the arrangement of $\mathcal{D}$ is at most l. Therefore, by Miller et al. [16], G has a separator of size $O(\sqrt{cn}) = O(\sqrt{n})$. Hence, by Theorem 7, we conclude that $|B| \ge (1 - \varepsilon)|R|$.

Remark. It is easy to see that one can replace "triangle" in the above result with any connected graph (i.e., structure) H with a constant number of vertices, and obtain a PTAS for maximum H-matching in unit disk graphs.

3.3 Guarding with Limited Visibility

We now demonstrate our generalization of the local search technique on a family of *covering* problems, or, more precisely, on a family of *guarding* problems. Throughout this section, $\mathcal{G}$ denotes a set of points, representing stationary guards, where each guard $g \in \mathcal{G}$ has its own range of sight r_g. Let D_g denote the disk of radius r_g centered at g, and set $D = \{D_g : g \in \mathcal{G}\}$. We assume that the depth of the arrangement of D is at most some constant l.

Guarding a polygon. Given a polygon P (possibly with holes) and a set $\mathcal{G} \subseteq P$, a *minimum guarding set* for P (with respect to $\mathcal{G}$) is a minimum-cardinality subset $\mathcal{G}'$ of $\mathcal{G}$, such that every $p \in P$ is guarded by $\mathcal{G}'$, i.e., for every $p \in P$, there exists $g \in \mathcal{G}'$, such that $|gp| \leq r_g$ and the segment $\overline{gp}$ is contained in P.

Eidenbenz et al. [9] proved that finding a minimum guarding set for a polygon, where the given set of guards is the polygon's set of vertices, is APX-hard, even if the polygon has no holes and there are no limits on the ranges. We show that our generalization of the local search technique enables us to apply local search to find a $(1 + \varepsilon)$-approximation of a minimum guarding set for P (with respect to $\mathcal{G}$), under the above assumption concerning the depth of the arrangement of D.

For a guard $g \in \mathcal{G}$, let VP_g denote its visibility polygon and let VR_g denote its visible region, where $VR_g = VP_g \cap D_g$. Notice that VR_g is not necessarily convex. For a subset $X \subseteq \mathcal{G}$, let D_X denote the set $\{D_g : g \in X\}$ and let VR_X denote the set $\{VR_g : g \in X\}$. Let $B \subseteq \mathcal{G}$ be the set of guards obtained by applying local search (to the set $VR_{\mathcal{G}}$) and let R be an optimal set. (We may assume that $B \cap R = \emptyset$, since otherwise, we can remove the guards that appear in both sets.) Consider the graph $G = (B \cup R, E)$, where $(g, g') \in E$ if and only if $VR_g \cap VR_{g'} \neq \emptyset$. We claim that G has a separator of size $\sqrt{ln} = O(\sqrt{n})$, since any separator of the intersection graph of $D_B \cup D_R$ is also a separator of G, and the former graph has a separator of size $\sqrt{ln} = O(\sqrt{n})$, by a result of Miller et al. [16]. Observe that the locality condition is satisfied, since for any $p \in P$, let $b \in B$ and $r \in R$ be two guards that guard p, then $VR_b \cap VR_r \neq \emptyset$ and hence the edge (b, r) is in G. Therefore, by Theorem 4, we conclude that $|B| \leq (1 + \varepsilon)|R|$.

Theorem 8. *There exists a PTAS for minimum guarding a (not necessarily simple) polygon with respect to a given set of guards, assuming that the depth of the corresponding arrangement of disks is bounded by a constant.*

It is possible to consider other versions of the problem in which the visible region of a guard $g \in \mathcal{G}$ is defined differently, as long as $VR_g \subseteq D_g$. The approximation analysis for these versions is the same as for the standard version. For example, consider the problem of guarding a polygon P, where each guard can see through at most some fixed number of walls (edges of the polygon P). Then, the visibility polygon of a guard $g \in \mathcal{G}$ is different than its visibility polygon in the standard version, but still its visible region is contained in D_g. For such problems, one needs to modify the verification procedure of the local search algorithm (i.e., whether a given set of guards consists of a solution). We thus obtain, for example, the following corollary.

Corollary 1. *There exists a PTAS for minimum guarding a polygon through walls, assuming that the depth of the corresponding arrangement of disks is bounded by a constant.*

Terrain guarding with limited visibility. Let T be a 1.5D terrain (i.e., an x-monotone polygonal chain), let $\mathcal{G} \subseteq T$ be a finite set of guards, and let $X \subseteq T$ be a finite set of points to be guarded. A guard $g \in \mathcal{G}$ sees a point $x \in X$ if $\overline{gx}$ does not cross any edge of T and $|gx| \leq r_g$. The goal is to find a minimum-cardinality subset $\mathcal{G}' \subseteq \mathcal{G}$, such that $\mathcal{G}'$ sees X (i.e., for each point $x \in X$, there exists a guard in $\mathcal{G}'$ that sees x). Assuming unlimited visibility (i.e., $r_g = \infty$, for each $g \in \mathcal{G}$), the problem is known to be NP-hard [7, 15], and there exists a local-search-based PTAS for it [13]. This PTAS relies heavily on the, so called, "order claim", which states that for any four points a, b, c, d on T (listed from left to right), if a sees c and b sees d then a also sees d. Unfortunately, this claim is false in the limited visibility version. However, this version is a simple variant of polygon guarding with limited visibility. Therefore, we have

Theorem 9. *There exists a PTAS for terrain guarding, assuming that the depth of the corresponding arrangement of disks is bounded by a constant.*

Consider the guarding problems above, but now set $r_g = 1$, for each $g \in \mathcal{G}$, and assume the requirement is to guard a finite set X of points in P (alt., on T). Then, one may replace the assumption that the depth of the arrangement of disks is bounded by a constant with the assumption that the points in X are not too crowded, i.e., the distance between any two points in X is at least some constant $\delta > 0$. This assumption is sometimes more convenient. Under this assumption, each guard sees at most some constant number $c = c(\delta)$ of points of X, and it is likely therefore that there are many redundant guards. (A guard is *redundant* if there exists another guard that sees the same subset of points of X.) We thus remove redundant guards from $\mathcal{G}$, one at a time, as long as there are such guards in $\mathcal{G}$. It is easy to see that the depth of the arrangement of disks corresponding to the set of remaining guards is bounded by a constant. We obtain, e.g., the following theorem.

Theorem 10. *There exists a PTAS for minimum guarding a sparse set of points X within a (not necessarily simple) polygon P with respect to a given set of guards of unit visibility range.*

4   Discrete Coverage of Points

We consider the following fundamental problem. Let P be a set of points in the plane and let D be a set of disks that covers P. (We assume that the centers of the disks in D are in general position.) Find a minimum-cardinality subset $D' \subseteq D$ that covers P.

If D is a set of unit disks, then this is the dual version of the problem known as discrete piercing of unit disks. Mustafa and Ray [17] presented a PTAS for the latter problem (even for the case of varying radii). We present a local-search-based PTAS for discrete coverage of points. Our proof is based on the framework developed in [17] and in Section 2, and is inspired by the work of Gibson and Pirwani [14].

Theorem 11. *There exists a PTAS for discrete coverage of points by disks (alt., by axis-parallel squares).*

Proof. We may assume that there do not exist two disks in D, such that one of them is contained in the other, since, if two such disks do exist, we can always prefer the larger one. Let B be the set of the centers of the disks returned by the local search algorithm, and let R be the set of the centers of the disks in an optimal solution. We may assume that $B \cap R = \emptyset$ (since otherwise, we can remove the centers that belong to both sets). For a center $c \in B \cup R$, let $D_c \in D$ denote the corresponding disk and r_c its radius. In order to prove that $|B| \leq (1 + \varepsilon)|R|$, we need to present a planar bipartite graph $G = (B \cup R, E)$ satisfying the following locality condition: For each $p \in P$ there exist $b \in B$ and $r \in R$, such that $p \in D_b$, $p \in D_r$, and $(b, r) \in E$.

Consider the additively weighted Voronoi diagram of $B \cup R$, constructed according to the distance function $\delta(p, c) = d(p, c) - r_c$. Observe first that for any center $c \in B \cup R$ and point $p \in \mathbb{R}^2$, if $\delta(p, c) \leq 0$, then $p \in D_c$ (and vice versa). Now, let $cell(c)$ denote the cell of the diagram corresponding to $c \in B \cup R$. It is well known that for each $c \in B \cup R$, $c \in cell(c)$ and $cell(c)$ is connected.

Our planar bipartite graph is the dual graph of the weighted Voronoi diagram defined above, without the monochromatic edges. That is, there is an edge between two centers c and c' in $B \cup R$, if and only if $c \in B$ and $c' \in R$, or vice versa, and their cells are adjacent to each other. We denote this graph by $G = (R \cup B, E)$. It is easy to see that G is planar. Indeed, by the diagram's properties, any edge $(c, c') \in E$ can be drawn such that it is contained in $cell(c) \cup cell(c')$, and within each cell $cell(c)$, it is easy to ensure that the edges do not cross each other.

We now show that G satisfies the locality condition. Let $p \in P$, and assume w.l.o.g. that $p \in cell(r)$, where $r \in R$. Let $b \in B$ be the closest center to p according to δ. That is, for any $b' \in B$, $\delta(p, b) \leq \delta(p, b')$. Notice that $\delta(p, b) \leq 0$, since there exists a center $b' \in B$ whose disk covers p, and therefore, $\delta(p, b') \leq 0$. We conclude by the observation above that $p \in D_b$. Consider the cells we visit when walking along the line segment $\overline{pb}$ from p to b. Since $p \notin cell(b)$ and $b \in cell(b)$, we must at some point enter $cell(b)$. Let $c \in B \cup R$ be the center for which $cell(c)$ is the last cell that we visit before entering $cell(b)$, and let q be the point on $\overline{pb}$, which is also on the boundaries of $cell(c)$ and $cell(b)$.

It remains to show that $c \in R$, implying that $(b, c) \in E$, and that $p \in D_c$. Using the triangle inequality and since the centers of the disks are in general position, we get that $\delta(p, c) = d(p, c) - r_c < d(p, q) + d(q, c) - r_c = d(p, q) + \delta(q, c)$. But, $d(p, q) + \delta(q, c) = d(p, q) + \delta(q, b) = d(p, q) + d(q, b) - r_b = d(p, b) - r_b = \delta(p, b)$, so we get that $\delta(p, c) < \delta(p, b)$. Now, since b is the closest center to p among the centers in B, we conclude that $c \in R$, and since $\delta(p, b) \leq 0$, we conclude that $p \in D_c$. The axis-parallel version is obtained by replacing the L_2 metric by the L_∞ metric.

4.1   Discrete Coverage of a Polygon

Consider now the following problem. Let Q be a polygon, and let D be a set of disks (alt., a set of axis-parallel rectangles). Find a minimum-cardinality subset of D that covers Q. The local search algorithm can be easily adapted to this setting. At each iteration of the algorithm, instead of checking whether all points are covered, one needs to check whether the entire polygon is covered. This can be done in polynomial time. The analysis is essentially the same as for discrete coverage of points, except that here p is any point in the polygon Q. We thus conclude that

Theorem 12. *There exists a PTAS for discrete coverage of a polygon by disks (alt., by axis-parallel squares).*

4.2   The Class Cover Problem

The class cover problem is defined as follows: Let B be a set of blue points and let R be a set of red points and set $n = |B| + |R|$. Find a minimum-cardinality set D of disks (alt., of axis-parallel squares) that covers the blue points, but does not cover any of the red points. That is, find a minimum-cardinality set D, such that $B \cap \cup_{d \in D} d = B$ and $R \cap \cup_{d \in D} d = \emptyset$. Bereg et al. [3] study several versions of the class cover problem with boxes. In particular, they prove that the class cover problem with axis-parallel squares is NP-hard and give an $O(1)$-approximation algorithm for this version. We show that the class cover problem with squares (resp., disks) is essentially equivalent to discrete coverage of points by squares (resp., disks), and therefore there exists a PTAS for both versions.

Indeed, consider any set D of disks that cover the points in B and does not cover any point in R. It is easy to see that one can replace each disk $d \in D$ with a "legal" disk d', such that $B \cap d \subseteq B \cap d'$ and either there are three points on d''s boundary, or there are exactly two points on d''s boundary and the line segment between them is a diameter of d'. Therefore, we can transform the class cover problem with disks to discrete coverage of points by disks, by first computing all $O(n^3)$ disks defined by either triplets or pairs of points and then removing those that are "illegal". Similarly, we can transform the class cover problem with squares to discrete coverage of points by squares. We thus obtain

Theorem 13. *There exists a PTAS for the class cover problem with disks (alt., with axis-parallel squares).*

Concluding Remark. Very recently it has been brought to our attention that Chan and Grant [5] observe that the PTAS of Mustafa and Ray [17] for discrete hitting set of half-spaces in $\mathbb{R}^3$ implies a PTAS for discrete coverage of points by disks, by a lifting transformation that maps disks to lower half-spaces in $\mathbb{R}^3$ and by duality between points and half-spaces. However, our PTAS above for discrete coverage of points by disks is direct and refrains from moving to $\mathbb{R}^3$.

References

1. Alon, N., Seymour, P., Thomas, R.: A separator theorem for graphs with an excluded minor and its applications. In: Proc. 22nd ACM Sympos. on Theory of Computing, pp. 293–299 (1990)
2. Baker, B.S.: Approximation algorithms for NP-complete problems on planar graphs. J. ACM 41(1), 153–180 (1994)
3. Bereg, S., Cabello, S., Díaz-Báñez, J.M., Pérez-Lantero, P., Seara, C., Ventura, I.: The class cover problem with boxes. Comput. Geom. Theory Appl. 45(7), 294–304 (2012)
4. Chan, T.M.: Polynomial-time approximation schemes for packing and piercing fat objects. J. Algorithms 46(2), 178–189 (2003)
5. Chan, T.M., Grant, E.: Exact algorithms and APX-hardness results for geometric packing and covering problems. Comput. Geom. Theory Appl. (in press, available online)
6. Chan, T.M., Har-Peled, S.: Approximation algorithms for maximum independent set of pseudo-disks. In: Proc. 25th ACM Sympos. on Computational Geometry, pp. 333–340 (2009)
7. Chen, D.Z., Estivill-Castro, V., Urrutia, J.: Optimal guarding of polygons and monotone chains. In: Proc. 7th Canadian Conf. on Computational Geometry, pp. 133–138 (1995)
8. Efrat, A., Har-Peled, S.: Guarding galleries and terrains. Inf. Process. Lett. 100, 238–245 (2006)
9. Eidenbenz, S., Stamm, C., Widmayer, P.: Inapproximability results for guarding polygons and terrains. Algorithmica 31(1), 79–113 (2001)
10. Fox, J., Pach, J.: Separator theorems and Turan-type results for planar intersection graphs. In: Advances in Mathematics, vol. 219, pp. 1070–1080 (2009)
11. Frederickson, G.N.: Fast algorithms for shortest paths in planar graphs. SIAM Journal on Computing 6, 1004–1022 (1987)
12. Ghosh, S.K.: Approximation algorithms for art gallery problems. In: Proc. Canadian Inform. Process. Soc. Congress, pp. 429–434 (1987)
13. Gibson, M., Kanade, G., Krohn, E., Varadarajan, K.: An Approximation Scheme for Terrain Guarding. In: Dinur, I., Jansen, K., Naor, J., Rolim, J. (eds.) APPROX and RANDOM 2009. LNCS, vol. 5687, pp. 140–148. Springer, Heidelberg (2009)
14. Gibson, M., Pirwani, I.A.: Algorithms for Dominating Set in Disk Graphs: Breaking the log n Barrier. In: de Berg, M., Meyer, U. (eds.) ESA 2010, Part I. LNCS, vol. 6346, pp. 243–254. Springer, Heidelberg (2010)
15. King, J., Krohn, E.: Terrain guarding is NP-hard. In: Proc. 21st ACM-SIAM Sympos. on Discrete Algorithms, pp. 1580–1593 (2010)
16. Miller, G.L., Teng, S.-H., Thurston, W.P., Vavasis, S.A.: Separators for sphere-packings and nearest neighbor graphs. J. ACM 44(1), 1–29 (1997)
17. Mustafa, N.H., Ray, S.: Improved results on geometric hitting set problems. Discrete & Computational Geometry 44(4), 883–895 (2010)
18. Smith, W.D., Wormald, N.C.: Geometric separator theorems & applications. In: Proc. IEEE 39th Sympos. on Foundations of Computer Science, pp. 232–243 (1998)

A Randomised Approximation Algorithm for the Hitting Set Problem

Mourad El Ouali[1], Helena Fohlin[2], and Anand Srivastav[1]

[1] Departement of Computer Science. University of Kiel. Germany
{meo,asr}@informatik.uni-kiel.de
[2] Department of Clinical and Experimental Medicine. Linköping University, Sweden
Helena.Fohlin@lio.se

Abstract. Let $\mathcal{H} = (V, \mathcal{E})$ be a hypergraph with vertex set V and edge set $\mathcal{E}$, where $n := |V|$ and $m := |\mathcal{E}|$. Let l be the maximum size of an edge and Δ be the maximum vertex degree. A hitting set (or vertex cover) in $\mathcal{H}$ is a set of vertices from V in which all edges are incident. The hitting set problem is to find a hitting set of minimum cardinality. It is known that an approximation ratio of l can be achieved easily. On the other side, for constant l, an approximation ratio better than l cannot be achieved in polynomial time under the unique games conjecture (Khot and Ragev 2008). Thus breaking the l-barrier for significant classes of hypergraphs is a complexity-theoretic and algorithmically interesting problem, which has been studied by several authors (Krivelevich (1997), Halperin (2000), Okun (2005)). We propose a randomised algorithm of hybrid type for the hitting set problem, which combines LP-based randomised rounding, graphs sparsening and greedy repairing and analyse it in different environments. For hypergraphs with $\Delta = O(n^{\frac{1}{4}})$ and $l = O(\sqrt{n})$ we achieve an approximation ratio of $l\left(1 - \frac{c}{\Delta}\right)$, for some constant $c > 0$, with constant probability. In the case of l-uniform hypergraphs, l and Δ being constants, we prove by analysing the expected size of the hitting set and using concentration inequalities, a ratio of $l\left(1 - \frac{l-1}{4\Delta}\right)$. Moreover, for quasi-regularisable hypergraphs, we achieve an approximation ratio of $l\left(1 - \frac{n}{8m}\right)$. We show how and when our results improve over the results of Krivelevich, Halperin and Okun.

Keywords: Approximation algorithms, probabilistic methods, randomised rounding, hitting set, vertex cover, greedy.

1 Introduction

A hypergraph $\mathcal{H} = (V, \mathcal{E})$ consists of a finite set V and a set $\mathcal{E}$ of subsets of V. We call the elements of V vertices and the elements of $\mathcal{E}$ (hyper-)edges. Further let $n := |V|$ and $m := |\mathcal{E}|$. A hitting set (or vertex cover) of a hypergraph $\mathcal{H}$ is a set C of vertices such that for every $E \in \mathcal{E}$ there exists a vertex $v \in E \cap C$. The hitting set problem in hypergraphs is the task of finding a hitting set of minimum cardinality. A set $\mathcal{S} \subseteq \mathcal{E}$ is called a set cover, if all vertices of $\mathcal{H}$ are contained in edges of $\mathcal{S}$, and the set cover problem is to find a set cover of minimum

S.K. Ghosh and T. Tokuyama (Eds.): WALCOM 2013, LNCS 7748, pp. 101–113, 2013.
© Springer-Verlag Berlin Heidelberg 2013

cardinality. Note that the hitting set problem in hypergraphs is equivalent to the set cover problem by changing the role of vertices and edges.

A number of inapproximability results are known. Lund and Yannakakis [20] proved for the set cover problem that for any $\alpha < \frac{1}{4}$, the existence of a polynomial-time $(\alpha \ln n)$-ratio approximation algorithm would imply that $\mathcal{NP}$ has a quasipolynomial, i.e., $n^{\mathcal{O}(\text{poly}(\ln n))}$ deterministic algorithm. This result was improved to $(1 - o(1)) \ln n$ by Feige [7]. A $c \cdot \ln n$-approximation under the assumption that $\mathcal{P} \neq \mathcal{NP}$ was established by Safra and Raz [24], where c is a constant. A similar result for larger values of c was proved by Alon, Moshkovitz and Safra [1].

The hitting set problem remains hard for many hypergraph classes. Most interesting are l-uniform hypergraphs with a constant l, because for them under the unique games conjecture (UGC), it is $\mathcal{NP}$-hard to approximate the problem within a factor of $l - \epsilon$, for any *fixed* $\epsilon > 0$, see [17], while an approximation ratio of l can be easily achieved by finding a maximal matching. Therefore, the problem of breaking the l-barrier for significant and interesting classes of hypergraphs received much attention.

Let us briefly give an overview of the known approximability results for the problem. The earliest published approximation algorithms for the hitting set problem achieve an approximation ratio of the order $\ln m + 1$ [6,16,19] by using a greedy heuristic. For l-uniform hypergraphs, several authors achieved the ratio of l using different techniques (see e.g. [3,11,13,14]). The first and important result breaking the barrier of l for l-uniform hypergraphs, is due to Krivelevich [18]. He proved an approximation ratio of $l(1 - cn^{\frac{1-l}{l}})$, for some constant $c > 0$, using a combination of the LP-based algorithm and the local ratio approach described by Bar-Yehuda and Even [4]. Later, for l-uniform hypergraphs with $l^3 = o(\frac{\ln \ln n}{\ln \ln \ln n})$ and $\Delta = O(n^{l-1})$, Halperin [12] presented a semidefinite programming based algorithm with an approximation ratio of $l - (1 - o(1))\frac{l \ln \ln n}{\ln n}$. Note that this condition enforces the doubly exponential bound, $n \geq 2^{2^{l^2}}$, and already for $l = 3$ the hypergraph is very large and is hardly suitable for practical purposes.

A further important class consists of hypergraphs with Δ and l being constants. In this case Krivelevich [18] gave an LP-based algorithm that provides an approximation ratio of $l(1 - c\Delta^{\frac{1}{1-l}})$ for some constant $c > 0$. An improved approximation ratio of $l - (1 - o(1))\frac{l(l-1)\ln \ln \Delta}{\ln \Delta}$ was presented by Halperin [12], provided that $l^3 = o(\frac{\ln \ln \Delta}{\ln \ln \ln \Delta})$. For hypergraphs which are not necessarily uniform, but with size of edges bounded from above by a constant l, an improvement of the result of Krivelevich was given by Okun [23]. He proved an approximation ratio of $l(1 - c(\beta, l)\Delta^{-\frac{1}{\beta l}})$ for $\beta \in (0, 1)$ and a constant $c(\beta, l) \in (0, 1)$ depending on β and l, by a modification of the algorithm presented in [18].

Our Results. We consider hypergraphs with maximum edge size l and maximum vertex degree Δ, at the moment not necessarily assumed to be constants. In Section 3 we present a randomised algorithm, combining LP-based randomised rounding, sparsening of the hypergraph and greedy repairing. Such a hybrid approach is frequently used in practice and it has been analysed for many problems, e.g., maximum graph bisection [9], maximum graph partitioning problems [8,15] and the vertex cover and partial vertex cover problem in graphs [11,12]. In Section 4.1 we show that our algorithm achieves for $l = O(\sqrt{n})$ and $\Delta = O(n^{\frac{1}{4}})$ an approximation ratio of $l\left(1 - \frac{c}{\Delta}\right)$, for some constant $c > 0$, with constant probability. In this case our result improves the result of Krivelevich, for any function $f(n)$ satisfying $f(n) = O(n^{\frac{1}{4}})$, since $n^{\frac{1}{4}} < n^{1-\frac{1}{l}}$ for $l \geq 2$, and the approximation is the better the smaller $f(n)$ becomes. For $\Delta \leq \frac{\ln n}{\ln \ln n}$ we obtain a better approximation than Halperin. In Section 5.1 we analyse the algorithm for the class of uniform, quasi-regularisable hypergraphs, which are known and useful in the combinatorics of hypergraphs (see Berge [5]). We prove an approximation ratio of $l\left(1 - \frac{n}{8m}\right)$ provided that $\Delta = O(n^{\frac{1}{3}})$. This result improves the approximation ratio given by Krivelevich and Halperin for sparse hypergraphs (roughly speaking sparseness means, $m \leq n^{\alpha}$, $\alpha \leq 2$, see section 5.1, page 9 for details). In Section 5.2 we consider l-uniform hypergraphs, where l and Δ are constants, and achieve a ratio of $l\left(1 - \frac{l-1}{4\Delta}\right)$. This improves over the result of Krivelevich for Δ smaller than $(l-1)^{1+\frac{1}{l-2}}$ and of Okun for Δ smaller than $(l-1)^{1+\frac{1}{3l-1}}$, respectively.

The paper is organised as follows: In Section 2 we give definitions and probabilistic tools. In Section 3 we present our randomised algorithm for the hitting set problem. In Section 4 we analyse the approximation ratio for hypergraphs with non-constant size of edges and non-constant vertex degree. In Section 5 we analyse the algorithm in a different way and prove an approximation ratio for the subclass of uniform quasi-regularisable hypergraphs (Section 5.1) and uniform hypergraphs with bounded vertex degree (Section 5.2). In Section 6 we comment on some future works.

2   Preliminaries and Definitions

Graph-theoretical Notions. Let $\mathcal{H} = (V, \mathcal{E})$ be a hypergraph, with V and $\mathcal{E}$ its set of vertices and edges. For $v \in V$ we define $d(v) = |\{E \in \mathcal{E}; v \in E\}|$ and $\Delta = \max_{v \in V}\{d(v)\}$. Here $d(v)$ is the vertex-degree of v and Δ is the maximum vertex degree of $\mathcal{H}$. Further for a set $X \subseteq V$ we denote by $\Gamma(X) := \{E \in \mathcal{E}; X \cap E \neq \emptyset\}$ the set of edges incident to the set X. Let $l, \Delta \in \mathbb{N}$ be two given constants. We call $\mathcal{H}$ l-uniform, if $|E| = l$ for all $E \in \mathcal{E}$, and with bounded degree Δ, if for every $v \in V$ it holds $d(v) \leq \Delta$. It is convenient to order the vertices and edges, i.e., $V = \{v_1, \ldots, v_n\}$ and $\mathcal{E} = \{E_1, \ldots, E_m\}$, and to identify the vertices and edges with their indices.

For an integer $k \geq 0$, multiplying the edge E_i by k means replacing the edge E_i in $\mathcal{H}$ by k identical copies of E_i. If $k = 0$, this operation is the deletion of the edge E_i. A hypergraph $\mathcal{H}$ is called *regularisable* if a regular hypergraph can be obtained from $\mathcal{H}$ by multiplying each edge E_i by an integer $k_i \geq 1$. Finally, a hypergraph $\mathcal{H}$ is called *quasi-regularisable* if a regular hypergraph is obtained by multiplying each edge E_i by an integer $k_i \geq 0$ where $\sum_i^m k_i > 0$. Regular implies regularisable and this implies quasi-regularisable (see [5]). Note that quasi-regularisable hypergraphs play an important role in the study of matching and covering in hypergraphs. e.g. [10].

Concentration Inequalities. For the one-sided deviation the following Chebychev-Cantelli inequality will be frequently used:

Theorem 1 ([2]). *Let X be a non-negative random variable with finite mean $\mathbb{E}(X)$ and variance $\mathrm{Var}(X)$. Then for any $a > 0$ we have*

$$\Pr(X \geq \mathbb{E}(X) + a) \leq \frac{\mathrm{Var}(X)}{\mathrm{Var}(X) + a^2}.$$

A further useful concentration result is the independent bounded differences inequality theorem:

Theorem 2 (see [21]). *Let $X = (X_1, X_2, ..., X_n)$ be a family of independent random variables with X_k taking values in a set A_k for each k. Suppose that the real-valued function f defined on $\Pi_{k=1}^n A_k$ satisfies $|f(x) - f(x')| \leq c_k$ if the vector x and x' differ only in the k-th coordinate. Let $\mathbb{E}(X)$ be the expected value of the random variable $f(X)$. Then for any $t > 0$ it holds*

$$\Pr(f(X) \leq \mathbb{E}(f(X)) - t) \leq \exp\left(\frac{-2t^2}{\sum_{k=1}^n c_k^2}\right).$$

The following estimate on the variance of a sum of dependent random variables can be proved as in the book of Alon and Spencer:

Lemma 1 (see [2]). *Let X be the sum of finitely many 0/1 random variables, i.e. $X = X_1 + \ldots + X_n$, and let $p_i = \mathbb{E}(X_i)$ for all $i = 1, \ldots, n$. For a pair $i, j \in \{1, \ldots, n\}$ we write $i \sim j$, if X_i and X_j are dependent. Let Γ be the set of all unordered dependent pairs i, j, i.e. 2-element sets $\{i, j\}$, and let $\gamma = \sum_{\{i,j\} \in \Gamma} \mathbb{E}(X_i X_j)$, then it holds: $\mathrm{Var}(X) \leq \mathbb{E}(X) + 2\gamma$.*

3   The Randomised Algorithm

An integer, linear programming formulation of the hitting set problem in a hypergraph $\mathcal{H}$ is the following.

$$(\text{ILP-VC}) \qquad \min \sum_{j=1}^n x_j$$

$$\sum_{j \in E} x_j \geq 1 \text{ for all } E \in \mathcal{E},$$

$$x_j \in \{0, 1\} \text{ for all } j \in [n] := \{1, \ldots, n\}.$$

Its linear programming relaxation, denoted by LP-VC, is given by relaxing the integrality constraints to $x_j \in [0, 1] \ \forall j \in [n]$. Let Opt and Opt* be the value of an optimal solution to ILP-VC and LP-VC, respectively. Clearly, Opt* $\leq$ Opt. Let x^* be an optimal solution of LP-VC. Let $\epsilon \in [0, 1]$ be a parameter that will be chosen based on the application, we set $\lambda = l(1 - \epsilon)$.

Algorithm 1. VC-$\mathcal{H}$

Input   : A hypergraph $\mathcal{H} = (V, \mathcal{E})$
Output: A hitting set C
1. Initialise $C := \emptyset$.
2. Solve the LP relaxation of ILP-VC
3. Set $S_0 := \{j \in [n] \mid x_j^* = 0\}$, $S_1 := \{j \in [n] \mid x_j^* = 1\}$,
   $S_\geq := \{j \in [n] \mid 1 \neq x_j^* \geq \frac{1}{\lambda}\}$ and $S_\leq := \{j \in [n] \mid 0 \neq x_j^* < \frac{1}{\lambda}\}$.
4. Delete the vertices in S_0 from $\mathcal{H}$, and set $V := V \setminus S_0$ and $\mathcal{E} := \{E \cap V \mid E \in \mathcal{E}\}$.
5. Take all vertices of S_1 and $S_\geq$ into the hitting set C.
   Set $V := V \setminus S_1$ and $\mathcal{E} := \mathcal{E} \setminus \Gamma(S_1)$.
6. (Randomised Rounding) For all vertices $j \in S_\leq$ include the vertex j
   in the hitting set C, independently for all such j, with probability $x_j^* \lambda$.
7. (Repairing) Repair the Hitting Set C (if necessary) as follows:
   a) If $|\{E \in \mathcal{E} \mid E \cap C \neq \emptyset\}| = |\mathcal{E}|$, then return C.
   b) If $|\{E \in \mathcal{E} \mid E \cap C \neq \emptyset\}| < |\mathcal{E}|$, then pick arbitrary at most $|\mathcal{E}| - |C|$
      additional vertices from not covered edges in the hitting set.
8. Return the hitting set C of $\mathcal{H}$

Let us briefly explain the ingredients of the algorithm. Usually, as in [8,9,11], the LP or semidefinite program is solved and randomised rounding or random hyperplane techniques are used followed by a repairing step. In our algorithm we thin out the hypergraph by removing vertices and edges corresponding to LP-variables with zero value, which will not be taken into the hitting set by randomised rounding (Step 4), *before* entering randomised rounding and repairing. This is an intuitively meaningful sparsening, and in fact will be necessary in Section 5 where we estimate the expected size of the repaired hitting set (one step analysis), while in Section 4 it is sufficient to analyse randomised rounding and repairing separately.

4   Two-Step Analysis of the Algorithm VC-$\mathcal{H}$

Let $X_1, ..., X_n$ be 0/1-random variables defined as follows: X_j is 1 if the vertex v_j was picked into the hitting set after the rounding step and 0 otherwise.
For all $i \in [m]$ we define the 0/1- random variables Z_i as follows: Z_i is 1 if the edge E_i is covered after the rounding step and 0 otherwise.
Then $Y := \sum_{j=1}^n X_j$ is the cardinality of the hitting set after the randomised rounding step in the algorithm and $W = \sum_{j=1}^m Z_j$ is the number of covered edges after this step.

For the expected size of the hitting set we have the following upper bound:

$$\mathbb{E}(|C|) \leq \mathbb{E}(Y) + \mathbb{E}(m - W). \tag{1}$$

For the computation of the expectation of W we need the following lemma (See Lemma 2.2 [22]).

Lemma 2. *For all $n \in \mathbb{N}$, $\lambda > 0$ and $x_1, \cdots, x_n, z \in [0,1]$ with $\sum_{i=1}^{n} x_i \geq z$ and $\lambda x_i < 1$ for all $i \in \mathbb{N}$, we have $\prod_{i=1}^{n}(1 - \lambda x_i) \leq (1 - \lambda \frac{z}{n})^n$, and this bound is the tight maximum.*

Lemma 3. *Let l and Δ be integers, not necessarily constant and let $\epsilon > 0$.*

(i) $\mathbb{E}(W) \geq (1 - \epsilon^2)m$.
(ii) $\text{Opt}^* \geq \frac{m}{\Delta}$.
(iii) *Let hypergraph $\mathcal{H} = (V, \mathcal{E})$ with $x_j^* > 0$ for all $j \in [n]$ it holds $\text{Opt}^* \geq \frac{n}{l}$, where l is the maximum size of a edge.*
(iv) $\text{Opt}^* \leq \mathbb{E}(Y) \leq \lambda \text{Opt}^*$.

Proof. (i) For this proof we consider an equivalent form of the LP relaxation of the problem given in section 2.

$$(LP-1) \quad \min \sum_{j=1}^{n} x_j$$

$$\sum_{j=1}^{n} a_{ij} x_j \geq z_i \text{ for all } i \in [m] := \{1, \ldots, m\}$$

$$\sum_{i=1}^{m} z_i \geq m$$

$$x_j, z_i \in [0,1] \text{ for all } i \in [m], j \in [n].$$

It is easy to show that an optimal solution of LP-1 is an optimal solution of LP and vice versa.

Let $i \in [m]$, $|E_i| = r$ and $z_i^* = \sum_{j \in E_i} x_j^*$. If there is a $j \in E_i$ with $\lambda x_j \geq 1$ then $\Pr(Z_i = 0) = 0$, else we have

$$\Pr(Z_i = 0) = \prod_{j \in E_i}(1 - \lambda x_j^*) \underset{\text{Lem 2}}{\leq} \left(1 - \frac{\lambda z_i^*}{r}\right)^r$$

$$\leq \left(1 - \frac{\lambda z_i^*}{l}\right)^r = (1 - (1 - \epsilon)z_i^*)^r$$

$$\leq (1 - (1 - \epsilon)z_i^*)^2 \leq 1 - z_i^*(1 - \epsilon^2)$$

and we get

$$\mathbb{E}(W) = \sum_{i=1}^{m} \Pr(Z_i = 1) = \sum_{i=1}^{m}(1 - \Pr(Z_i = 0))$$

$$\geq \sum_{i=1}^{m}(1 - (1 - z_i^*(1 - \epsilon^2))) = \sum_{i=1}^{m} z_i^*(1 - \epsilon^2) = (1 - \epsilon^2)\sum_{i=1}^{m} z_i^*$$

$$\geq (1 - \epsilon^2)m.$$

(ii) Let $d(v_j)$ the degree of the vertex v_j. With the ILP constraints we have

$$m \leq \sum_{i=1}^{m} z_i^* \leq \sum_{i=1}^{m}\sum_{j \in E_i} x_j^* = \sum_{j=1}^{n} d(v_j)x_j^* \leq \Delta \sum_{j=1}^{n} x_j^* = \Delta \cdot \text{Opt}^*$$

(iii) Let consider the LP problem dual to the hitting set LP problem

$$\text{(D-VC)} \qquad \max \sum_{j \in \mathcal{E}} y_j$$

$$\sum_{j \in \mathcal{E},\, i \in j} y_j \leq 1 \ \text{ for every } i \in V,$$

$$y_j \in [0, 1] \quad \text{ for all } j \in \mathcal{E}.$$

Let $(y_j^*)_{j \in [m]}$ resp. $\text{Opt}^*(D)$ be an optimal solution of D-VC resp. the value of the optimal solution, than the duality Theorem of Linear Programming applied to the (LP-VC) and (D-VC) implies:

(a) $\text{Opt}^* = \text{Opt}^*(D)$
(b) If $x_i^* > 0 \Rightarrow \sum_{j \in \mathcal{E},\, i \in j} y_j = 1$.

Therefore, we have
$$n = \sum_{i \in V} 1 = \sum_{i \in V} \sum_{j \in \mathcal{E},\, i \in j} y_j^* = \sum_{j \in \mathcal{E}} y_j^* |j \cap V| \leq l \sum_{j \in \mathcal{E}} y_j^* \underset{(a)}{=} l\text{Opt}^*.$$

(iv) By using the LP relaxation and the definition of the sets S_1, $S_\geq$ and $S_\leq$, and since $\lambda \geq 1$, we get

$$\text{Opt}^* \leq \overbrace{\underbrace{|S_1|}_{\text{Opt}^*(S_1)} + \underbrace{|S_\geq|}_{\leq \lambda \text{Opt}^*(S_\geq)}}^{=\mathbb{E}(Y)} + \lambda \text{Opt}^*(S_\leq) \leq \lambda \text{Opt}^*.$$

$\square$

4.1 Hypergraphs with Non-constant l, Δ

In this section we will analyse the algorithm for hypergraphs with maximum degree and maximum edge size that are not constant but may be given as functions of n. The main result in this section is:

Theorem 3. *Let $\mathcal{H}$ be a hypergraph with maximum edge size $l = O(\sqrt{n})$ and maximum vertex degree $\Delta = O(n^{\frac{1}{4}})$. The algorithm VC-$\mathcal{H}$ returns a hitting set C such that, $|C| \leq l \left(1 - \frac{\sqrt{2}-1}{4\sqrt{2}\Delta}\right) \text{Opt}$ with probability at least $\frac{3}{5}$.*

Proof. *Case 1 :* $S_0 = \emptyset$.
Let

$$\epsilon := \frac{l\text{Opt}^*(1+\beta)}{4m} \quad \text{for} \quad \beta = \frac{\sqrt{2l}}{\sqrt{n}}. \tag{2}$$

We can assume that

$$\epsilon \leq \frac{1+\beta}{4-\eta}, \quad \text{for all} \quad \eta \in (0, 1), \tag{3}$$

because otherwise it follows from the definition of ϵ in (2) that $l\mathrm{Opt}^* \geq \frac{4m}{4-\eta}$, hence $l(1 - \frac{\eta}{4})\mathrm{Opt}^* \geq m$. Since a hitting set of size m can be trivially found by picking m arbitrary edges and taking one vertex from each of them pairwise distinct, we can get a $l(1 - \frac{\eta}{4})$-approximation —i.e. a constant factor strictly better than l— in this case.

It is straightforward to check that (3) implies $\epsilon \leq \frac{2}{3}$, so $\lambda = l(1 - \epsilon) \underset{l \geq 3}{>} 1$.

Claim 1. $\Pr\left(W \leq m(1 - \epsilon^2) - \sqrt{\sum_{i=1}^n d^2(v_i)}\right) \leq \frac{1}{5}$.

Proof of Claim 1. First we consider the function: $f(X_1, ..., X_n) = \sum_{j=1}^m Z_j$.
f satisfies: $|f(X_1, .., X_k, .., X_n) - f(X_1, .., X_k', .., X_n)| \leq d(v_k)$, with $X_k' \in \{0, 1\}$ and $X_k \neq X_k'$.

Since the $X_1, ..., X_n$ are chosen independently at random, by Theorem 2 we get for any $t > 0$

$$\Pr(f(X) - \mathbb{E}(f(X)) \leq -t) \leq \exp\left(\frac{-2t^2}{\sum_{i=1}^n d^2(v_i)}\right). \qquad (4)$$

Let us choose $t = \sqrt{\sum_{i=1}^n d^2(v_i)}$. By Lemma 3 (ii)

$$\Pr\left(W \leq m(1 - \epsilon^2) - \sqrt{\sum_{i=1}^n d^2(v_i)}\right) \leq \Pr\left(W \leq \mathbb{E}(W) - \sqrt{\sum_{i=1}^n d^2(v_i)}\right)$$

$$\underset{\mathrm{Ineq}\,(4)}{\leq} \exp\left(\frac{-2\sum_{i=1}^n d^2(v_i)}{\sum_{i=1}^n d^2(v_i)}\right) < \frac{1}{5}.$$

This concludes the proof of Claim 1.

Claim 2. For $\beta = \frac{\sqrt{2l}}{\sqrt{n}}$ it holds that $\Pr\left(Y \geq l \cdot \mathrm{Opt}^*(1 - \epsilon)(1 + \beta)\right) < \frac{1}{5}$.

Proof of Claim 2. The random variables $X_1, ..., X_n$ in the rounding step are independent. Moreover, since $l \leq \sqrt{2n}$ we have $\beta \in (0, 1)$. Thus the Angluin-Valliant form of Chernoff bound ([21], Theorem 2.3, p. 200) shows

$$\Pr\left(Y \geq l(1 - \epsilon)(1 + \beta)\mathrm{Opt}^*\right) \underset{\mathrm{Lem}\,3(iv)}{\leq} \Pr\left(Y \geq \mathbb{E}(Y)(1+\beta)\right) \leq \exp\left(-\frac{\beta^2\mathbb{E}(Y)}{3}\right).$$

On the other hand we have: $\frac{\mathbb{E}(Y)\beta^2}{3} \underset{\mathrm{Lem}\,3(iv)}{\geq} \frac{\mathrm{Opt}^*\beta^2}{3} \underset{\mathrm{Lem}\,3(iii)}{\geq} \frac{n\beta^2}{3l} \geq \frac{2l^2n}{3ln} \underset{l \geq 3}{\geq} 2.$

Finally we get: $\Pr\left(Y \geq l(1 - \epsilon)(1 + \beta)\mathrm{Opt}^*\right) \leq \exp\left(-2\right) < \frac{1}{5}.$
This concludes the proof of Claim 2.

By Claims 1 and 2 we get with probability at least $1 - (\frac{1}{5} + \frac{1}{5}) \geq \frac{3}{5}$ an upper bound for the final hitting set:

$$|C| \leq \underbrace{l(1 - \epsilon)(1 + \beta)\mathrm{Opt}^* + m\epsilon^2}_{(*)} + \underbrace{\sqrt{\sum_{i=1}^n d^2(v_i)}}_{(**)}.$$

By Lemma 3(iii) and the condition $\Delta \leq \frac{1}{32}n^{\frac{1}{4}}$ it holds:

$$(**) = \sqrt{\sum_{i=1}^{n} d^2(v_i)} \leq \Delta\sqrt{n} \leq \sqrt{\frac{n}{l}}\sqrt{l}\Delta \leq l\sqrt{Opt^*}\sqrt{Opt^*}\frac{1}{4\sqrt{2}\Delta} \leq lOpt^*\frac{1}{4\sqrt{2}\Delta}.$$

Furthermore we have

$$(*) \underset{Eq\ (2)}{=} l\left((1+\beta)(1-\epsilon) + \frac{lOpt^*(1+\beta)}{16m}\right)Opt^* = l(1+\beta)\left(1 - \frac{3lOpt^*(1+\beta)}{16m}\right)Opt^*$$

$$\underset{Lem\ 3(ii)}{\leq} l(1+\beta)\left(1 - \frac{3l(1+\beta)}{16\Delta}\right)Opt^* = l\left(1 + \beta - \frac{3l(1+\beta)^2}{16\Delta}\right)Opt^*.$$

On the other hand we can easily check, that $\frac{3l(1+\beta)^2}{16\Delta} - \beta \geq \frac{1}{4\Delta}$, therefore

$$l(1-\epsilon)(1+\beta)Opt^* + m\epsilon^2 \leq l\left(1 - \frac{1}{4\Delta}\right)Opt^*.$$

Finally $(*) + (**) \leq l\left(1 - \frac{1}{4\Delta} + \frac{1}{4\sqrt{2}\Delta}\right)Opt^* \leq l\left(1 - \frac{\sqrt{2}-1}{4\sqrt{2}\Delta}\right)Opt^*.$

The randomised algorithm returns with probability at least $\frac{3}{5}$ a hitting set C with cardinality at most $l\left(1 - \frac{c}{\Delta}\right)Opt^*$, where $c = \frac{1}{4}(1 - \frac{1}{\sqrt{2}})$.

Case 2: If S_0 is not empty, we can consider the sub-hypergraph $\mathcal{H}$ constructed in step 4 of algorithm VC-$\mathcal{H}$. Let $\tilde{\Delta}$ resp. $\tilde{l}$ be the maximum vertex degree resp. the maximum edge size of this sub-hypergraph. Now for this hypergraph we have $S_0 = \emptyset$ and with Case 1 we get a hitting set of cardinality at most $\tilde{l}(1 - \frac{c}{\tilde{\Delta}})Opt$. Since $\tilde{l} \leq l$ and $\tilde{\Delta} \leq \Delta$, the assertion of Theorem 3 holds.   □

Remark 1. For hypergraphs addressed in Theorem 3 we have an improvement over the result of Krivelevich [18], for any function $f(n)$ satisfying $f(n) = O(n^{\frac{1}{4}})$, since $n^{\frac{1}{4}} < n^{1-\frac{1}{l}}$ for $l \geq 2$, and our approximation is the better the smaller $f(n)$ becomes. For $\Delta \leq \frac{\ln(n)}{\ln\ln(n)}$ we obtain a better approximation than Halperin [12].

5   One-Step Analysis of the Algorithm VC-$\mathcal{H}$

Instead bounding the error probability of the randomised rounding step and the repairing step separately as above, in this section we analyse the expected size of the hitting set including repairing, and then use concentration inequalities.

For a set $S \subset \{1, ..., n\}$ let $Opt^*(S) := \sum_{j \in S} x_j^*$. By (1) it holds

$$\mathbb{E}(|C|) \leq Opt^*(S_1) + l(1-\epsilon)(Opt^*(S_\geq) + Opt^*(S_\leq)) + m\epsilon^2 \qquad (5)$$

Let us choose:

$$\epsilon = \frac{l(Opt^*(S_\geq) + Opt^*(S_\leq))}{2m}. \qquad (6)$$

We can assume that $\dfrac{l\left(\text{Opt}^*(S_\geq)+\text{Opt}^*(S_\leq)\right)}{2m} \in [0,1]$. Otherwise, if $\dfrac{l\left(\text{Opt}^*(S_\geq)+\text{Opt}^*(S_\leq)\right)}{2m} > 1$ then $\frac{l}{2}\text{Opt}^* \geq \frac{l}{2}\left((\text{Opt}^*(S_\geq) + \text{Opt}^*(S_\leq))\right) > m$.
Since any hitting set of cardinality m can be found trivially, this approximates the optimum within a factor of $\frac{l}{2} < l$.
Let $S_f := S_\geq \cup S_\leq \backslash \{j \in [n] | x_j^* = 0\}$. Plugging in ϵ from (6) into (5), we get

$$\mathbb{E}(|C|) \leq \text{Opt}^*(S_1) + l\left(1 - \frac{l\text{Opt}^*(S_f)}{4m}\right)\text{Opt}^*(S_f). \tag{7}$$

We observe here that the LP-based sparsening of the instance becomes relevant. At next we compute the variance of the size of the hitting set. We get,

Lemma 4. *Let* $X_1,\ldots,X_n$ *be the 0/1-random variables returned by algorithm VC-$\mathcal{H}$. Then we have* $\text{Var}(|C|) \leq l\Delta\mathbb{E}(|C|)$.

Proof. Let Γ and γ like in Lemma 1. Furthermore for every $v_i, v_j \in V, X_i, X_j$ are dependent iff they belong to the same edge. Thus, for a fixed v_i, there are at the most $(l-1)d(v_i)$ random variables X_j depending on X_i. Furthermore it holds for every $v_i, v_j \in V$:

$$\mathbb{E}(X_iX_j) = \Pr(X_i = 1 \wedge X_j = 1) \leq \min\{\Pr(X_i = 1), \Pr(X_j = 1)\}$$
$$\leq \frac{\Pr(X_i = 1) + \Pr(X_j = 1)}{2}.$$

Moreover
$$\gamma = \sum_{\{v_i,v_j\}\in\Gamma} \mathbb{E}(X_iX_j) \leq \sum_{\{v_i,v_j\}\in\Gamma} \frac{\Pr(X_i = 1) + \Pr(X_j = 1)}{2}$$
$$\leq \sum_{i=1}^{n} \frac{(l-1)d(v_i)}{2}\Pr(X_i = 1) = \frac{(l-1)d(v_i)}{2}\mathbb{E}(|C|)$$

by Lemma 1 we have: $\text{Var}(|C|) \leq \mathbb{E}(|C|) + (l-1)d(v_i)\mathbb{E}(|C|) \leq l\Delta\mathbb{E}(|C|)$.   □

5.1   Quasi-Regularisable l-Uniform Hypergraphs

Recall that S_1 is the set $S_1 = \{j \in [n] \mid x_j^* = 1\}$, containing those vertices for which the LP-optimal solution is tight (see algorithm VC-$\mathcal{H}$, step 3).
The next theorem is the main result of this section and it is proved using the above stated estimation (7) of $\mathbb{E}(|C|)$ and the Chebychev-Cantelli inequality.

Theorem 4. *Let* $\mathcal{H}$ *be a l-uniform, quasi-regularisable hypergraph with arbitrary l and maximum vertex degree* $\Delta = O(n^{\frac{1}{3}})$, *then the algorithm VC-$\mathcal{H}$ returns a hitting set C such that,* $|C| \leq l\left(1 - \frac{n}{8m}\right)\text{Opt}^*$ *with probability at least* $\frac{3}{4}$.

We need the following theorem of Berge [5].

Theorem 5. *For an l-uniform hypergraph* $\mathcal{H}$, *the following properties are equivalent:*

1. $\mathcal{H}$ is quasi-regularisable;
2. $\text{Opt}^* = \frac{n}{l}$ (i.e. the vector $x^* = (\frac{1}{l}, ..., \frac{1}{l})$ is an optimal solution for the LP relaxation and l is the size of the edges).

By this theorem, the condition $S_1 = \emptyset$ becomes a graph-theoretical meaning.

Proof of Theorem 4. By (7) and Theorem 5 we get for quasi-regularisable l-uniform hypergraphs with arbitrary l and bounded degree Δ the approximation

$$\mathbb{E}(|C|) \leq l \left(1 - \frac{n}{4m}\right) \text{Opt}^*. \tag{8}$$

Hence

$$\Pr\left(|C| \geq l \left(1 - \frac{n}{8m}\right) \text{Opt}^*\right) \underset{\text{Ineq (8)}}{\leq} \Pr\left(|C| \geq \mathbb{E}(|C|) + \frac{nl\text{Opt}^*}{8m}\right) \underset{\text{Th 1}}{\leq} \frac{1}{4}.$$

Namely for $n \geq 8^2\Delta^3$ we get with a straightforward calculation that $\frac{\left(\frac{ln\text{Opt}^*}{8m}\right)^2}{\text{Var}(|C|)} \geq l \geq 3$. So we obtain a hitting set C of size at most $l \left(1 - \frac{n}{8m}\right) \text{Opt}^*$ with probability at least $\frac{3}{4}$.   □

Remark 2. In Theorem 4, we can assume that $n < 8m$, because otherwise we have $\text{Opt}^* = \frac{n}{l} \geq \frac{8m}{l}$ thus $m \leq \frac{l}{8}\text{Opt}^*$. By taking one vertex for each edge we obtain a hitting set of cardinality $\frac{l}{8}\text{Opt}^*$, which gives already an approximation ratio of $l/8$. For hypergraphs addressed in Theorem 4 we have an improvement over the ratio of Krivelevich if $m \leq cn^{\frac{2l-1}{l}}$ and the ratio of Halperin if $m \leq \frac{(1-o(1))^{-1}\ln(n)n}{\ln\ln(n)}$.

5.2 l-Uniform Hypergraphs with Bounded Vertex Degree

In this section l and Δ are constants and $\mathcal{H}$ is an l-uniform hypergraph. Let $\tilde{\mathcal{H}} = (\tilde{V}, \tilde{\mathcal{E}})$ be the sub-hypergraph of $\mathcal{H}$ constructed in step 5 of the algorithm VC-$\mathcal{H}$ with $|\tilde{V}| = \tilde{n}$ and $|\tilde{\mathcal{E}}| = \tilde{m}$. We denote by $\tilde{l}$ and $\tilde{\Delta}$ the maximum size of all edges and the maximum vertex degree in $\tilde{\mathcal{H}}$. We consider the LP relaxation of the ILP formulation of the hitting set problem in $\tilde{\mathcal{H}}$ which we denote by $\text{LP}(\tilde{\mathcal{H}})$. By $\text{Opt}^*(\tilde{\mathcal{H}})$ we denote the value of the optimal solution of $\text{LP}(\tilde{\mathcal{H}})$. The optimal LP solution for $\mathcal{H}$ is Opt^*. Then the following holds.

Lemma 5. $\text{Opt}^*(\tilde{\mathcal{H}}) = \text{Opt}^* - |S_1|$ and $\mathbb{E}(|C|) \leq |S_1| + \mathbb{E}(|\tilde{C}|)$.

Lemma 6. Let l and Δ be constants, and let $\mathcal{H}$ be a l-uniform hypergraph with maximum vertex degree Δ. Then: $\mathbb{E}(|C|) \leq l \left(1 - \frac{l}{4\Delta}\right) \text{Opt}^*$.

Proof. Since there is no tight $\text{LP}(\tilde{\mathcal{H}})$-variable, because there are no 1's in the solution $(\tilde{x}_1, ..., \tilde{x}_{\tilde{n}})$, we get using (5)

$$\mathbb{E}(|\tilde{C}|) \leq \tilde{l} \left(1 - \frac{\tilde{l}\text{Opt}^*(\tilde{\mathcal{H}})}{4\tilde{m}}\right) \text{Opt}^*(\tilde{\mathcal{H}}) \underset{\text{Lem 3}(ii)}{\leq} \tilde{l} \left(1 - \frac{\tilde{l}}{4\tilde{\Delta}}\right) \text{Opt}^*(\tilde{\mathcal{H}}).$$

Furthermore,

$$
\begin{aligned}
\mathbb{E}(|C|) \quad &\leq \quad |S_1| + \mathbb{E}(|\tilde{C}|) \leq |S_1| + \tilde{l}\left(1 - \frac{\tilde{l}}{4\tilde{\Delta}}\right) \mathrm{Opt}^*(\tilde{\mathcal{H}}) \\
&\underset{\lambda \geq 1}{\leq} \quad \tilde{l}\left(1 - \frac{\tilde{l}\,\mathrm{Opt}^*(\tilde{\mathcal{H}})}{4\tilde{m}}\right)|S_1| + \tilde{l}\left(1 - \frac{\tilde{l}}{4\tilde{\Delta}}\right)\mathrm{Opt}^*(\tilde{\mathcal{H}}) \\
&\underset{\mathrm{Lem}\,3(ii)}{\leq} \quad \tilde{l}\left(1 - \frac{\tilde{l}}{4\tilde{\Delta}}\right)\left(\mathrm{Opt}^*(\tilde{\mathcal{H}}) + |S_1|\right) \underset{\mathrm{Lem}\,5}{=} \tilde{l}\left(1 - \frac{\tilde{l}}{4\tilde{\Delta}}\right)\mathrm{Opt}^*.
\end{aligned}
$$

and because $\mathcal{H}$ is uniform and $\Delta \geq \tilde{\Delta}$ we have: $\mathbb{E}(|C|) \leq l\left(1 - \frac{l}{4\Delta}\right)\mathrm{Opt}^*$ □

Lemma 6 and Lemma 4 imply the following theorem using the Chebyshev-Cantelli inequality and standard calculations.

Theorem 6. *Let $\mathcal{H}$ be an l-uniform hypergraph with bounded vertex degree, then the algorithm VC-$\mathcal{H}$ returns a hitting set C such that, $|C| \leq l\left(1 - \frac{l-1}{4\Delta}\right)\mathrm{Opt}^*$ with probability at least $\frac{3}{4}$.*

Proof. Assuming that $m \geq 16\Delta^5$ the proof is similar to the proof of Theorem 4. □

This improves over the result of Krivelevich [18] for Δ smaller then $(l-1)^{1+\frac{1}{l-2}}$ and of Okun [23] for Δ smaller then $(l-1)^{1+\frac{1}{\beta l-1}}$. The approximation ratio in this result is little weaker than the ratio of Halperin [12]. But the advantage here is that l and Δ are not coupled anymore, so a significantly larger class of hypergraphs than in [12] is covered.

6   Further Work

We believe that the analysis presented in this paper can incorporate other hypergraph parameters in a natural way, like bounded VC-dimension, uncrowdnedness, or exclusion of subgraphs. We hope that this may lead to new and better approximation results for the hitting set problem in such hypergraphs.

References

1. Alon, N., Moshkovitz, D., Safra, S.: Algorithmic construction of sets for k-restrictions. ACM Trans. Algorithms (ACM) 2, 153–177 (2006)
2. Alon, N., Spencer, J.: The probabilistic method, 2nd edn. Wiley Interscience (2000)
3. Bar-Yehuda, R., Even, S.: A linear-time approximation algorithm for the weighted vertex cover problem. Journal of Algorithms 2, 198–203 (1981)
4. Bar-Yehuda, R., Even, S.: A local ratio theorem for approximating weighted vertex cover problem. In: Ausiello, G., Lucertini, M. (eds.) Analysis and Design of Algorithms for Combinatorial Problems. Annals of Discrete Math., vol. 25, pp. 27–46. Elsevier, Amsterdam (1985)

5. Berge, C.: Hypergraphs-combinatorics of finite sets. North Holland Mathematical Library (1989)
6. Chvatal, V.: A greedy heuristic for the set covering problem. Math. Oper. Res. 4(3), 233–235 (1979)
7. Feige, U.: A treshold of ln n for approximating set cover. Journal of the ACM 45(4), 634–652 (1998)
8. Feige, U., Langberg, M.: Approximation algorithms for maximization problems arising in graph partitioning. Journal of Algorithms 41(2), 174–201 (2001)
9. Frieze, A., Jerrum, M.: Improved approximation algorithms for max k-cut and max bisection. Algorithmica 18, 67–81 (1997)
10. Füredi, Z.: Matchings and covers in hypergraphs. Graphs and Combinatorics 4(1), 115–206 (1988)
11. Gandhi, R., Khuller, S., Srinivasan, A.: Approximation Algorithms for Partial Covering Problems. J. Algorithms 53(1), 55–84 (2004)
12. Halperin, E.: Improved approximation algorithms for the vertex cover problem in graphs and hypergraphs. In: ACM-SIAM Symposium on Discrete Algorithms, vol. 11, pp. 329–337 (2000)
13. Hochbaum, D.S.: Approximation algorithms for the set covering and vertex cover problems. SIAM J. Computation 11(3), 555–556 (1982)
14. Hall, N.G., Hochbaum, D.S.: A fast approximation for the multicovering problem. Discrete Appl. Math. 15, 35–40 (1986)
15. Jäger, G., Srivastav, A.: Improved approximation algorithms for maximum graph partitioning problems. Journal of Combinatorial Optimization 10(2), 133–167 (2005)
16. Johnson, D.S.: Approximation algorithms for combinatorial problems. J. Comput. System Sci. 9, 256–278 (1974)
17. Khot, S., Regev, O.: Vertex cover might be hard to approximate to within 2-epsilon. J. Comput. Syst. Sci. 74(3), 335–349 (2008)
18. Krivelevich, J.: Approximate set covering in uniform hypergraphs. J. Algorithms 25(1), 118–143 (1997)
19. Lovász, L.: On the ratio of optimal integral and fractional covers. Discrete Math. 13, 383–390 (1975)
20. Lund, C., Yannakakis, M.: On the hardness of approximating minimization problems. J. Assoc. Comput. Mach. 41, 960–981 (1994)
21. McDiarmid, C.: Concentration. In: Habib, M., McDiarmid, C., Ramirez-Alfonsin, J., Reed, B. (eds.) Probabilistic Methods for Algorithmic Discrete Mathematics, pp. 195–248. Springer, Berlin (1998)
22. Peleg, D., Schechtman, G., Wool, A.: Randomized approximation of bounded multicovering problems. Algorithmica 18(1), 44–66 (1997)
23. Okun, M.: On approximation of the vertex cover problem in hypergraphs. Discrete Optimization (DISOPT) 2(1), 101–111 (2005)
24. Raz, R., Safra, S.: A sub-constant error-probability low-degree test, and a sub-constant error-probability PCP characterization of NP. In: Proc. 29th ACM Symp. on Theory of Computing, pp. 475–484 (1997)

Exact and Approximation Algorithms for Densest k-Subgraph[*]

(Extended Abstract)

Nicolas Bourgeois[1], Aristotelis Giannakos[2], Giorgio Lucarelli[3], Ioannis Milis[4], and Vangelis Th. Paschos[2,5]

[1] ESSEC, France
nbourgeo@phare.normalesup.org
[2] PSL Research University, Université Paris-Dauphine, LAMSADE, CNRS UMR 7243, Paris, France
{giannako,paschos}@lamsade.dauphine.fr
[3] Université Pierre et Marie Curie, LIP6, Paris, France
Giorgio.Lucarelli@lip6.fr
[4] Athens University of Economics and Business, Dept. of Informatics, Athens, Greece
milis@aueb.gr
[5] Institut Universitaire de France

Abstract. The DENSEST k-SUBGRAPH problem is a generalization of the maximum clique problem, in which we are given a graph G and a positive integer k, and we search among the subsets of k vertices of G one inducing a maximum number of edges. In this paper, we present algorithms for finding exact solutions of DENSEST k-SUBGRAPH improving the trivial exponential time complexity of $O^*(2^n)$ and using polynomial space. Two FPT algorithms are also proposed; the first considers as parameter the treewidth of the input graph and uses exponential space, while the second is parameterized by the size of the minimum vertex cover and uses polynomial space. Finally, we propose several approximation algorithms running in moderately exponential or parameterized time.

1 Introduction and Preliminaries

In the DENSEST k-SUBGRAPH problem we are given a graph $G = (V, E)$, $|V| = n$, $|E| = m$, and an integer $k \in \mathbb{N}^+$, and we ask for a subset $A \subseteq V$ of k vertices such that the number of edges induced by A is maximized. This problem belongs to a known class of problems, called *fixed cardinality problems*, most of which are generalizations of well-known combinatorial optimization problems. For instance, this is the case for DENSEST k-SUBGRAPH with respect to the MAX CLIQUE problem that is NP-hard [21]. Furthermore, it is NP-complete even to decide if there is a solution with at least $k^{1+\epsilon}$ edges, for any $\epsilon > 0$ [2].

In this paper, we present (sub)exponential and parameterized algorithms that compute optimal or approximate solutions for the DENSEST k-SUBGRAPH

---

[*] Research supported by the French Agency for Research under the DEFIS program TODO, ANR-09-EMER-010.

problem. In Section 2 we propose exact algorithms for finding an optimal solution to DENSEST k-SUBGRAPH. These algorithms improve the trivial complexity $O^*(2^n)$ for the problem (throughout the paper we use notation $O^*(\cdot)$ that ignores polynomial factors in the complexity expressions). In contrast to the algorithm presented in [12], they need only polynomial space. In this direction, we first present a general decomposition schema which, depending on the way the graph is decomposed, leads to different time complexities for finding an optimal solution. Let us note that this schema is quite general and can be applied to solve a lot of graph-problems, in particular problems whose feasible solutions are subsets of the vertex-set of the input graph. The interesting algorithmic point of this technique is that it can avoid a complete enumeration of k-element vertex subsets, when it can be restricted to subsets of the vertex-set whose complement induces a graph of maximum degree at most 2. Next, in Section 2.2, we propose a branch-and-cut algorithm and we analyze its complexity using the "measure and conquer" and the "bottom-up" techniques. An interesting point in this section is that we initiate a kind of multivariate analysis within moderately exponential algorithms, by getting running time upper-bounds depending not only on the order of the input graph but also on the maximum degree, or the chromatic number or the diameter of the graph, or ... This kind of analysis gives complexity expressions that represent more "tightly" the running time of an algorithm. In Section 3, we present algorithms of parameterized complexity for DENSEST k-SUBGRAPH. We first propose an algorithm of complexity exponential in the treewidth tw of the input graph, supposing that a tree decomposition is given. However, this algorithm uses exponential space. In order to fix this, we show that DENSEST k-SUBGRAPH is FPT with respect to the size τ of a minimum vertex cover of the input graph. Note that $tw \leqslant \tau$, but the later algorithm uses polynomial space. In Section 4, we first present two **XP**-approximation schemata for DENSEST k-SUBGRAPH whose approximation ratios depend on their complexity (see [17] for a formal definition of the problem-class **XP**). We also give approximation algorithms that run in moderately exponential or parameterized time. The omitted proofs can be found in [10].

In what follows, we denote by $\delta(G)$, $\Delta(G)$ and $\bar{\Delta}(G)$ (or simply δ, Δ and $\bar{\Delta}$) the minimum, maximum and average degree, respectively, of a graph G. The diameter $\mathcal{D}(G)$ of a graph G is the length of the largest shortest path between any two vertices of the graph. For a graph G, we denote by $tw(G)$ and $\chi(G)$ its treewidth and chromatic number, respectively. Given two sets of vertices $A, B \subseteq V$, $G[A]$ denotes the subgraph induced by A, $E(A)$ the set of edges induced by $G[A]$ and $E(A, B)$ the set of edges with their one endpoint in A and the other in B.

The approximability of DENSEST k-SUBGRAPH has been studied in several papers. For instance, an approximation algorithm achieving ratio $\frac{8k}{9n}$ has been proposed in [4]. In [19], three procedures are used in order to obtain a $O(n^{-1/3})$-approximation ratio, while the best known approximation algorithm achieves a ratio of $O(n^{-(1/4+\epsilon)})$ within $n^{O(1/\epsilon)}$ time, for any $\epsilon > 0$ [3]. From a negative point of view, it is known that DENSEST k-SUBGRAPH in general graphs does not admit a PTAS [22].

DENSEST k-SUBGRAPH can be solved in time $O^*(kn^{\omega \lfloor k/3 \rfloor + 1 + k \mod 3})$ where $\omega < 2.376$, by the exact algorithm proposed in [12]. Notice, however, that this algorithm requires exponential space.

A problem is fixed-parameter tractable (FPT) with respect to a parameter t if it can be solved (to optimality) with time-complexity $O(f(t)p(n))$ where f is a function that depends on the parameter t and p is a polynomial in the size n of the instance. Cai in [12] proved that DENSEST k-SUBGRAPH is $\mathbf{W}[1]$-hard, with respect to k even for regular graphs. This result immediately implies also that DENSEST k-SUBGRAPH is $\mathbf{W}[1]$-hard with respect to the size ℓ of the solution, as any solution cannot contain more than $k(k-1)/2$ edges.

2   Exact Algorithms

2.1   A Decomposition Technique

A general idea for finding an exact solution for the DENSEST k-SUBGRAPH problem in a graph $G = (V, E)$ is to split the vertex set V into two subsets V_1 and V_2. Then, for each j, $0 \leqslant j \leqslant k$, and each subset $A_1 \subseteq V_1$ with $|A_1| = j$, we search for a subset $A_2 \subseteq V_2$, $|A_2| = k - j$, such that the number of edges in $G[A_1 \cup A_2]$ is maximized. Clearly, the complexity of this algorithm depends on:

- the size of set V_1, as we create all subsets of V_1;
- the complexity of determining, given the set A_1, the appropriate set $A_2 \subseteq V_2$.

Hence, it is required for $V_2 = V \setminus V_1$ to be of non-trivial size and to have some specific property that allows A_2 to be determined in polynomial time. We will show that the property $\Delta(G[V_2]) \leqslant 2$ is an application of this idea.

This general idea can be applied to many problems especially to those where feasible solutions are subsets of V satisfying some specific property. As we will see in what follows, this method provides also a general framework for the complexity analysis of several algorithms (depending on the way V_1 is chosen and on its size), and uses polynomial space. Therefore, it allows to achieve non-trivial bounds to running time (using polynomial space), in particular for problems where no bounds better than $O^*(2^n)$ are known.

GENERIC(V_1, V_2) is a procedure that takes as input a partition of the vertex set (V_1, V_2) and returns an optimal DENSEST k-SUBGRAPH in G through exhaustive search.

Whenever $\Delta(G[V_2]) \leqslant 2$, the following proposition states that A_2 is found in polynomial time.

Proposition 1. *Consider a graph $G = (V, E)$, some partition of the vertex set V into two subsets V_1 and V_2 such that $\Delta(G[V_2]) \leqslant 2$, and a subset $A_1 \subseteq V_1$, $|A_1| \leqslant k$. A solution $A = A_1 \cup A_2$ for the DENSEST k-SUBGRAPH problem in G such that $A_2 \subseteq V_2$, $|A_2| = k - |A_1|$, and $|E(A)|$ is maximized, can be found in $O(nk^2)$ time.*

Generic(V_1, V_2)
1: **for** $j = 0$ to k **do**
2:   **for** any subset $A_1 \subseteq V_1$, $|A_1| = j$ **do**
3:     find a solution $A = A_1 \cup A_2$ for the DENSEST k-SUBGRAPH problem in G such that $A_2 \subseteq V_2$, $|A_2| = k - j$, and $|E(A)|$ is maximized;
4: **return** the best among the solutions found in Line 3;

Note that, if $\Delta(G[V_2]) = 0$, i.e., V_2 is an independent set, then the set A_2 can be found in $O(n \log k)$ time, by selecting the $k - |A_1|$ vertices of V_2 with the largest degree to A_1.

Proposition 2. GENERIC(V_1, V_2) *returns an optimal* DENSEST k-SUBGRAPH-*solution on* $G[V_1 \cup V_2]$ *in* $O^* \left(2^{|V_1|}\right)$ *time, whenever* $A_2 \subseteq V_2$ *can be computed in polynomial time.*

Note that, if $k \leqslant \frac{|V_1|}{2}$ then the term $O^* \left(\binom{|V_1|}{k}\right)$ is a better expression for the complexity.

The following theorem handles four decompositions (V_1, V_2) of G, each one determined by the way V_1 is obtained. Other decompositions based on specific structural properties of the set V_1 can be also used to obtain different complexities.

Theorem 1. GENERIC(V_1, V_2) *leads to a polynomial space algorithm for* DENSEST k-SUBGRAPH *of time complexity: (i)* $O^* \left(2^{\left(1 - (5/8)^{\Delta - 2}\right)n}\right)$, *if* V_1 *is obtained by repeated excavations of minimum dominating sets; (ii)* $O^* \left(2^{\frac{\chi - 1}{\chi}n}\right)$ *or* $O^* \left(2^{\frac{\Delta - 1}{\Delta}n}\right)$, *if* V_1 *is a minimum vertex cover; (iii)* $O^* \left(2^{\frac{\Delta - 2}{\Delta - 1}n}\right)$, *for any* $\Delta \geqslant 3$, *if* V_1 *is obtained by repeated excavations of minimum independent dominating sets; (iv)* $O^* \left(2^{n - \mathcal{D}(G)}\right)$, *for any* $\Delta \geqslant 3$, *if* V_1 *is the complement of the vertices of a longest path of the graph.*

We prove here Items (ii) and (iii). The whole proof of the theorem can be found in [10].

Proof of Item (ii). Consider Algorithm 1. A minimum vertex cover B, can be found in time $O^* (1.2738^\tau)$ [14]. The set $V \setminus B$ is a maximum independent set of size at least $\frac{n}{\chi}$, since the vertex set of the input graph can be partitioned into χ independent sets. Hence, by Proposition 2 the first part of Item (ii) of the theorem holds. If the input graph is a clique or an odd cycle then the DENSEST

Algorithm 1. Decomposition by Minimum Vertex Cover

1: find a minimum vertex cover B, of G;
2: **return** GENERIC$(B, V \setminus B)$;

---

k-SUBGRAPH problem is polynomial. Otherwise, $\chi \leqslant \Delta$ and the second part of Item (ii) of the theorem holds.

Proof of Item (iii). Consider Algorithm 2. Note that, if there exits a D_i such that $|D_i| \geqslant \frac{n}{\Delta-1}$, then by Line 5 of the algorithm we have that $|D| \leqslant n - \frac{n}{\Delta-1} = \frac{\Delta-2}{\Delta-1}n$. Otherwise, for each i, $3 \leqslant i \leqslant \Delta$, it holds that $|D_i| \leqslant \frac{n}{\Delta-1}$, and hence, $|D| \leqslant (\Delta-2)\frac{n}{\Delta-1}$. Since in both cases $G[V \setminus D]$ is a graph of maximum degree 2, we can apply Proposition 2, completing the proof of Item (iii).   □

Algorithm 2. Decomposition by Minimum Independent Dominating Set

---

1: $V_\Delta = V$; $D = \emptyset$;
2: **for** $i = \Delta$ to 3 **do**
3:    find an independent dominating set D_i on $G[V_i]$;
4:    **if** $|D_i| \geqslant \frac{n}{\Delta-1}$ **then**
5:        $D = V \setminus D_i$; Go to Line 8;
6:    **else**
7:        $D = D \cup D_i$; $V_{i-1} = V_i \setminus D_i$;
8: **return** GENERIC$(D, V \setminus D)$;

The complexity for optimally solving DENSEST k-SUBGRAPH in bipartite graphs can be further improved. Observe that, given a bipartite graph $G = (U \cup V, E)$, we can apply GENERIC(U, V) getting an algorithm with running time $O^*\left(\binom{n/2}{k}\right)$ (or $2^{n/2}$ if $k \geqslant n/2$). We now show how to improve this result, by considering the balance of the vertices among the two independent sets in an optimal solution. In what follows, we define $\phi(k, n)$ to be the worst-case complexity of our algorithm running on general instances of DENSEST k-SUBGRAPH.

We now show that DENSEST k-SUBGRAPH *can be solved on bipartite graphs in time* $O^*(\phi(k, n))$, where the behavior of $\phi(k, n)$ for different ratios k/n is illustrated in the next table.

k/n	1/100	1/20	1/10	1/6	1/4	1/3
$\phi(k, n)$	1.029^n	1.105^n	1.177^n	1.253^n	1.325^n	1.375^n
$\sum_{i \leqslant k} \binom{n/2}{i}$	1.051^n	1.177^n	1.285^n	1.375^n	$\sqrt{2}^n$	$\sqrt{2}^n$

Indeed, w.l.o.g., assume that $|U| \leqslant n/2$ and let $\lambda = |U|/n \leqslant 1/2$. GENERIC$(U, V)$ solves DENSEST k-SUBGRAPH in $O^*(\phi(k, \lambda n))$ time, while GENERIC(V, U) solves it in time $O^*(\phi(k, (1 - \lambda)n))$. We fix some scalar $\nu(\lambda) \leqslant 1/2$. Notice that either V contains at most νk vertices from an optimal solution, or U contains less than $(1 - \nu)k$ of them. Hence, we only need to consider small subsets: $T(n) \leqslant \max_\lambda \min_\nu \{\phi(\nu k, (1 - \lambda)n) + \phi((1 - \nu)k, \lambda n)\}$. Since the second term in the previous expression involves an increasing and a decreasing function, it is easy to find the solution of this minimization problem for a given set of parameters (k, n). However, it would be very tedious to try to give an exact formula, especially considering all the specific cases when k is close to n. As a consequence, we prefer to give a sample of values for the function ϕ.

2.2   Branch-and-Cut Algorithms

In this section we propose two slightly different branching algorithms for DEN-SEST k-SUBGRAPH and we prove upper bounds on their time complexity. For the analysis of the first algorithm we use the well known technique of *measure and conquer* introduced in [20]. For the analysis of the second algorithm we use the *bottom-up* technique which has been developed in [8] as a technique for finely measuring the progression of a branching algorithm. This method has led to the best known worst-case complexity for the independent set problem [8], and it has been also used in [9].

Let us first consider a simple branch-and-cut algorithm that branches on a vertex of maximum degree. The branching tree is pruned whenever the remaining graph is of maximum degree 2. In this case, a solution for the whole graph can be obtained by extending the solution implied by the particular path of the search tree as stated in Proposition 1.

Theorem 2. *Using measure and conquer, the basic branching algorithm solves* DENSEST k-SUBGRAPH *in time* $O^*\left(2^{\frac{\Delta-1}{\Delta+1}n}\right)$.

We now slightly modify the previous basic branching algorithm by proceeding to search tree cutting whenever the remaining graph has average degree three. The analysis of this modified branching algorithm is based on the bottom-up method. The following Lemma 1 settles the case where the average degree of the graph is at most 3, while Lemma 2 handles the complexity of finding a DEN-SEST k-SUBGRAPH on graphs with average degree at least $d - 1$, given that the complexity of finding a DENSEST k-SUBGRAPH for graphs with average degree at most $d - 1$ is known.

Lemma 1. DENSEST k-SUBGRAPH *can be solved on graphs of average degree* $\bar{\Delta} \leqslant 3$ *with running time* $O^*(2^{21n/46})$.

Lemma 2. *Assume* DENSEST k-SUBGRAPH *can be computed in graphs with average degree at most* $d - 1$, *in time* $O^*(2^{\alpha_d n})$ *for a given* $\alpha_d \geqslant 1/2$, $d \in \mathbb{N}$. *Then, its computation time in graphs with average degree at least* $d - 1$ *is* $O^*(2^{\alpha_d n + \beta_d(m-(d-1)n/2)})$, *where* $\beta_d = \frac{2(1-\alpha_d)}{d+1}$. *Furthermore, in graphs with average degree at most* d, *this time is* $O^*(2^{\alpha_{d+1}n})$, *where* $\alpha_{d+1} = \frac{d\alpha_d+1}{d+1}$.

Theorem 3. DENSEST k-SUBGRAPH *can be solved on graphs of average degree* $\bar{\Delta} \leqslant d$ *with running time* $O^*(2^{\frac{d-27/23}{d+1}n})$, *for any* $d \in \mathbb{N}, d \geqslant 3$.

Proof. For $d = 3$ the result follows by Lemma 1. Assume that it is true for $\bar{\Delta} \leqslant d - 1$. Then, by Lemma 2, we can find a solution when $\bar{\Delta} \leqslant d$ with running time $O^*(2^{\alpha_{d+1}n})$, where $\alpha_{d+1} = \frac{d\alpha_d+1}{d+1} = \frac{d\frac{d-1-27/23}{d}+1}{d+1} = \frac{d-27/23}{d+1}$. Thus, the statement holds by induction on d. □

3   Parameterized Algorithms

Given a graph $G = (V, E)$, a tree decomposition is a pair (X, T), where $X = \{X_1, X_2, \ldots, X_{|X|}\}$, $X_i \subseteq V$, and $T = (X, F)$ is a tree such that: (i) $\bigcup X_i = V$, (ii) for each $e = (u, v) \in E$ there is a X_i where $u, v \in X_i$, and (iii) for each X_i, X_j, X_l such that X_j appears on the path between X_i and X_l it holds that $X_i \cap X_l \subseteq X_j$. The treewidth, tw, of such a decomposition is defined as $tw = \max\{|X_i|, 1 \leqslant i \leqslant |X|\} - 1$. It is known that finding a minimum treewidth decomposition of a given graph is NP-hard [1]. However, deciding whether there is a tree decomposition of a graph of a fixed treewidth is polynomial [6]. A similar approach to that of [25] for MAX k-COVER, can be used for DENSEST k-SUBGRAPH, deriving the following result.

Theorem 4. *There exist an algorithm for* DENSEST k-SUBGRAPH *that runs in time* $O^*(2^{tw} \cdot k \cdot (tw^2 + k) \cdot |X|)$ *and uses space exponential in* tw. *Consequently,* DENSEST k-SUBGRAPH *parameterized by the treewidth of the input graph is FPT.*

Let us note that the graph $G[V_2]$ in Proposition 1 has bounded treewidth. However, observe that, although the completion of A_1 is done by vertices of V_2, the DENSEST k-SUBGRAPH problem itself is not solved in $G[V_2]$. So, Proposition 1 cannot be substituted by Theorem 4.

For the size of the minimum vertex cover τ of the input graph it holds that $tw \leqslant \tau$. So, Theorem 4 implies that DENSEST k-SUBGRAPH is FPT with respect to τ too. In what follows, we present another application of GENERIC and we restate Item (ii) of Theorem 1 in order to obtain the following parameterized result, which implies only polynomial space. The proof of the following theorem follows directly from the proof of Theorem 1, Item (ii) given earlier.

Theorem 5. *There exists an* $O^*(2^\tau)$-*time algorithm for* DENSEST k-SUBGRAPH *that uses polynomial space.*

We now improve the analysis of Theorem 5 and prove that, informally, the instances of DENSEST k-SUBGRAPH that are not fixed-parameter tractable (with respect to k) are those solved with running time better than $O^*(2^\tau)$.

Theorem 6. DENSEST k-SUBGRAPH *can be solved in* $O^*(\max\{\gamma^\tau, c^k\})$, *for two related constants* $\gamma < 2$ *and* $c > 4$, *and with polynomial space.*

Proof. Note that by the proof of Theorem 5 the running time of GENERIC is $O^*(\sum_{i=1}^k \binom{\tau}{i})$. If $\tau \leqslant k$, it follows that DENSEST k-SUBGRAPH can be solved in $O^*(2^\tau) = O^*(2^k)$ time. Hence, we can assume that $k < \tau$.

We will prove that, for any $0 < \lambda < 1/2$, we can determine constants $\gamma = \gamma(\lambda) < 2$ and $c = c(\lambda) > 4$ such that DENSEST k-SUBGRAPH can be solved in time $O^*(\max\{\gamma^\tau, c^k\})$. We distinguish the following two cases: $\tau > k \geqslant \lambda\tau$ and $k < \lambda\tau$.

If $k \geqslant \lambda\tau$, then using the fact that $k < \frac{k/\lambda}{2}$ and Stirling's formula we get

$$\sum_{i=1}^k \binom{\tau}{i} \leqslant k\binom{k/\lambda}{k} \sim k\left(\frac{(\lambda^{-1})^{\lambda^{-1}}}{(\lambda^{-1}-1)^{(\lambda^{-1}-1)}}\right)^k = O^*\left(c^k\right), \text{ for some constant } c \text{ that}$$

depends on λ.

If $k < \lambda\tau$, then $k < \tau/2$ and hence $\sum_{i=1}^{k} \binom{\tau}{i} \leqslant k\binom{\tau}{k} \leqslant k\left(\frac{1}{\lambda^{\lambda}(1-\lambda)^{(1-\lambda)}}\right)^{\tau} = O^*(\gamma^{\tau})$, for some constant $\gamma < 2$ that depends on λ. $\qquad\square$

The table below contains the values of c and γ for some values of λ.

λ	0.01	0.05	0.1	0.15	0.2	0.25	0.3	0.35	0.40	0.45	0.49
$c = \dfrac{\lambda^{\frac{1}{\lambda}}}{\left(\frac{1}{\lambda}-1\right)^{\left(\frac{1}{\lambda}-1\right)}}$	270.47	53.00	25.81	16.74	12.21	9.48	7.66	6.36	5.38	4.61	4.11
$\gamma = \dfrac{1}{\lambda^{\lambda}(1-\lambda)^{1-\lambda}}$	1.06	1.22	1.38	1.53	1.65	1.75	1.84	1.91	1.96	1.99	1.9996

4   Approximation Algorithms

Up to now no constant factor approximation algorithm for DENSEST k-SUB-GRAPH that runs in polynomial time is known. In this section, by relaxing the demand of polynomiality, we present approximation algorithms that run in time exponential, but faster than the time needed for computing an exact solution. This approach has already been considered for several other paradigmatic problems such as MINIMUM SET COVER [15], MIN COLORING [5], MAX INDEPENDENT SET and MIN VERTEX COVER [7], MIN BANDWIDTH [16], etc. Note that, the $O(n^{-(1/4+\epsilon)})$-approximation algorithm with complexity $n^{O(1/\epsilon)}$ presented in [3], can be considered as an approximation algorithm in this context, since whenever ϵ is chosen to be of the form $\log_n c$, where c is a constant, a constant factor approximation ratio is achieved in subexponential time. Note finally that similar issues arise in the field of FPT algorithms, where approximation notions have been introduced, for instance, in [11,13,18,24].

For better readability, we partition the results of this section into two parts. In the first part, we give approximation algorithms with complexity of the form $O^*(n^{ck})$, with $0 < c \leqslant 1$. In the second part, we present approximation algorithms either with complexity of the form $O^*(c^n)$, with $1 < c \leqslant 2$, or parameterized complexity.

4.1   XP-Approximation Algorithms

A general idea for the design of an exponential time approximation algorithm is to construct a "good" subgraph of ρk vertices in exponential time and select the remaining $(1-\rho)k$ vertices in a greedy way. In this vein, the following proposition gives a property of such a good subgraph.

Proposition 3. *For an optimal solution A^* for* DENSEST k-SUBGRAPH *and a rational ρ such that $0 < \rho \leqslant 1$, there exists a partition of the vertices of A^* into two subsets A_1^*, $|A_1^*| = \rho k$, and A_2^*, $|A_2^*| = (1-\rho)k$, such that $|E(A_1^*)| \geqslant \frac{\rho}{1-\rho} \cdot |E(A_2^*)|$.*

Theorem 7. *For any ρ, $0 < \rho \leqslant 1$, Algorithm 3 achieves a ρ-approximation ratio in $O^*(n^{\rho k})$ time.*

Algorithm 3. Create all subsets

1: **for** each of the $\binom{n}{\rho k}$ subsets of vertices $A_1 \subseteq V$, $|A_1| = \rho k$, **do**
2:    find the set of vertices $A_2 \in V \setminus A_1$, $|A_2| = (1 - \rho)k$, which have the highest
      degree to A_1;
3: **return** the maximum solution $A_1 \cup A_2$ found;

Another way to construct a good subgraph of ρk vertices is to run an exact
algorithm for DENSEST ρk-SUBGRAPH and to complete the solution with $(1 - \rho)k$
arbitrary selected vertices. The following lemma deals with the density of an
induced subgraph and will be used to count the number of edges induced by
such an optimal DENSEST ρk-SUBGRAPH.

Lemma 3. *Consider a graph $G = (V, E)$ of density $\varpi = 2|E|/|V|(|V| - 1)$. For
any p, $2 \leqslant p \leqslant |V|$, there exists a set of vertices $V_p \subseteq V$, $|V_p| = p$, such that the
induced subgraph $G_p(V_p, E(V_p))$ has density at most ϖ.*

In the following theorem, we assume that an algorithm of complexity $\phi(k, t)$ is
known for finding a DENSEST k-SUBGRAPH, where t is some parameter of the
instance, e.g., $t = \Delta, \tau, \ell, n$. This algorithm is used in order to obtain an optimal
solution of size ρk for the problem, where $0 < \rho \leqslant 1$.

Theorem 8. *Let $\mathcal{A}$ be an exact algorithm of complexity $\phi(k, t)$ for finding a
DENSEST k-SUBGRAPH, where t is a parameter of the instance. For any ρ such
that $0 < \rho \leqslant 1$, it is possible to find a ρ^2-approximation for DENSEST k-SUB-
GRAPH in G with running time $O^*(\phi(\rho k, t))$.*

In Theorem 8, we count only the edges induced by the DENSEST $(\lceil \rho k \rceil + 1)$-
SUBGRAPH, as the remaining vertices are selected arbitrarily. In Algorithm 4,
we replace this greedy step by searching for successive DENSEST $(\lceil \rho k \rceil + 1)$-
SUBGRAPHS.

Algorithm 4. Approximate subsets

1: $A = \emptyset$; $i = 1$; $G_i = G$;
2: **while** $|A| < k$ **do**
3:    compute a DENSEST $(\lceil \rho k \rceil + 1)$-SUBGRAPH in G_i;
4:    let A_i be the set of vertices of this subgraph;
5:    **if** $|A \cup A_i| \leqslant k$ **then**
6:        create the graph G_{i+1} by removing from G_i the edges of $E(A_i)$;
7:        $A = A \cup A_i$;
8:        **if** the vertices of G_{i+1} induce an independent set **then**
9:            arbitrarily complete A with vertices in $V \setminus A$ such that $|A| = k$;
10:   **else**
11:       arbitrarily complete A with vertices in $V \setminus A$ such that $|A| = k$;
12:       **return** A;
13:   $i = i + 1$;
14: **return** A;

Theorem 9. *Let $\mathcal{A}$ be an exact algorithm of complexity $\phi(k,t)$ for finding a* DENSEST k-SUBGRAPH. *For any ρ such that $0 < \rho \leqslant 1$, Algorithm 4 achieves a $\rho(1 - 3\rho/2)$-approximation for* DENSEST k-SUBGRAPH *in G with running time $O^*(\phi(\rho k, t))$.*

Proof. Let λ be the number of iterations of Algorithm 4. As at the beginning of each iteration there exists at least one edge in G_i, there exists also a vertex $v \in A_i$ such that $v \notin A$. Moreover, in each iteration at most ρk new vertices are added in the solution. Thus, it holds that $1/\rho \leqslant \lambda \leqslant k(1 - \rho)$. Therefore, the running time of the algorithm is bounded by $O^*(\phi(\rho k, t))$.

At the beginning of iteration $i + 1$, $i \geqslant 1$, the current graph G_{i+1} contains $|E(A^*)| - |E_i|$ edges, where $|E_i| = \sum_{j=1}^{i} |E(A_j)|$ and A^* is an optimal solution for the DENSEST k-SUBGRAPH problem. Thus, there exists a subgraph of G_{i+1} with size ρk that contains at least $\rho^2(|E(A^*)| - |E_i|)$ edges. We prove by induction on i that $|E_i| \geqslant \rho^2\left(i - \frac{i(i-1)}{2}\rho^2\right)|E(A^*)|$. For $i = 1$, by Theorem 8 the inequality holds. Assume that it is true for $i - 1$. Then:

$$|E_i| \geqslant \rho^2|E(A^*)| + (1 - \rho^2)|E_{i-1}|$$
$$\geqslant \rho^2\left(1 + (1 - \rho^2)\left(i - 1 - \frac{(i-1)(i-2)}{2}\rho^2\right)\right)|E(A^*)|$$
$$\geqslant \rho^2\left(i - \frac{i(i-1)}{2}\rho^2\right)|E(A^*)|$$

Let $E(A)$ be the set of edges of the final solution obtained by the algorithm. As Algorithm 4 iterates at least $\left(\frac{1}{\rho} - 1\right)$ times, we have

$$|E(A)| \geqslant \rho^2\left(\left(\frac{1}{\rho} - 1\right) - \frac{\left(\frac{1}{\rho} - 1\right)\left(\frac{1}{\rho} - 2\right)}{2}\rho^2\right)|E(A^*)| \geqslant \rho\left(1 - \frac{3\rho}{2}\right)|E(A^*)|. \qquad \square$$

In general, Algorithm 4 performs better than Algorithm 3 for small values of ρ, since in that case $\rho(1 - 3\rho/2)$ is close to ρ and $\mathcal{A}$ runs faster than exhaustive enumeration. Algorithm 3 outperforms Algorithm 4 whenever ρ is close to 1.

4.2 Parameterized and Moderately Exponential Approximation

As already mentioned, DENSEST k-SUBGRAPH is not fixed parameter tractable with respect to k [12], and hence, neither with respect to the size of the solution ℓ. However, in this section we show that there is an approximation algorithm for DENSEST k-SUBGRAPH achieving non-trivial approximation ratios (though non-constants) unattainable in polynomial time, with complexity parameterized by k (and hence by ℓ).

Theorem 10. DENSEST k-SUBGRAPH *is approximable within any ratio $R(n)$, where R is any strictly increasing function, in parameterized time w.r.t. k.*

Proof. If $k \leqslant R(n)$, then we arbitrarily select $k/2$ edges. In this case, the solution consists of the vertices incident to these edges, while we arbitrarily add some vertices if necessary in order to have size exactly k. In general, it holds that

$\ell \leqslant k(k-1)/2$ and hence $\ell \leqslant R(n)(k-1)/2$. Therefore, the algorithm achieves $R(n)$-approximation ratio in polynomial time.

If $k > R(n)$, then let R^{-1} be the inverse function of R. We consider all possible subgraphs of size k and return the densest one. In this case, the algorithm finds an exact solution with running time $O^*(2^n) = O^*(2^{R^{-1}(k)})$.   □

In the two last algorithms, we use again the idea of splitting the vertex set.

Algorithm 5. Decomposition by Vertex Cover

1: find a minimum vertex cover V^* ($|V^*| = \tau$);
2: consider a partition of V into V_1 and V_2 s.t. $V_1 \subseteq V^*$ and $|V_1| = |V_2 \cap V^*| = \tau/2$;
3: solve DENSEST k-SUBGRAPH on $G[V_1]$ (let A_1 be the solution);
4: solve DENSEST k-SUBGRAPH on $G[V_2]$ (let A_2 be the solution);
5: solve DENSEST k-SUBGRAPH on the bipartite graph $B = (V_1, V_2; E')$ obtained by removing the edges in $E(V_1)$ and $E(V_2)$ (let A_3 be the solution);
6: **return** the best of A_1, A_2 and A_3;

Theorem 11. *Algorithm 5 achieves a $1/3$-approximation ratio for* DENSEST k-SUBGRAPH *in time* $O^*(2^{\tau/2})$.

Proof. By construction $E = E(V_1) \cup E(V_2) \cup E'$. Thus, the approximation ratio of Algorithm 5 is $1/3$, since optimal densest k-subgraphs are built for $G[V_1]$, $G[V_2]$ and B, and one of them contains at least one third times the optimum number of edges. In Line 1, a minimum vertex cover can be computed as in [14]. As $|V_1| = \tau/2$, Line 3 runs in $O^*(2^{\tau/2})$. In Line 4, use GENERIC($V_2 \cap V^*, V \setminus V^*$) which, by Proposition 2, runs in $O^*(2^{\tau/2})$, since $|V_2 \cap V^*| = \tau/2$ and $V \setminus V^*$ is an independent set. Finally, as B is a bipartite graph, Line 5 runs in $O^*(2^{\min\{|V_1|,|V_2|\}}) = O^*(2^{|V_1|}) = O^*(2^{\tau/2})$.   □

Using similar arguments as in the proof of Theorem 11, the following theorem can be proved.

Theorem 12. DENSEST k-SUBGRAPH *is approximable within ratio $1/2$ in time* $O^*(2^{n/2})$.

References

1. Arnborg, S., Corneil, D.G., Proskurowski, A.: Complexity of finding embeddings in a k-tree. SIAM Journal on Algebraic and Discrete Methods 8, 277–284 (1987)
2. Asahiro, Y., Hassin, R., Iwama, K.: Complexity of finding dense subgraphs. Discrete Applied Mathematics 121, 15–26 (2002)
3. Bhaskara, A., Charikar, M., Chlamtac, E., Feige, U., Vijayaraghavan, A.: Detecting high log-densities: An $O(n^{1/4})$ approximation for densest k-subgraph. In: STOC 2010, pp. 201–210 (2010)
4. Billionnet, A., Roupin, F.: A deterministic approximation algorithm for the densest k-subgraph problem. International Journal of Operational Research 3, 301–314 (2008)

5. Björklund, A., Husfeldt, T., Koivisto, M.: Set partitioning via inclusion-exclusion. SIAM Journal of Computing 39(2), 546–563 (2009)
6. Bodlaender, H.L.: A linear-time algorithm for finding tree-decompositions of small treewidth. SIAM Journal on Computing 25, 1305–1317 (1996)
7. Bourgeois, N., Escoffier, B., Paschos, V.T.: Approximation of MAX INDEPENDENT SET, MIN VERTEX COVER and related problems by moderately exponential algorithms. Discrete Applied Mathematics 159(17), 1954–1970 (2011)
8. Bourgeois, N., Escoffier, B., Paschos, V.T., van Rooij, J.M.M.: Fast algorithms for MAX INDEPENDENT SET. Algorithmica 62, 382–415 (2012)
9. Bourgeois, N., Giannakos, A., Lucarelli, G., Milis, I., Paschos, V.T., Pottié, O.: The MAX QUASI-INDEPENDENT SET PROBLEM. Journal of Combinatorial Optimization 23, 94–117 (2012)
10. Bourgeois, N., Giannakos, A., Lucarelli, G., Milis, I., Paschos, V.T.: The Exact and approximation algorithms for DENSEST k-SUBGRAPH. Cahiers du LAMSADE (324) (2012)
11. Brankovic, L., Fernau, H.: Combining Two Worlds: Parameterised Approximation for Vertex Cover. In: Cheong, O., Chwa, K.-Y., Park, K. (eds.) ISAAC 2010, Part I. LNCS, vol. 6506, pp. 390–402. Springer, Heidelberg (2010)
12. Cai, L.: Parameterized complexity of cardinality constrained optimization problems. The Computer Journal 51, 102–121 (2007)
13. Cai, L., Huang, X.: Fixed-Parameter Approximation: Conceptual Framework and Approximability Results. In: Bodlaender, H.L., Langston, M.A. (eds.) IWPEC 2006. LNCS, vol. 4169, pp. 96–108. Springer, Heidelberg (2006)
14. Chen, J., Kanj, I.A., Xia, G.: Improved upper bounds for vertex cover. Theoretical Computer Science 411, 3736–3756 (2010)
15. Cygan, M., Kowalik, L., Wykurz, M.: Exponential-time approximation of weighted set cover. Information Processing Letters 109(16), 957–961 (2009)
16. Cygan, M., Pilipczuk, M.: Exact and approximate bandwidth. Theoretical Computer Science 411(40-42), 3701–3713 (2010)
17. Downey, R.G., Fellows, M.R.: Parameterized complexity. Monographs in Computer Science. Springer, New York (1999)
18. Downey, R.G., Fellows, M.R., McCartin, C.: Parameterized Approximation Problems. In: Bodlaender, H.L., Langston, M.A. (eds.) IWPEC 2006. LNCS, vol. 4169, pp. 121–129. Springer, Heidelberg (2006)
19. Feige, U., Kortsarz, G., Peleg, D.: The dense k-subgraph problem. Algorithmica 29, 410–421 (2001)
20. Fomin, F.V., Grandoni, F., Kratsch, D.: A measure & conquer approach for the analysis of exact algorithms. Journal of the ACM 56 (2009)
21. Garey, M.R., Johnson, D.S.: Computers and intractability: A guide to the theory of NP-completeness. Freeman, San Francisco (1979)
22. Khot, S.: Ruling out PTAS for graph min-bisection, densest subgraph and bipartite clique. In: FOCS 2004, pp. 136–145 (2004)
23. Kloks, T.: Treewidth. LNCS, vol. 842. Springer, Heidelberg (1994)
24. Marx, D.: Parameterized complexity and approximation algorithms. The Computer Journal 51(1), 60–78 (2008)
25. Moser, H.: Exact algorithms for generalizations of vertex cover. PhD thesis, Friedrich-Schiller-Universität Jena (2005)

Linear-Time Constant-Ratio Approximation Algorithm and Tight Bounds for the Contiguity of Cographs

Christophe Crespelle[1] and Philippe Gambette[2]

[1] Université Claude Bernard Lyon 1, DNET/INRIA, LIP UMR CNRS 5668, ENS de Lyon, Université de Lyon
`christophe.crespelle@inria.fr`
[2] Université Paris-Est, LIGM UMR CNRS 8049, Université Paris-Est Marne-la-Vallée, 5 boulevard Descartes, 77420 Champs-sur-Marne, France
`philippe.gambette@univ-mlv.fr`

Abstract. In this paper we consider a graph parameter called *contiguity* which aims at encoding a graph by a linear ordering of its vertices. We prove that the contiguity of cographs is unbounded but is always dominated by $O(\log n)$, where n is the number of vertices of the graph. And we prove that this bound is tight in the sense that there exists a family of cographs on n vertices whose contiguity is $\Omega(\log n)$. In addition to these results on the worst-case contiguity of cographs, we design a linear-time constant-ratio approximation algorithm for computing the contiguity of an arbitrary cograph, which constitutes our main result. As a by-product of our proofs, we obtain a min-max theorem, which is worth of interest in itself, stating equality between the rank of a tree and the minimum height its path partitions.

Introduction

In many contexts, such as genomics, biology, physics, linguistics, computer science and transportation for examples, industrials and academics are led to algorithmically treat large dataset organised in the form of networks or graphs. The algorithms used to do so generally make extensive use of *neighborhood queries*, which, given a vertex x of a graph G, ask for the list of neighbors of x in G. Therefore, as pointed out by [1], due to the huge size of the graphs considered, finding compact representations of a graph providing optimal-time neighborhood queries is a crucial issue in practice.

One possible way to achieve this goal is to find an order σ on the vertices of G such that the neighborhood of each vertex x of G is an interval in σ. In this way, one can store the list of vertices of the graph in the order defined by σ and assign two pointers to each vertex: one toward its first neighbor in σ and one toward its last neighbor in σ. Therefore, one can answer adjacency queries on vertex x simply by listing the vertices appearing in σ between its first and last pointer. It must be clear that such an order on the vertices of G does not

S.K. Ghosh and T. Tokuyama (Eds.): WALCOM 2013, LNCS 7748, pp. 126–136, 2013.

exist for all graphs G. Nevertheless, this idea turns out to be quite efficient in practice and some compression techniques are precisely based on it [2,3]: they try to find orders of the vertices that group the neighborhoods together, as much as possible.

When one relaxes the constraints of the initial problem by rather asking for the minimum k such that there exists an order σ on the vertices of G where the neighborhood of each vertex is split in at most k intervals, this gives rise to a graph parameter called the *contiguity* of G [4]. This parameter was originally introduced in the broader context of binary matrices under the name k-consecutive-ones property [5]. It is worth to note that there are two variants of the parameter, respectively called *open contiguity* and *closed contiguity*, depending on whether one considers open neighborhoods (excluding the vertex itself) or closed neighborhoods (always containing the considered vertex). But this distinction is not fundamental as the two parameters always differ by at most one.

Here, we are interested in determining what is the worst-case contiguity for the cographs on n vertices, which are the graphs having no induced P_4 (path on 4 vertices). We also design an approximation algorithm that computes the contiguity of any cograph up to a constant ratio, in linear time with regard to the size of the input.

Related Works. Only very little is known about the contiguity of graphs. Only the class of graphs having open contiguity 1 and the class of graphs having closed contiguity 1 have been characterized: the former are biconvex graphs [6] and the latter are proper interval graphs [7]. But the classes of graphs having contiguity at most k, where k is an integer greater than 1, have not been characterized, even for $k = 2$. Actually, closed contiguity has initially been studied in the context of $0 - 1$ matrices and [5,8,9] showed that deciding whether an arbitrary graph has closed contiguity at most k is NP-complete for any fixed $k \geq 2$. For arbitrary graphs again, [10] (Corollary 3.4) gave an upper bound on the value of closed contiguity which is $n/4 + O(\sqrt{n \log n})$, whose interest lies in the constant $1/4$ since it is clear that the contiguity of a graph is always less than $n/2$ (where n is the number of vertices of the graph). Finally, let us mention that [4] showed that the contiguity is unbounded for interval graphs as well as for permutation graphs, and that it can be up to $\Omega(\log n / \log \log n)$.

Our Results. In this paper, we show that even for cographs, the contiguity is unbounded, but dominated by $O(\log n)$ for a cograph on n vertices. To this purpose, we show in Section 4 that the contiguity of a cograph G is mathematically equivalent the maximum height of a complete binary tree included (as a minor) in the cotree of G. This also allows us to exhibit a family of cographs $(G_n)_{n \in \mathbb{N}}$ on n vertices whose asymptotic contiguity is $\Omega(\log n)$, which implies that our $O(\log n)$ bound is tight. We give in Section 5 a constant-ratio approximation algorithm that computes the contiguity of an arbitrary cograph (up to a multiplicative constant) in linear time wrt. the size of the input, that is $O(n)$

time provided that the input cograph is given by its cotree (see Section 1 for a definition). In addition, our algorithm can also provide a linear ordering σ of the vertices of G, together with the pointers from each vertex to the at most k intervals partitioning its neighborhood, where k is in a constant ratio from the optimal one, i.e. the contiguity of G. In this case, the complexity of our algorithm is linear wrt. the size of the output, that is $O(kn)$ time.

As a by-product of our proofs, we also establish in Section 2 a min-max theorem which is worth of interest in itself: the maximum height of a complete binary tree included (as a minor) in a tree T (known as the *rank* of tree T [11,12]) is equal to the minimum height of a partition of T into vertex-disjoint paths.

1   Preliminaries

All graphs considered here are finite, undirected, simple and loopless. In the following, G denotes a graph, V (or $V(G)$ to avoid ambiguity) denotes its vertex set and E (or $E(G)$) its edge set. We use the notation $G = (V, E)$ and we denote $|V| = n$. The set of subsets of V is denoted by 2^V. An edge between vertices x and y will be arbitrarily denoted by xy or yx. The (open) neighborhood of x is denoted by $N(x)$ (or $N_G(x)$ to avoid ambiguity) and its closed neighborhood by $N[x] = N(x) \cup \{x\}$. The subgraph of G induced by the subset of vertices $X \subseteq V$ is denoted by $G[X] = (X, \{xy \in E \mid x, y \in X\})$.

Cographs are the graphs that do not have any P_4 (path on 4 vertices) as induced subgraph. They are also known to be the graphs G admitting a *cotree*, i.e. a rooted tree T whose leaves are the vertices of G, and whose internal nodes are labelled *series* or *parallel* with the following property: any two vertices x and y of G are adjacent iff the least common ancestor u of leaves x and y in T is a series node. Otherwise, if u is a parallel node, x and y are not adjacent.

For a rooted tree T and a node $u \in T$, the depth of u in T is the number of edges in the path from the root to u (the root has depth 0). The *height* of T, denoted by $height(T)$ or simply $h(T)$, is the greatest depth of its leaves. For a rooted tree T, the subtree of T rooted at u, denoted by T_u, is the tree induced by node u and all its descendants in T. We now give a simplified definition of *minors* of rooted trees, which is a special case of minors of graphs (see e.g. [13]).

Definition 1.1. *The* contraction of edge uv in a rooted tree T, where u is the parent of v, consists in removing v from T and assigning its children (if any) to node u.

A rooted tree T' is a minor *of a rooted tree T if it can be obtained from T by a sequence of edge contractions.*

Let us now formally define the contiguity of a graph.

Definition 1.2. *A* closed *p-interval-model (resp.* open *p-interval-model) of a graph $G = (V, E)$ is a linear order σ on V such that $\forall v \in V, \exists (I_1, \ldots, I_p) \in (2^V)^p$ such that $\forall i \in [\![1, p]\!]$, I_i is an interval of σ and $N[v] = \bigcup_{1 \leq i \leq p} I_i$ (resp. $N(v) = \bigcup_{1 \leq i \leq p} I_i$).*

The closed contiguity *(resp.* open contiguity*) of G, denoted by cc(G) (resp.*
oc(G)), is the minimum integer p such that there exists a closed p-interval-model
(resp. open p-interval-model) of G.

It is worth to note that the closed and open contiguity never differ by more than
one. Indeed, given a closed k-interval model of a graph G, we directly get an
open $k+1$-interval model for G by simply splitting, for each vertex x, the interval
containing x into (at most) two intervals. Conversely, adding (at most) one trivial
interval $\{x\}$ for each vertex x in an open k-interval-model results in a closed $(k+
1)$-interval model. Therefore, from now on, we only consider closed contiguity, but
all our results also hold for open contiguity. We will abusively extend the notion
of contiguity to cotrees referring to the contiguity of the associated cograph.

2   Some General Results on the Rank of Trees

In this section, we give two general results on trees which will play a key role
in the rest of the paper. The first result links the rank of a tree T with the
minimum height of a partition of vertices of T into paths.

The *rank* [11,12] of a tree T is the maximal height of a complete binary
tree obtained from T by edge contractions, that is: $rank(T) = \max\{h(T') \mid
T'$ complete binary tree, minor of $T\}$. A *path partition* of a tree T is a partition
$\{P_1, \ldots, P_k\}$ of $V(T)$ such that for any i, the subgraph $T[P_i]$ of T induced by
P_i is a path, as shown in Fig. 1(a). The *partition tree* of a path partition $\mathcal{P}$,
denoted by $T_p(\mathcal{P})$, is the tree whose nodes are P_i's and where the node of $T_p(\mathcal{P})$
corresponding to P_i is the parent of the node corresponding to P_j iff some node
of P_i is the parent in T of the root of P_j (see Fig. 1(b)). The height of a path
partition $\mathcal{P}$ of a tree T, denoted by $h(\mathcal{P})$, is the height $h(T_p(\mathcal{P}))$ of its partition
tree. The *path-height* of T, $ph(T)$, is the minimal height of the path partitions
of T.

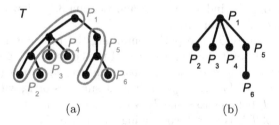

(a)   (b)

Fig. 1. A tree T, a path partition $\mathcal{P} = \{P_1, P_2, P_3, P_4, P_5, P_6\}$ of T (a), as well as the
partition tree of $\mathcal{P}$ (b)

Theorem 2.1. *For any rooted tree T, $rank(T) = ph(T)$.*

Sketch of Proof. It is not difficult to show that the path-height of a tree T
is at least the path-height of any tree T' included in T as a minor. On the
other hand, a simple recursion on the height of a complete binary tree shows

that its path-height is at least its height. It follows that the path-height of any tree T is at least the maximum height of a complete binary tree T' included as a minor in T, that is $ph(T) \geq rank(T)$. The converse inequality, namely $ph(T) \leq rank(T)$, can be shown by induction on $rank(T)$. Indeed, consider a tree T such that $rank(T) = k+1$. The nodes u of T such that $rank(T_u) = k+1$ form a path P containing the root of T. By definition, any node $v \notin P$ is such that $rank(T_v) \leq k$. Then, by the induction hypothesis, $ph(T_v) \leq rank(T_v) \leq k$. And it follows that $ph(T) \leq max_{v \notin P} ph(T_v) + 1 \leq k+1 = rank(T)$. $\qquad \square$

We now consider *bicolored trees*, i.e. trees whose nodes are colored either black or white. We define the *black rank* (resp. *white rank*), denoted $r_B(T)$ (resp. $r_W(t)$), of a bicolored tree T as the maximum height of an entirely black (resp. entirely white) complete binary tree being a minor of T.

Theorem 2.2. *For any bicolored complete binary tree T, $r_B(T) + r_W(T) \geq h(T) - 1$.*

Sketch of Proof. The proof is by induction on $h(T)$. Consider a complete binary bicolored tree T of height $k+1$, whose root is colored black wlog. and has two children denoted by u_1 and u_2. If $r_B(T_{u_1}) = r_B(T_{u_2})$, then $r_B(T) = r_B(T_{u_1}) + 1$. And since $r_W(T) \geq r_W(T_{u_1})$, by the induction hypothesis, we obtain $r_B(T) + r_W(T) \geq h(T_{u_1}) + 1 - 1 = h(T) - 1$. On the other hand, if $r_B(T_{u_1}) \neq r_B(T_{u_2})$, then assume wlog. that $r_B(T_{u_1}) > r_B(T_{u_2})$. Then, we have $r_B(T) + r_W(T) \geq r_B(T_{u_1}) + r_W(T_{u_2}) \geq r_B(T_{u_2}) + 1 + r_W(T_{u_2})$, and by the induction hypothesis, we obtain the desired inequality for T. $\qquad \square$

3   An Upper Bound for the Contiguity of Cographs

We now prove that the contiguity of any cograph is linearly bounded by the rank of its cotree T (Theorem 3.1 below) by using a path partition of T of minimal height h. Our proof is by induction on h and Lemma 3.1 below constitutes the recursive encoding step of our proof.

In a path partition of T, the path containing the root naturally induces a partition of the leaves of T, i.e. the vertices of the corresponding cograph G, as described in the following definition. A *root-path decomposition* (see Fig. 2) of a rooted tree T is a set $\{T_1, \dots, T_p\}$ of disjoint subtrees of T, with $p \geq 2$, such that every leaf of T belongs to some T_i, with $i \in [1..p]$, and the sets of parents in T of the roots of T_i's is a path containing the root of T.

Lemma 3.1 (Caterpillar Composition Lemma). *Given a cograph $G = (V, E)$ and a root-path decomposition $\{T_i\}_{1 \leq i \leq p}$ of its cotree, where X_i is the set of leaves of T_i, $cc(G) \leq 2 + \max_{i \in [1..p]} cc(G[X_i])$.*

Sketch of Proof. It is straightforward to check that for any $i \in [1..p]$ and for any vertex $x \in X_i$, the neighbors of x that are not in X_i are split in at most two intervals in the order σ given on Fig. 3. Therefore, by choosing for σ an order

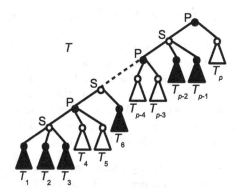

Fig. 2. The root-path decomposition $\{T_1, \ldots, T_p\}$ of a rooted tree T

on each X_i that realizes the contiguity of $G[X_i]$, we obtain that in σ, the neighborhood of any vertex $x \in X_i$ of G is split in at most $2 + cc(G[X_i])$ intervals, which proves the lemma. □

Fig. 3. The general structure of the order σ used in Lemma 3.1 for the root-path decomposition of Fig 2

From this, we deduce an upper bound on the contiguity of a cograph depending on the rank of its cotree.

Theorem 3.1. *For any cograph G with cotree T, $cc(G) \leq 2\,rank(T) + 1$.*

Sketch of Proof. The proof is by induction on $rank(T)$. The inequality holds for $rank(T) = 0$. Consider a cograph G whose cotree T is of rank $k + 1$, with $k \geq 0$. The nodes u of T such that $rank(T_u) = k + 1$ form a path containing only internal nodes of T and containing its root. Therefore, this path induces a root-path decomposition $\{T_1, \ldots, T_p\}$ of T as shown on Fig. 2. By definition, this root-path decomposition is such that for any $i \in [1..p]$, tree T_i is of rank at most k. Then, by the induction hypothesis, $cc(G[X_i]) \leq 2\,rank(T_i) + 1$. From Lemma 3.1, we have that $cc(G) \leq 2 + \max_{i \in [1..p]} cc(G[X_i])$. It follows that $cc(G) \leq 2 + \max_{i \in [1..p]} \{2\,rank(T_i) + 1\} \leq 2 + 2k + 1 = 2(k + 1) + 1$, which ends the induction and the proof of Theorem 3.1. □

As the rank of a tree T is bounded by the logarithm of its number of leaves, we directly obtain the following corollary.

Theorem 3.2. *If $G = (V, E)$ is a cograph, then $cc(G) \leq 2\log_2 |V| + 1$.*

4   A Lower Bound for the Contiguity of Cographs

In this section, we show that the rank of a cotree also provides a lower bound on the contiguity of its associated cograph. Together with the result of the previous section, this give that the contiguity of a cograph and the rank of its cotree are mathematicaly equivalent functions, which is at the core of the approximation algorithm we present in next section. We also use the lower bound to exhibit a family of cographs on n vertices whose contiguity is at least $\Omega(\log n)$, showing that the upper bound of the previous section is tight.

Lemma 4.1. *Let G_k be the underlying undirected graph of the transitive closure of the directed rooted complete ternary tree T^k of depth $k \geq 0$. Then, we have $cc(G_k) \geq k + 1$.*

Proof: We prove it by induction on k. We obviously have $cc(G_0) \geq 1$.

Now, suppose that $cc(G_k) \geq k + 1$ for some $k \geq 0$, and let us show that $cc(G_{k+1}) \geq k+2$. Consider a $cc(G_{k+1})$-interval-model of G_{k+1}, and denote by σ the corresponding order on the vertices of G. We denote by r the root of T^{k+1}, and by v_1, v_2 and v_3 its three children in T^{k+1}. Since there are at most two vertices of T^{k+1} that are next to r in σ, then there exists $i \in [\![1,3]\!]$ such that no vertex of $T^{k+1}_{v_i}$ is next to r in σ. Denoting by G_{v_i} the subgraph of G_{k+1} induced by the vertices of $T^{k+1}_{v_i}$, the induction hypothesis gives that $cc(G_{v_i}) \geq k + 1$. In particular, this implies that in the restriction of σ to the vertices of G_{v_i}, there exists some $v \in G_{v_i}$ such that its neighborhood in G_{v_i} is split into at least $k + 1$ intervals. Since r is adjacent to v in G_{k+1}, and since the at most two vertices next to r in σ are not in the neighborhood of v, it follows that the neighborhood of v requires one more interval for r in σ. Thus, $cc(G_{k+1}) \geq k + 2$.   □

Note that G_k is a cograph whose cotree can be recursively built as follows. The cotree of G_1 (the *claw* graph) is made of one series root having two children: a leaf and a parallel node having three leaf children. Replacing these three leaves by three copies of the cotree of G_k, for any $k \geq 1$, results in the cotree of G_{k+1}. From now on, in order to use the result of Theorem 2.2, we consider cotrees bicolored as explained in the following definition. The *bicolored cotree* of a cograph G is its cotree where the parallel nodes are colored black and the series nodes are colored white.

Lemma 4.2. *For any cograph G, whose bicolored cotree has a white root and black rank at least $2k$, we have $cc(G) \geq k + 1$.*

Sketch of Proof. We prove by induction on k that G contains G_k as an induced subgraph, which has contiguity at least $k + 1$ from Lemma 4.1. Since G has black rank at least $2k$, its cotree T contains, as a minor, an entirely black complete binary tree T'. Consider the two nodes at depth 1 in T', they have at least four different children in T. For three of them u_1, u_2 and u_3, take one leaf of their subcotree in T: since the nodes of T' are all black (i.e. parallel), this gives three independent vertices denoted by x_1, x_2 and x_3. Now recall that the root of T is white (i.e. series) and must have a child which is not an ancestor of the root of T'. Take a leaf y being a descendant of this child: y, x_1, x_2 and x_3 induce the claw graph, i.e. the graph G_1 of Lemma 4.1. Finally, note that for any $u \in \{u_1, u_2, u_3\}$, u is necessarily white (as a parallel node of a cotree has only series, or leaf, children) and the cotree T_u has black rank at least $2k - 2$. Then, from the induction hypothesis, the cograph associated to T_u contains G_{k-1} as an induced subgraph. And from the recursive construction of the cotree of G_k that precedes Lemma 4.2, we conclude that G contains G_k as an induced subgraph. □

Theorem 4.1. *For any cograph G and its cotree T, $cc(G) \geq (rank(T) - 7)/4$.*

Proof: Consider the bicolored cotree T of G. Then, from Lemma 2.2, $r_B(T) + r_W(T) \geq rank(T) - 1$, so either $r_B(T) \geq (rank(T) - 1)/2$, or $r_W(T) \geq (rank(T) - 1)/2$.

In the first case, $r_B(T) \geq 2((rank(T) - 3)/4) + 1$, so T has a black complete binary tree of height $2((rank(T) - 3)/4)$, below a white node, as a minor, so Lemma 4.2 implies that $cc(G) \geq (rank(T) - 3)/4 + 1 \geq (rank(T) - 7)/4$.

In the second case, the bicolored cotree $T_{\bar{G}}$ of the complement of G is simply the bicolored cotree of G where series and parallel nodes, and so black and white nodes, have been exchanged. Then, we have $r_B(T_{\bar{G}}) = r_W(T) \geq (rank(T)-1)/2$. Lemma 4.2 implies similarly as above that $cc(\bar{G}) \geq (rank(T) - 3)/4 + 1$. We can easily check that $cc(\bar{G}) \leq cc(G) + 2$ (the closed neighborhood of any vertex x in $\bar{G}$ is composed by at most $cc(G) + 1$ intervals surrounding the intervals of the closed neighborhood of x in G, plus possibly one interval for vertex x), so $cc(G) \geq (rank(T) - 3)/4 + 1 - 2 = (rank(T) - 7)/4$. □

From Theorem 4.1 we obtain that the contiguity of cographs is unbounded. Moreover, since the rank of a complete binary tree is, by definition, its height, it follows that the family of cographs having a complete binary cotree reaches the $O(\log n)$ upper bound of Theorem 3.2, showing that this bound is tight.

Corollary 4.1. *For any cograph G whose cotree is a complete binary tree, $cc(G) = \Omega(\log n)$.*

As a direct consequence of Theorems 4.1 and 3.1, we obtain the equivalence between the contiguity of a cograph and the rank of its cotree, which is at the core of the approximation algorithm presented in next section.

Corollary 4.2. *For any cograph G with cotree T, $cc(G) = \Omega(rank(T))$.*

5   An Approximation Algorithm for the Contiguity of Cographs

We now provide an algorithm that outputs an integer k and a k-interval-model of the cograph G given as input, such that k is in a constant ratio from the contiguity of G. Our algorithm takes as input a cograph G given by its cotree, which takes $O(n)$ space, and outputs a k-interval model of G in $O(kn)$ time, i.e. linear wrt. the size of the output. Alternatively, if only the value of k is needed, the algorithm runs in $O(n)$ time, which is linear wrt. the size of the input.

First Step: Approximation of the Contiguity of G. The first step of our algorithm computes an integer k which is in a constant ratio from the contiguity of G. To this purpose, we simply compute recursively the rank of T_u for any node $u \in T$ by a bottom-up process: for a node u we have either i) $rank(T_u) = max_{v \text{ child of } u} rank(T_v)$ if this maximum is reached by only one child of u, or ii) $rank(T_u) = 1 + max_{v \text{ child of } u} rank(T_v)$ otherwise. Then, our algorithm outputs the value $k = 2\,rank(T_r) + 1$, where r is the root of T, and so $rank(T_r) = rank(T)$. This clearly takes $O(n)$ time.

Furthermore, the approximation ratio ρ of this algorithm is constant. Using Theorem 4.1 and the fact that $cc(G) \geq 1$, we deduce that $\rho = (2\,rank(T) + 1)/cc(G) \leq (2\,rank(T) + 1)/max(1, (rank(T) - 7)/4)$. This function reaches its maximum of 23 for $rank(T) = 11$, and then the algorithm performs a 23-approximation of the closed contiguity.

Second Step: Building a k-Interval Model of G. We follow the construction provided by Theorem 3.1 and Lemma 3.1 (see Fig. 3) in order to output a k-interval-model of G, where $k = 2\,rank(T) + 1$. During this process, we build the order σ on the vertices as well as a table $Neighborhoods$ of n tables of $2k$ pointers to the bounds of the intervals of each vertex in order σ. To this purpose, we call the recursively defined routine $Build(u)$, where u is a node of T, on the root r of T. $Build(u)$ outputs an order σ_u on the subset X_u of vertices of G being the leaves of T_u and table $Neighborhoods_u$ containing the pointers of the vertices of X_u toward σ_u. Routine $Build(u)$ proceeds in three steps as follows.

i) thanks to the ranks computed in the first step of the algorithm, $Build(u)$ finds the subset P_u of nodes $\tilde{u}$ of T_u such that $rank(T_{\tilde{u}}) = rank(T_u)$, and the subset C_u of children in T of nodes $\tilde{u} \in P_u$.

ii) $Build(u)$ recursively calls $Build(v)$ for all nodes $v \in C_u$.

iii) $Build(u)$ builds σ_u by concatenating all the orders σ_v returned by the recursive calls as shown on Fig. 3, and builds $Neighborhoods_u$ by merging all the tables $Neighborhoods_v$ returned by the recursive calls. Then, for each $v \in C_u$ and for each vertex $x \in X_v$, we add to $Neighborhoods_u$ the pointers of x toward the at most two intervals of σ_u formed by its neighbors that are not in X_v, as explained in the proof of Lemma 3.1.

The terminal case of Routine $Build(u)$ is when u is a leaf of T, for which the computation of σ_u and $Neighborhoods_u$ is trivial and takes constant time. The fact that a call to $Build(r)$ indeed gives the desired k-interval model of G comes from the fact that the routine follows the constructive proof of Theorem 3.1. Let us analyze its complexity. Step i), Step ii) and the construction of σ_u in Step iii) take $O(|C_u| + |P_u|)$, that is $O(n)$ for the whole process on tree T. In Step iii), the merge of table $Neighborhoods$ and the addition of the pointers to the two new intervals of each vertex take $O(X_u)$ time. It turns out that, during the whole process on T, a vertex x will be involved in at most h different sets X_u, where h is the height of the path partition of T defined by the set of paths P_u computed along the process. From Theorem 3.1, $h = rank(T) = (k - 1)/2$. Thus, the total computation time of the k-interval model output by our algorithm is $O(kn)$ time.

Conclusion

We showed that the contiguity of a cograph is equivalent to the maximum height of a complete binary tree contained in its cotree as a minor. From this, we obtained a tight $O(\log n)$ upper bound on the maximum contiguity of a cograph on n vertices. Even more interesting, this allowed us to design a linear time algorithm that does not only compute an approximation of the contiguity of a cograph G but also provides a k-interval model realizing a k which is in a constant-ratio to the optimal one, i.e. the contiguity of G. Then, the first question raised by our work is whether it is possible to compute efficiently the exact value of the contiguity of a cograph and to provide a model realizing this optimal value. Another key perspective is to extend our results to larger classes of graphs, such as permutation graphs (which are a proper generalization of cographs) and interval graphs. Does the $O(\log n)$ upper bound still hold for those graphs? Is it possible to compute efficiently an exact or approximated value of their contiguity?

Acknowledgments. The authors thank Pierre Charbit and Stéphan Thomassé for useful discussions on the subject, as well as George Oreste Manoussakis for proofreading a draft of this article. This work was partially supported by the PEPS-C1P CNRS project.

References

1. Turan, G.: On the succinct representation of graphs. Discr. Appl. Math. 8, 289–294 (1984)
2. Boldi, P., Vigna, S.: The webgraph framework I: compression techniques. In: WWW 2004, pp. 595–602. ACM (2004)
3. Boldi, P., Vigna, S.: Codes for the world wide web. Internet Mathematics 2(4), 407–429 (2005)
4. Crespelle, C., Gambette, P.: Efficient Neighborhood Encoding for Interval Graphs and Permutation Graphs and $O(n)$ Breadth-First Search. In: Fiala, J., Kratochvíl, J., Miller, M. (eds.) IWOCA 2009. LNCS, vol. 5874, pp. 146–157. Springer, Heidelberg (2009)
5. Goldberg, P., Golumbic, M., Kaplan, H., Shamir, R.: Four strikes against physical mapping of DNA. Journal of Computational Biology 2(1), 139–152 (1995)
6. Brandstädt, A., Le, V., Spinrad, J.: Graph Classes: a Survey. SIAM Monographs on Discrete Mathematics and Applications (1999)
7. Roberts, F.: Representations of indifference relations, Ph.D. thesis, Stanford University (1968)
8. Johnson, D., Krishnan, S., Chhugani, J., Kumar, S., Venkatasubramanian, S.: Compressing large boolean matrices using reordering techniques. In: Proceedings of the Thirtieth International Conference on Very Large Data Bases, VLDB 2004, vol. 30, pp. 13–23 (2004)
9. Wang, R., Lau, F., Zhao, Y.: Hamiltonicity of regular graphs and blocks of consecutive ones in symmetric matrices. Discr. Appl. Math. 155(17), 2312–2320 (2007)

10. Gavoille, C., Peleg, D.: The compactness of interval routing. SIAM Journal on Discrete Mathematics 12(4), 459–473 (1999)
11. Ehrenfeucht, D.H.A.: Learning decision trees from random examples. Information and Computation 82(3), 231–246 (1989)
12. Gavaldà, R., Thérien, D.: Algebraic Characterizations of Small Classes of Boolean Functions. In: Alt, H., Habib, M. (eds.) STACS 2003. LNCS, vol. 2607, pp. 331–342. Springer, Heidelberg (2003)
13. Lováz, L.: Graph minor theory. Bulletin of the American Mathematical Society 43(1), 75–86 (2006)

Approximation Algorithms for the Partition Vertex Cover Problem

Suman Kalyan Bera[1], Shalmoli Gupta[2], Amit Kumar[3], and Sambuddha Roy[1]

[1] IBM India Research Lab, New Delhi, India
[2] University of Illinois at Urbana-Champaign, USA
[3] Indian Institute of Technology Delhi, India

Abstract. We consider a natural generalization of the `Partial Vertex Cover` problem. Here an instance consists of a graph $G = (V, E)$, a cost function $c : V \to \mathbb{Z}^+$, a partition $P_1, \ldots, P_r$ of the edge set E, and a parameter k_i for each partition P_i. The goal is to find a minimum cost set of vertices which cover at least k_i edges from the partition P_i. We call this the `Partition-VC` problem. In this paper, we give matching upper and lower bound on the approximability of this problem. Our algorithm is based on a novel LP relaxation for this problem. This LP relaxation is obtained by adding knapsack cover inequalities to a natural LP relaxation of the problem. We show that this LP has integrality gap of $O(\log r)$, where r is the number of sets in the partition of the edge set. We also extend our result to more general settings.

1 Introduction

The `Vertex Cover` problem is one of the most fundamental NP-hard problems and has been widely studied in the context of approximation algorithms [1, 2]. In this problem, we are given an undirected graph $G = (V, E)$ and a cost function $c : V \to \mathbb{Z}^+$. The goal is to find a minimum cost set of vertices which *cover* all the edges in E : a set of vertices S covers an edge e if S contains at least one of the end-points of e. Several 2-approximation algorithms are known for this problem [3, 4]. The `Partial Vertex Cover` problem is a generalization of the `Vertex Cover` problem, where we are also given a parameter k. The goal is to find a minimum cost set of vertices which cover at least k edges. This problem was proposed by Bshouty and Burroughs [5], and they gave a 2-approximation for this problem using LP-rounding. Since then, many different techniques have been shown to give 2-approximation algorithm for this problem ([6, 7, 8]).

In this paper, we consider a natural generalization of the `Partial Vertex Cover` problem. Here an instance consists of a graph $G = (V, E)$, a cost function $c : V \to \mathbb{Z}^+$, a partition $P_1, \ldots, P_r$ of the edge set E, and a parameter k_i for each partition P_i. The goal is to find a minimum cost set of vertices which cover at least k_i edges from the partition P_i. We call this the `Partition-VC` problem. In this paper, we give matching upper and lower bound on the approximability of this problem. We give an $O(\log r)$-approximation algorithm for the `Partition-VC` problem, and show that unless P=NP, we cannot do better. Recall that r denotes

S.K. Ghosh and T. Tokuyama (Eds.): WALCOM 2013, LNCS 7748, pp. 137–145, 2013.

the number of sets in the partition of the edge set E. Note that for constant number of partitions this gives a constant approximation. Our results also extend to the slightly more general problem where edges have weights, and we would like to pick a minimum cost set of vertices which cover edges of total weight at least Π_i for each partition P_i. We call this the Knapsack Partition Vertex Cover problem.

Our Techniques: The hardness result for the Partition-VC problem follows by an approximation preserving reduction from the Set Cover problem. The approximation algorithm uses a novel LP relaxation – this is the main contribution of the paper, and we expect this idea to have more applications. The natural LP relaxation for even the Partial Vertex Cover turns out to be unbounded. Indeed, consider the following example : the graph is a star – there is a vertex v of degree D and all its neighbors are leaves. All vertex costs are 1, and the parameter $k = 1$. Clearly, any optimal solution must cost at least 1 unit. But a fractional solution will pick the vertex v to an extent of $\frac{1}{D}$, and hence will cover all the D edges fractionally to an extent of $\frac{1}{D}$. So, the fractional solution pays only $\frac{1}{D}$. One way of getting around this problem is to augment the LP with more information. Here, we can guess the most expensive vertex an optimal solution will pick, and can remove all vertices with cost more than the cost of this vertex. Further, the cost of this vertex is also a lower bound on the optimal value. This idea was used by [8] to give a 2-approximation for the Partial Vertex Cover problem. However, applying such an idea to the Partition-VC problem turns out to be non-trivial. We cannot guess the most expensive vertex in each of the partitions – this will take time exponential in r. Our approach is to strengthen the natural LP relaxation such that no guesswork is required. We show how to do this using knapsack cover inequalities [9]. Armed with this stronger relaxation, we show that one can carefully use randomized rounding based techniques to get the approximation algorithm.

Related Work: There has been recent work on partial covering versions of several covering problems. For the set cover problem, the partial covering version namely the partial set cover problem was first studied by Kearns [10], who proved that the approximation ratio of the greedy algorithm is at most $2H(n)+3$, where n is the size of the ground set and $H(n)$ is the nth harmonic number. Later Slavík [11] showed that it is actually bounded by $H(k)$ where k is the number of elements to be covered. This natural greedy approach when extended for the Partition-VC problem gives only a $H(|V|)$-approximation, which is much worse than the lower bound of $O(\log r)$ that we have proved for this problem.

The partial vertex cover problem has also been widely studied in the literature. Bshouty and Burroughs [5] were the first to give a polynomial time 2-approximation algorithm for it. Subsequently, several other algorithms based on Lagrangian Relaxation, local-ratio, primal-dual techniques with the same approximation guarantee were proposed [6, 7, 8, 12]. Mestre's [12] primal-dual technique can also be used to get a 2-approximation for a more general version of the problem, the partial capacitated vertex cover problem. Bar-Yehuda et al. [13] gave constant factor approximation algorithms for several variants of this problem using the local-ratio technique. Partial versions have also been studied for the Facility Location problem and its variants. Charikar et al. [14] explored the outlier or robust version of the uncapacitated facility location problem and k-center problem where only a fraction of the clients need to be serviced. For both the problems they gave constant approximation algorithms. Apart from these there are several other partial covering problems: e.g. k-median with outliers [15], k-MST problem [16] and k-multicut problem [17]. However, these approaches do not seem to work for the `Partition-VC` problem.

The set of partitions of the edge set in the `Partition-VC` problem are a special case of matroids. There has been significant work on maximizing a submodular function under matroid constraints [18, 19], but none of these results apply to the `Partition-VC` problem.

The rest of the paper is organized as follows. We present the hardness of the `Partition-VC` problem in Section 2. Our rounding algorithm and its analysis is presented in Section 3. Finally concluding remarks are made in Section 4.

2   Hardness of the `Partition-VC` Problem

In this section, we prove that it is NP-hard to get better than $O(\log r)$-approximation for the `Partition-VC` problem. We give a reduction from the `Set Cover` problem. Recall that an instance $\mathcal{I}$ of the set cover problem consists of a set X containing r elements, and a set $\mathcal{S}$ of subsets $S_1, \ldots, S_m$ of X. The goal is to find minimum number of sets in $\mathcal{S}$ such that their union is X. This problem is known to be NP-hard, and in fact, it is known that unless $NP \subset DTIME(n^{O(\log \log n)})$, any polynomial time algorithm for the `Set Cover` problem must have approximation ratio of $\Omega(\log r)$ [20].

We now describe the reduction in detail. Let $\mathcal{I}$ be an instance of the `Set Cover` problem as described above. We now construct an instance $\mathcal{I}'$ of the `Partition-VC` problem. The graph G' in $\mathcal{I}'$ is a bipartite graph. The vertices on left side, V_L' are defined as follows : for each set $S_i \in \mathcal{S}$, we add a vertex s_i' to V_L'. All these vertices have unit cost. The vertices on the right side, V_R' are as follows : for every element $u \in X$, we have a corresponding vertex $u' \in V_R'$. Each of these vertices has infinite cost. Now, we define the set of edges E' and the partition $P_1, \ldots, P_r$ (note that the number of sets in the partition is same as the size of the set X in $\mathcal{I}$). For a vertex $s_i' \in V_L'$ and $u' \in V_R'$, we have an edge between them in E' iff $u \in S_i$ in the instance $\mathcal{I}$. We partition the set E' as follows : for every $u' \in V_R'$, define $P_{u'}$ as the set of edges incident to u'. The partition of E' is $\{P_{u'} : u' \in V_R'\}$. Further, the quantities $k_{u'}$, which tell how many of the edges in the set $P_{u'}$ need to be covered, are 1. This completes the description of the instance $\mathcal{I}'$. The following lemma is now easy to see.

Lemma 1. *There is a solution to $\mathcal{I}$ of cost C iff there is a solution to $\mathcal{I}'$ of cost C. Hence, unless P=NP, any polynomial time algorithm for the `Partition-VC` problem must have approximation ratio of $\Omega(\log r)$.*

Proof. Any solution to $\mathcal{I}$ picks a some subsets $S_{i_1}, \ldots, S_{i_l}$ in $\mathcal{S}$. Then we can have a solution of the same cost for $\mathcal{I}'$ in which we pick the corresponding vertices in V'_L. Similarly, consider a solution to $\mathcal{I}'$. None of the vertices picked by this solution can be in V'_R (because of infinite cost). Thus, we can look at the corresponding subsets in $\mathcal{S}$, and these subsets will form a set cover in $\mathcal{I}$.   □

3   Approximation Algorithm for the Partition-VC Problem

In this section, we give an $O(\log r)$-approximation algorithm for this problem. We begin with the natural LP relaxation, and then strengthen it by adding knapsack cover inequalities. Fix an instance $\mathcal{I}$ consisting of a graph $G = (V, E)$, partition $P_1, \ldots, P_r$ of E, and parameters $k_1, \ldots, k_r$. The natural LP relaxation is described below. For every vertex $v \in V$, we have a variable x_v which should be 1 if we pick this vertex, and 0 otherwise. For an edge e, we have a variable y_e, which should be 1 if e gets covered by the solution, 0 otherwise.

$$\min \sum_{v \in V} c_v x_v \tag{LP1}$$

$$x_u + x_v \geq y_e \qquad \text{for all edges } e = (u, v) \in E$$

$$\sum_{e \in P_i} y_e \geq k_i \qquad \text{for every partition } P_i$$

$$0 \leq x_v \leq 1 \qquad \text{for all vertices } v \in V$$

$$0 \leq y_e \leq 1 \qquad \text{for all edges } e \in E$$

As explained in the introduction, the integrality gap of this LP relaxation is unbounded even for the Partial Vertex Cover problem.

To improve the integrality gap, we add knapsack cover inequalities as follows. Consider a subset of vertices A. Suppose we select all the vertices in A. Now, A will cover some of the edges in each of the partitions P_i. Define $k_i(A)$ as the number of edges we still need to cover from P_i (after having picked A). So, we must choose enough vertices from $V \setminus A$ such that the remaining covering constraint in every partition is met. In other words, the following conditions must be satisfied

$$x_u + x_v \geq y_e \qquad \forall e = (u, v) \in E \text{ and } u, v \notin A$$

$$\sum_{\substack{e = (u, v) \in P_i \\ u, v \notin A}} y_e \geq k_i(A) \qquad \text{for every partition } P_i$$

If we replace the variable y_e in the second inequality above by using the first inequality, we get that for every partition P_i,

$$\sum_{\substack{e=(u,v)\in P_i \\ u,v\notin A}} (x_u + x_v) \geq k_i(A)$$

$$\text{i.e., } \sum_{v\notin A} \deg_i(v, A)x_v \geq k_i(A),$$

where $\deg_i(v, A)$ denotes the degree of v in the subgraph of G considering edges in P_i only and removing the vertices of A. Since x_v have value lying the range $[0, 1]$, we can further strengthen the above by truncating the values $\deg_i(v, A)$ to $\min(\deg_i(v, A), k_i(A))$. Thus we get the following strengthened LP relaxation:

$$\min \sum_{v\in V} c_v x_v \qquad\qquad \text{(PVC-LP)}$$

$$\sum_{v\notin A} \min(k_i(A), \deg_i(v, A))x_v \geq k_i(A) \qquad \text{for all partitions } P_i, \text{ subsets } A \subseteq V$$

$$\tag{1}$$

$$0 \leq x_v \leq 1 \qquad \text{for all vertices } v \in V$$

We shall show that the integrality gap of the above LP relaxation is $O(\log r)$. We first present the rounding algorithm, and then we will discuss how to solve this LP. So assume that we have a solution $x^\star$ to the LP above. The rounding algorithm is described below.

Algorithm 1. Rounding a solution to the PVC-LP
___
    Given a solution $x^\star$.
    Let $\hat{x}$ be the integral solution that we will build. Initially, $\hat{x}_v = 0$ for all v.
    for $\forall v \in V$ **do**
      if $x_v^\star \geq 1/6$ **then**
        Set $\hat{x}_v \leftarrow 1$
      else
        Set $\hat{x}_v \leftarrow 1$ with probability $6x_v^\star$.
      end if
    end for
    Pick the set of vertices v in the solution for which $\hat{x}_v = 1$

Analysis of the Rounding Algorithm

We begin by showing that the solution constructed above is good for any partition P_i with constant probability.

Theorem 1. *For any partition P_i, the solution $\hat{x}$ covers at least k_i edges of P_i with probability at least $5/8$.*

Proof. Define A to be the set $\{v \in V : x_v^* \geq 1/6\}$. Note that our algorithm picks all the vertices in A. Therefore, we just need to show that the vertices picked from $V \setminus A$ cover at least $k_i(A)$ edges from P_i after we remove the vertices in A. Let $\beta_i(v, A)$ denote $\frac{\min(\deg_i(v,A), k_i(A))}{k_i(A)}$. Then the constraint (1) applied to this particular set A implies that

$$\sum_{v \notin A} \beta_i(v, A) x_v^* \geq 1. \tag{2}$$

Lemma 2. *For any partition P_i,*

$$\sum_{v \notin A} \beta_i(v, A)\hat{x}_v < 2,$$

happens with probability at most 3/8.

Proof. The proof is a simple application of Chebychev's inequality and uses the fact that the quantities $\beta_i(v, A)$ are at most 1. For a vertex $v \notin A$, let Y_v be an indicator random variable which is 1 if v is included in the solution (i.e., $\hat{x}_v = 1$), 0 otherwise. Let Z_i denote $\sum_{v \notin A} \beta_i(v, A)Y_v$. The expected value $E[Z_i]$ can be expressed as

$$E[Z_i] = \sum_{v \notin A} 6\beta_i(v, A)x_v^* \geq 6,$$

using the inequality (2). We now bound the variance $Var(Z_i)$ of Z_i.

$$Var[Z_i] = \sum_{v \notin A} \beta_i(v, A)^2 Var(Y_v)$$

$$= \sum_{v \notin A} \beta_i(v, A)^2 \cdot 6x_v^*(1 - 6x_v^*)$$

$$\leq 6 \sum_{v \notin A} \beta_i(v, A)x_v^* \quad \text{because } \beta_i(v, A) \leq 1$$

The claim now follows from Chebychev's inequality. Indeed, we want to bound the probability $\Pr[Z_i < 2]$. This can be done as follows :

$$\Pr[Z_i < 2] \leq \Pr[Z_i < E[Z_i]/3] \leq \Pr[|Z_i - E[Z_i]| \geq \frac{2E[Z_i]}{3}]$$

$$\leq \frac{9}{4} \cdot \frac{Var[Z_i]}{E[Z_i]^2}$$

$$\leq \frac{9}{4} \cdot \frac{6\sum_{v \notin A} \beta_i(v, A)x_v^*}{\left(\sum_{v \notin A} 6\beta_i(v, A)x_v^*\right)^2} \leq \frac{3}{8},$$

where the last inequality uses (2). $\qquad\qquad\square$

Now, suppose the solution $\hat{x}$ does not cover at least k_i edges of P_i. Then, restricted to the sub-graph of G where we include edges in P_i only, and remove all vertices in A, the total degree of the vertices picked by our algorithm (in $V \setminus A$) will be less than $2k_i(A)$. In other words,

$$\sum_{v \notin A} \beta_i(v, A)\hat{x}_v < 2.$$

But lemma 2 shows that the probability of this event is at most $3/8$. This proves the theorem.   □

Thus, in expectation, more than half of the partitions get satisfied. To satisfy all the partitions, we just repeat our algorithm $O(\log r)$ times. So, our final algorithm is : repeat Algorithm 1 $c \log r$ times, where c is a large constant. We output the union of all the vertices chosen in each such round. The following theorem now shows that our algorithm is an $O(\log r)$-approximation algorithm.

Theorem 2. *With high probability, the algorithm outputs a feasible solution and its cost is $O(\log r) \cdot \sum_{v \in V} c_v x_v^\star$.*

Proof. Lemma 2 shows that in any particular round, we cover at least k_i edges of P_i with probability at least $5/8$. So, the probability that we do not satisfy the constraint for P_i in any of the rounds is at most $1/r^{c'}$ for some large constant c', and hence, by union bound, our algorithm outputs a feasible solution with high probability. Also, the expected cost of each round is at most $6 \sum_{v \in V} c_v x_v^\star$. This proves the theorem.   □

Solving the LP Relaxation. Finally, we show how we can get a solution $x^\star$ for the PVC-LP. We first guess the value of the optimal solution – call it Δ (we can always do this up to any constant precision by binary search). We convert the LP to a feasibility LP by removing the objective function, and adding a constraint

$$\sum_v c_v x_v \leq \Delta.$$

Now, we use the ellipsoid method to solve the LP. Given a candidate solution x, we first check if it satisfies the above constraint – if not, we can just return this violated constraint. Otherwise, we define $A = \{v : x_v \geq 1/6\}$. We check the constraint (1) for this set A, and again, if this is not satisfied, we can return this as a violated constraint. Now, notice that our rounding algorithm just requires the solution $x^\star$ to satisfy these two inequalities, and we need not even check all the (exponentially many) constraints (1).

3.1   Extensions

We now show that our result can be extended to more general settings.

The Knapsack Partition Vertex Cover Problem : Recall that in this problem, we have weights w_e associated with each edge e. Again given a partition $P_1, \ldots, P_r$, and parameters Π_i, we would like to pick a minimum cost subset of vertices such that they cover edges of cost at least Π_i from the set P_i for each i. Our algorithm and analysis extend in straightforward way to this setting as well.

The Sets P_i Need Not Be Disjoint : Our analysis does not require these sets to be disjoint. The same algorithm works here as well. Note that our hardness results holds in the stronger setting where we want these sets to be disjoint.

4   Conclusion

We have presented algorithms for the `Partition-VC` problem. For this problem using primal-dual schema similar to the one described by Tim Carnes & David Shmoys [21] we can obtain an $O(f)$-approximation algorithm, where f is the maximum number of edges in a partition P_i. The proof is quite straight forward. This result is analogous to the f-approximation result for the `Set Cover` problem [22, 3]. It will be interesting to extend our techniques to partition versions of other partial covering problems. One natural related problem is the `Partition Set Cover` problem. The `Partition Set Cover` problem can be seen as a generalization of the `Partial Set Cover` problem where $P_1, \ldots, P_r$ forms partition of the element set, and the goal is to find a minimum cost sub-collection of sets such that atleast k_i elements are covered from partition P_i. For that we can get a $H(\sum_{P_i} k_i)$-approximation by directly extending Slavík's [11] greedy approach, and unless P=NP we cannot do any better.

References

[1] Vazirani, V.V.: Approximation algorithms. Springer-Verlag New York, Inc., New York (2001)

[2] Williamson, D.P., Shmoys, D.B.: The Design of Approximation Algorithms. Cambridge University Press (2010)

[3] Bar-Yehuda, R., Even, S.: A linear time approximation algorithm for approximating the weighted vertex cover (1981)

[4] Hochbaum, D.S. (ed.): Approximation algorithms for NP-hard problems. PWS Publishing Co., Boston (1997)

[5] Bshouty, N.H., Burroughs, L.: Massaging a Linear Programming Solution to Give a 2-Approximation for a Generalization of the Vertex Cover Problem. In: Meinel, C., Morvan, M. (eds.) STACS 1998. LNCS, vol. 1373, pp. 298–308. Springer, Heidelberg (1998)

[6] Hochbaum, D.S.: The t-Vertex Cover Problem: Extending the Half Integrality Framework with Budget Constraints. In: Jansen, K., Rolim, J.D.P. (eds.) APPROX 1998. LNCS, vol. 1444, pp. 111–122. Springer, Heidelberg (1998)

[7] Bar-Yehuda, R.: Using homogenous weights for approximating the partial cover problem. In: Proceedings of the Tenth Annual ACM-SIAM Symposium on Discrete Algorithms, SODA 1999, pp. 71–75. Society for Industrial and Applied Mathematics, Philadelphia (1999)

[8] Gandhi, R., Khuller, S., Srinivasan, A.: Approximation algorithms for partial covering problems. J. Algorithms 53(1), 55–84 (2004)

[9] Carr, R.D., Fleischer, L.K., Leung, V.J., Phillips, C.A.: Strengthening integrality gaps for capacitated network design and covering problems. In: Proceedings of the Eleventh Annual ACM-SIAM Symposium on Discrete Algorithms, SODA 2000, pp. 106–115. Society for Industrial and Applied Mathematics, Philadelphia (2000)

[10] Kearns, M.J.: The computational complexity of machine learning (1990)

[11] Slavík, P.: Improved performance of the greedy algorithm for partial cover. Inf. Process. Lett. 64(5), 251–254 (1997)

[12] Mestre, J.: A primal-dual approximation algorithm for partial vertex cover: Making educated guesses. Algorithmica 55(1), 227–239 (2009)

[13] Bar-Yehuda, R., Flysher, G., Mestre, J., Rawitz, D.: Approximation of Partial Capacitated Vertex Cover. In: Arge, L., Hoffmann, M., Welzl, E. (eds.) ESA 2007. LNCS, vol. 4698, pp. 335–346. Springer, Heidelberg (2007)

[14] Charikar, M., Khuller, S., Mount, D.M., Narasimhan, G.: Algorithms for facility location problems with outliers. In: Proceedings of the Twelfth Annual ACM-SIAM Symposium on Discrete Algorithms, SODA 2001, pp. 642–651. Society for Industrial and Applied Mathematics, Philadelphia (2001)

[15] Chen, K.: A constant factor approximation algorithm for k-median clustering with outliers. In: Proceedings of the Nineteenth Annual ACM-SIAM Symposium on Discrete Algorithms, SODA 2008, pp. 826–835. Society for Industrial and Applied Mathematics, Philadelphia (2008)

[16] Garg, N.: Saving an epsilon: a 2-approximation for the k-mst problem in graphs. In: Proceedings of the Thirty-Seventh Annual ACM Symposium on Theory of Computing, STOC 2005, pp. 396–402. ACM, New York (2005)

[17] Golovin, D., Nagarajan, V., Singh, M.: Approximating the k-multicut problem. In: Proceedings of the Seventeenth Annual ACM-SIAM Symposium on Discrete Algorithm, SODA 2006, pp. 621–630. ACM, New York (2006)

[18] Vondrák, J., Chekuri, C., Zenklusen, R.: Submodular function maximization via the multilinear relaxation and contention resolution schemes. In: STOC, pp. 783–792 (2011)

[19] Călinescu, G., Chekuri, C., Pál, M., Vondrák, J.: Maximizing a monotone submodular function subject to a matroid constraint. SIAM J. Comput. 40(6), 1740–1766 (2011)

[20] Feige, U.: A threshold of ln n for approximating set cover. J. ACM 45(4), 634–652 (1998)

[21] Carnes, T., Shmoys, D.: Primal-Dual Schema for Capacitated Covering Problems. In: Lodi, A., Panconesi, A., Rinaldi, G. (eds.) IPCO 2008. LNCS, vol. 5035, pp. 288–302. Springer, Heidelberg (2008)

[22] Hochbaum, D.S.: Approximation algorithm for the weighted set covering and node cover problems (1980) (unpublished manuscript)

Daemon Conversions in Distributed Self-stabilizing Algorithms

Wayne Goddard and Pradip K. Srimani

School of Computing, Clemson University, Clemson, SC 29634–0974

Abstract. We consider protocols to transform a self-stabilizing algorithm for one daemon to one that can run under a different daemon. In the literature, there are several daemons, and several possible attributes of those daemons, and it is customary to detail the choice of daemon one is using in designing a specific self-stabilizing algorithm. The choice of daemon plays an important role in designing self-stabilizing algorithm in terms of correctness and convergence time analysis; techniques and complexity vary widely with the type of daemons used. In order to simplify algorithm development in a systematic way, it would be useful to have to consider only one "canonical" daemon and then to use a relatively mechanical procedure to convert the algorithm to any other daemon when needed. We give the first (full) proof that, provided there are IDs, any algorithm that self-stabilizes only under a fair central daemon can be converted to one that self-stabilizes under an unfair read/write daemon.

1 Introduction

A self-stabilizing algorithm is a distributed algorithm that is designed to converge to a desired state without coordination or initialization [1]. Each node participates in the distributed algorithm based on local knowledge: its own state and the states of its immediate neighbors. The objective is to achieve some global objective – a predicate defined on the local states of all the nodes in the network – based on local actions where individual nodes have no global knowledge about the network. In order to analyze the correctness of the algorithm and its time complexity, a *daemon* is assumed: the daemon plays the role of both scheduler and adversary. In the literature, there are several daemons, and several possible attributes of those daemons, and it is customary to detail the choice of daemon one is using in designing a specific self-stabilizing algorithm. Indeed, the choice of daemon plays an important role in designing self-stabilizing algorithm in terms of correctness and convergence time analysis; techniques and complexity vary widely with the type of daemons used.

In order to simplify algorithm development in a systematic way, it would be useful to have to consider only one "canonical" daemon and then to use a relatively mechanical procedure to convert the algorithm to any other daemon if need be. One of the goals of this paper is to show that, for a price (either slowdown of the convergence time or additional storage requirement at the nodes),

S.K. Ghosh and T. Tokuyama (Eds.): WALCOM 2013, LNCS 7748, pp. 146–157, 2013.

one can restrict one's attention to a specific daemon, at least when the nodes in the network are assigned unique IDs. In general, we consider the problem of converting self-stabilizing algorithms that run under one daemon to work under another daemon; these are called *transforms* [2–5]. Our purpose in this paper is to do a systematic investigation of all possible transforms between different daemons. We provide new transforms and improve upon some of the existing schemes. We give the first (full) proof that, provided there are IDs, any algorithm that self-stabilizes only under a fair central daemon can be converted to one that self-stabilizes under an unfair read/write daemon.

1.1 Definitions and Terminology

For this paper, we work in the *shared-variable* or *state-reading* model in which a node can directly read its neighbors' variables. We restrict attention to undirected, bidirectional links. All computation is deterministic unless otherwise stated. One can look at [6, 7] for a general overview of the paradigm of self-stabilization and its requirements.

A self-stabilizing algorithm is usually written as a collection of production *rules* at each node: each rule specifies a condition and an action. The *condition* is a boolean predicate on the state of the node and the states of its neighbors; the action or *move* is a change in the state of the node executing the action. A node is *privileged* at a particular time if the condition of one or more of its rules is satisfied. Note that a node might stop being privileged if a neighbor moves.

The *central daemon* (sometimes called a serial daemon) chooses or *taps* exactly one privileged node to move at each step. In contrast, the *distributed daemon* taps a nonempty subset of the privileged nodes to move at each step. These daemons are considered adversarial. A special case of the distributed daemon is the *synchronous daemon*: under this scheduler, at every step every privileged node moves.

One can also consider the granularity of the computation at a node. In the *coarse* daemon, all computation by a node is completed in one atomic step. This is the most common assumption in self-stabilization. Dolev et al. [8] proposed a self-stabilizing version of *read/write atomicity*. In this, a node's action is broken down into two atomic steps: (i) reading the state of all its neighbors, and (ii) updating its state. That is, each rule executes in a two-phase fashion: the read phase and the write phase. Whenever a node is tapped by the daemon, it reads the states of its neighbors and remembers the information; when it is tapped for the second time it executes the rule and changes its own variables. Since the daemon can choose an arbitrary gap between the read- and write-steps, the update might be made on stale information. The read/write daemon is defined to be a distributed daemon with read/write atomicity. See, e.g. [9] or [8].

In another direction, there is the concept of a fair and unfair daemon. For a (weakly) *fair daemon*, every node that is continuously privileged is tapped eventually. For an *unfair* daemon, there is no such restriction. See [6] for details. An algorithm running under fair daemon is thus expected to converge faster than when running under an unfair daemon: an unfair daemon represents a worst-case analysis of the algorithm. While it is to some extent a theoretical device, an unfair daemon does cover a situation where some node is "frozen" because of conditions external to the program.

In yet another direction of classification of the daemons, one may assume that the nodes in the network have unique *IDs*. If this is not assumed we say that the algorithm is *anonymous*. Note that in case of anonymous algorithms, no node can explicitly use any node ID to make any decision or to take any action; we use node identifiers for reference purpose only. A self-stabilizing algorithm is said to be *silent* if it terminates; that is, it is guaranteed to reach a state where no node is privileged. See [10]. Some problems such as token circulation and clock synchronization are inherently not silent. Other problems such as leader election and maximal independent set may be silent, though some of the solutions to leader election are not (e.g., [11]).

We measure the running times of the algorithms in terms of the maximum number of *steps* needed for the algorithm to converge to a stable legitimate state in the worst case. Another notion used is that of *round*: this is a minimal interval during which every continuously privileged node is tapped.

2   Main Results

In this section we lay out the results of the paper. The proofs are provided in Section 3. We consider first the case that IDs are not available. Randomness can clearly be used to create IDs (see for example [12]) – and most daemon-conversion results assume IDs or create them. However, IDs require $\Omega(\log n)$ space, where n denotes the number of nodes. We have previously shown [13] that that one can convert a central-daemon algorithm to a distributed-daemon algorithm using only $O(1)$ extra storage per node:

Theorem 1. *[13] Any algorithm that self-stabilizes under an unfair central daemon can be converted to a randomized one that self-stabilizes under an unfair distributed daemon, using constant extra space, without IDs, and with at most $O(n^3)$ expected slowdown, while preserving silentness.*

We provide a simple algorithm for the subproblem of *ensuring fairness*. As a consequence we have a simpler proof of the following result which can be extracted from the work of Beauquier et al. [14]:

Theorem 2. *[14] Any algorithm that self-stabilizes only under a fair central daemon can be converted to one that self-stabilizes under an unfair central daemon, without using IDs.*

This algorithm uses bounded memory, provided an upper bound on the size of the network is known. A significant open question is whether one can achieve this transform in bounded memory in an anonymous network *without any assumptions*. We give our alternative proof of Theorem 2 in Section 3.2.

It is essentially known, though to our knowledge not explicitly proven in the literature, that any algorithm that self-stabilizes under the central daemon using coarse atomicity can be converted to one that stabilizes under a read/write daemon provided there is symmetry breaking such as IDs or randomness. (See the next subsection for a discussion of the literature.) The problem that remains is to determine the price one pays in terms of slowdown in convergence time and/or additional storage need at the nodes. Our first result in this regard is that the price for conversion from central to distributed daemon is small.

Theorem 3. *Provided there are IDs, any algorithm that self-stabilizes under an unfair central daemon can be converted to one that self-stabilizes under an unfair distributed daemon, preserving silentness. For the general distributed daemon, the slowdown is linear (in the number of nodes). For the synchronous daemon, the slowdown is constant.*

Thus, for example, one immediately obtains linear-move synchronous algorithms for several problems like maximal matching and maximal independent set; linear-move algorithms under the central daemon are given in [15]. We prove Theorem 3 in Section 3.3. Our next result provides a conversion from central to read/write daemon, but the price seems higher.

Theorem 4. *Provided there are IDs, any algorithm that self-stabilizes under an unfair central daemon can be converted to one that self-stabilizes under an unfair read/write daemon, while preserving silentness.*

The **cross-over composition** idea of Beauquier et al. [16] is more powerful and general, but the resulting scheduling/synchronizing algorithm is not silent. Theorem 4 is proved in Section 3.4. Putting Theorems 2 and 4 together, we immediately obtain the most general result:

Theorem 5. *Provided there are IDs, any algorithm that self-stabilizes only under a fair central daemon can be converted to one that self-stabilizes under an unfair read/write daemon.*

2.1   Related Work

The concept of daemon can be thought of as two parts: The central daemon promises exclusivity, while the fair daemon promises each processor gets its turn. So to solve the problem of daemon conversion, one must show how to ensure (1) that no two adjacent nodes are ever simultaneously enabled for the original algorithm, and (2) that any node gets its turn after a while. The first of these is called local mutual exclusion, and the second is called fairness.

One solution to mutual exclusion is given by Lamport [17] (the bakery paper), and the ULME algorithm of [14] is largely the natural extension to general networks. Most of the papers on self-stabilizing mutual exclusion (see for example [9] or [6]) are essentially token-passing algorithms.

Fairness can also be thought of as a clock synchronization problem, and as such was investigated by Awerbuch et al. [18] and others. Their results hold in an anonymous network. If the processors have no global knowledge of the network, then these results use unbounded memory. If the processors have global knowledge (e.g. a bound on the order n or diameter d), then Awerbuch et al. provide a time-optimal (meaning $O(d)$ rounds given a fair daemon) if one also has IDs. Karaata [19] discusses ways of fairness enhancement.

There are several important related daemon-conversion results which address both fairness and mutual exclusion. Indeed, several of these results are deeper and more powerful than ours, but do not quite answer the questions we consider here. These include:

- Gouda and Haddix [20] provided an alternator which can be used to convert a central daemon algorithm to run under an unfair distributed daemon using IDs.
- Mizuno and Kakugawa [21] provided tools to convert a central daemon algorithm to run in a real distributed environment. They later reported on case studies [22]
- Beauquier et al. [14] provided a solution to local mutual exclusion which uses IDs and which also handles fairness, provided the initial algorithm can already handle a read/write daemon.
- Nesterenko and Arora [23] provided several results about conversion from one daemon to another, provided the daemon is fair. These results built on their solution to the dining philosophers problem.
- In the book [6], Dolev provides a conversion from central to read/write daemon, provided the daemon is fair and there is a distinguished processor.

Shukla et al. [24] provided a method using randomness that can be used to convert some central daemon algorithms to run under a distributed daemon.

3   The Transforms

We now prove Theorems 1–4. In each case we provide a transform from an algorithm S to an algorithm S'. The algorithms are presented as the code for a node i. We define the boolean predicate

$$p_S(i) \stackrel{\text{def}}{=} \begin{cases} 1 & \text{node } i \text{ is privileged for algorithm } S \text{ in a given system state} \\ 0 & \text{otherwise.} \end{cases}$$

The notation $N(i)$ denotes the set of neighbors of node i.

3.1   Central to Distributed without IDs

It has been shown [13] that randomization can be used to ensure local mutual exclusion between neighboring nodes, and so to convert a central daemon self-stabilizing algorithm to its distributed daemon equivalent. We briefly describe the algorithm and the result for the sake of completeness.

Let S be a self-stabilizing algorithm that works for an unfair, central, coarse daemon, but does not use IDs. In order to design the new algorithm S' to work for a distributed daemon, we add to each node i a boolean flag $b(i)$ in addition to the S-variables. This flag is designed to be true if the node is privileged for the underlying algorithm S and if the node is the only node in its neighborhood that has its flag set. When two adjacent nodes are simultaneously S-privileged and have their flag bits set, the nodes randomly determine a new value of their flag bits. A node can only execute the underlying algorithm if it is indeed privileged for S and is the only node in its neighborhood that has its flag set. The new algorithm S' is shown as Algorithm 1.

Algorithm 1. *Using randomness for exclusivity*

Variables: binary $b(i)$ (and variables needed for S)
BitClear: if $b(i) = 1$ and not $p_S(i)$
    then set $b(i) = 0$
BitSet: if $p_S(i)$ and $b(i) = 0$ and $\forall j \in N(i) : b(j) = 0$
    then set $b(i) = 1$
BitToss: if $p_S(i)$ and $b(i) = 1$ and $\exists j \in N(i)$ with $b(j) = 1$
    then set $b(i) = \text{RANDOM}$ (toss a fair coin to determine the new value of $b(i)$)
Step: if $p_S(i)$ and $b(i) = 1$ and $\forall j \in N(i) : b(j) = 0$
    then execute one step of S at i

Under a distributed daemon, Algorithm 1 (i) achieves local exclusivity for S, i.e., no two adjacent nodes execute the underlying algorithm S concurrently; (ii) cannot terminate while there is an S-privileged node.

3.2   Fair to Unfair Central Daemon

Beauquier et al. [14] showed how to enforce fairness using bounded variables, provided the nodes know a bound on the number of nodes in the graph. We give an alternative algorithm. Our result achieves fairness without having to wait for a period of stabilization: it is immediately self-stable. As in [14], the resulting algorithm is not silent. The algorithm is also similar to the alternator of Gouda and Haddix [20]. However, Algorithm 2 does not use IDs and achieves only fairness.

Given a self-stabilizing algorithm S that works for a fair central daemon, we define a new algorithm S' as follows. Each node maintains an additional *counter* variable $c(i)$ in the range $0 \ldots n$ with wrap around (where n is the number of nodes in the network). All arithmetic is modulo $n + 1$. We define the Boolean predicate $p_S(i)$ as before. A node is privileged for algorithm S' provided there does not exist a neighbor with a counter with the next higher value. If it is tapped, then a node executes S if S-privileged, and in any event, it increments its counter. The new algorithm S' is shown as Algorithm 2.

Algorithm 2. *Bounded counters for fairness*

Variables: counter $c(i)$ in the range $0 \ldots n$ inclusive (and variables needed for $\mathcal{S}$)
Update: if $\not\exists j \in N(i)$ with $c(j) = c(i) + 1$
    then (a) if $p_{\mathcal{S}}(i)$ then execute one step of $\mathcal{S}$ at i; and

            (b) in any event, $c(i) ++$ (modulo $n + 1$)

Lemma 1. *Algorithm 2 is always live and ensures fairness for $\mathcal{S}$.*

Proof. Suppose the algorithm terminates. Consider any node i; say $c(i) = 0$. It must have a neighbor, say j, with counter $c(j) = 1$. That node in turn must have a neighbor, say k, with counter $c(k) = 2$. And so on. But this implies $n + 1$ different values of the counters, and hence $n + 1$ different nodes, which is impossible. Thus, at every step there exists at least one node that is privileged for Algorithm 2.

As for fairness, consider a period of time when a node's counter goes from 0 to 0. There cannot be a neighbor that has not moved in that period, since each move precludes a neighbor with the next higher value. Indeed, if $M_i(t)$ denotes the number of moves of node i up until time t, then it always holds that $|M_i(t) - M_j(t)| \leq n$ for any pair i and j of adjacent nodes. Overall, there can be at most $O(n^2)$ other moves between two consecutive moves of the same node. Thus, as long as a node is $\mathcal{S}$-privileged, it will get to move in at most $O(n^2)$ steps.

This establishes Theorem 2:

Theorem 2. *[14] Any algorithm that self-stabilizes only under a fair central daemon can be converted to one that self-stabilizes under an unfair central daemon, without using IDs, with at most $O(n^2)$ slowdown.*

Actually, nowhere in the proof did we use the fact that the daemon was central. So one can use this to ensure fairness for other daemons such as the distributed daemon. By combining the conversions of this subsection and the previous subsection, one can obtain a new local mutual exclusion algorithm; we omit the details.

3.3   Central to Distributed with IDs

We next consider the conversion to distributed daemon without introducing randomness. Note that we do not require fairness. The central daemon as a scheduler enforces that neighboring nodes do not move concurrently, that is, local mutual exclusion. So it suffices for us to achieve this under the distributed daemon. We use the standard idea of interpreting the IDs as priorities such as in work on philosopher problems or the MIS algorithm of Kakugawa et al. [21].

Let $\mathcal{S}$ be a self-stabilizing algorithm that works for an unfair, central, coarse daemon. We define a new algorithm $\mathcal{S}'$ as follows. We add to each node i a

---

Algorithm 3. *Using IDs for exclusivity*

Variables: binary $b(i)$ (and variables needed for $\mathcal{S}$)
BitUpdate: if $b(i) \neq p_{\mathcal{S}}(i)$
    then $b(i) = p_{\mathcal{S}}(i)$
Step: if $b(i) = 1$ **and** $p_{\mathcal{S}}(i)$ **and** $\not\exists \, j \in N(i)$ with $b(j) = 1$ and $j < i$
    then execute one step of $\mathcal{S}$ at i, and set $b(i) = 0$

---

boolean flag $b(i)$. Algorithm 3 assumes that nodes are assigned unique IDs and has two rules as shown below.

The effect of Rule BitUpdate is that $b(i)$ contains the correct value of $p_{\mathcal{S}}(i)$. The effect of Rule Step is that a node i can enter execution of $\mathcal{S}$ only if its bit $b(i)$ is set and its ID is a local minimum amongst those neighbors which have their respective bits set.

Lemma 2. *Algorithm 3 under a distributed daemon achieves exclusivity for $\mathcal{S}$, i.e., no two adjacent nodes execute the underlying algorithm $\mathcal{S}$ concurrently.*

Proof. Consider two adjacent nodes i and j with $i < j$. For a node i to be able to execute Rule Step when it is tapped by the distributed daemon, at the point it reads the variables, it must have its b-bit set, and none of its smaller neighbors can have their b-bit set. Thus if node i is privileged for Algorithm 3 for Rule Step in a given system state, since $j > i$, the neighbor j, cannot be privileged for Algorithm 3 for Rule Step. Thus two adjacent nodes cannot execute Rule Step simultaneously.

Lemma 3. *If $\mathcal{S}'$ (Algorithm 3) terminates, then no node is privileged for $\mathcal{S}$.*

Proof. Assume that Algorithm 3 terminates. Then no node is privileged for Rule BitUpdate. So the bit $b(i) = p_{\mathcal{S}}(i)$ for each node i. The fact that no node is privileged for Rule Step then means that no node is $\mathcal{S}$-privileged (consider the smallest $\mathcal{S}$-privileged node; it must be privileged by Rule Step). Thus $\mathcal{S}$ has finished executing.

Lemma 4. *Algorithm 3 achieves progress in the original algorithm $\mathcal{S}$. In particular, the slowdown under the general distributed daemon is $O(n)$, and under the synchronous daemon is constant.*

Proof. Consider first the distributed daemon. The first rule can execute at most once per node between executions of Rule Step. Hence the slowdown is at most linear.

For the synchronous daemon, there cannot be two successive steps without Rule Step being executed. Hence the slowdown is at most constant. Thus,

Theorem 3. *Provided there are IDs, any algorithm that self-stabilizes under an unfair central daemon can be converted to one that self-stabilizes under an unfair distributed daemon, preserving silentness. For the general distributed daemon, the slowdown is at most $O(n)$. For the synchronous daemon, the slowdown is constant.*

3.4   Central to Read/Write with IDs

Algorithm 3 does not quite work for read/write atomicity, but we can modify the idea to handle read/write atomicity. In the worst case, this transform might lead to quadratic slowdown.

As before, let $\mathcal{S}$ be a self-stabilizing algorithm that works for an unfair, central, coarse daemon. In order to design the new algorithm $\mathcal{S}'$ to work for a read/write daemon, we add to each node i a boolean flag $b(i)$. This flag is designed to be true if the node is privileged for the underlying algorithm $\mathcal{S}$ and has higher priority (lower ID) than its waiting $\mathcal{S}$-privileged neighbors. The node can only execute the underlying algorithm if it is indeed privileged for $\mathcal{S}$ and is the only node in its neighborhood that has its flag set. The new algorithm $\mathcal{S}'$ is shown as Algorithm 4.

Algorithm 4. *Using IDs for exclusivity under read/write*

Variables: binary $b(i)$ (and variables for $\mathcal{S}$)
Withdraw: if $b(i) = 1$ and not $p_S(i)$
   then set $b(i) = 0$
Concede: if $b(i) = 1$ and $\exists j \in N(i)$ with $b(j) = 1$ and $j < i$
   then set $b(i) = 0$
Assert: if $p_S(i)$ and $b(i) = 0$ and $\forall j \in N(i) : b(j) = 0$
   then set $b(i) = 1$
Step: if $p_S(i)$ and $b(i) = 1$ and $\forall j \in N(i) : b(j) = 0$
   then execute one step of $\mathcal{S}$ at i

Remark 1. The daemon can tap a node to execute Rule **Step** immediately after Rule **Assert**, but we need the operation in two steps for coping with the read/write atomicity. Essentially, at the read-tap of the Rule **Step**, the node double-checks that every neighbor has bit clear. Even if one is in the process of setting its flag (write-tap for Rule **Assert**), that neighbor has to double-check (read-tap of Rule **Step**) before it executes write-phase of Rule **Step** and will be blocked.

Lemma 5. *If Algorithm 4 terminates, then no node is privileged for $\mathcal{S}$.*

Proof. Suppose at termination of Algorithm 4 there is a node i with $b(i) = 1$. Since node i is not privileged for Rule **Withdraw**, it is $\mathcal{S}$-privileged. Further, since no node is privileged for Rule **Concede**, all neighbors of i have b-bits zero (if there is a node j, $j > i$ and $b(j) = 1$, node j must be privileged by Rule **Concede**). Then node i is $\mathcal{S}$-privileged and all its neighbors have their b-bits zero; hence node i is privileged for Rule **Step**, a contradiction. Hence, at termination of Algorithm 4, each node i has $b(i) = 0$.

Since no node is privileged for Rule **Assert**, it follows that there is no $\mathcal{S}$-privileged node. That is, if $\mathcal{S}'$ terminates, then $\mathcal{S}$ has also terminated.

Lemma 6. *Algorithm 4 achieves local mutual exclusion among neighbor nodes and faithfully executes the original algorithm S.*

Proof. Consider a node i that executes Rule Step. We claim that between the read- and the write-steps for this move, no neighbor can start or finish Rule Step.

When node i performs the read-phase for Rule Step, it must have that $b(i) = 1$ and for all $j \in N(i)$, $b(j) = 0$; i.e., no neighbor j is currently executing Rule Step. Also, until node i completes the write-phase of Rule Step, the flag $b(i)$ remains set at 1 and hence no neighbor j of node i can be privileged for Rule Step. This also means that the step of S made by node i is based on correct (current) data.

Lemma 7. *Algorithm 4 achieves progress in the original algorithm S; when S stabilizes, Algorithm 4 also stabilizes.*

Proof. Consider an interval T (series of moves made by Algorithm 4) during which the write-phase of Rule Step is not executed. Then the S-variables at each node i remain constant during this interval T, and so whether a node is S-privileged remains constant. (Predicate $p_S(i)$ for each node i remains constant but $b(i)$ may change since some node(s) may still have stale read-data from before T started.)

We observe that during this interval T, each node i can execute either Rule Withdraw or Rule Assert at most once (since $p_S(i)$ does not change); thus, there can be at most n moves during this interval T without any node executing Rule Concede. When a node i executes Rule Concede, the node i cannot execute Rule Concede again until the b-bit of node i or any of its neighbors is changed by execution of Rule Withdraw or Rule Assert. Also, when a node i executes Rule Concede, and no node changes its b-bit, none of the neighbors of node i can execute Rule Concede. Thus, when neither Rule Withdraw nor Rule Assert is executed by any node, the maximum number of times Rule Concede can be executed during the interval T is at most the size of the maximum independent set of the network graph; that is at most n, the number of nodes.

Hence, the maximum number of moves possible during the interval T is $O(n^2)$. Hence, if Rule Step is not executed—meaning the underlying algorithm S has stabilized—then the algorithm S' will eventually terminate.

The above three lemmas establish Theorem 4:

Theorem 4. *Provided there are IDs, any algorithm that self-stabilizes under an unfair central daemon can be converted to one that self-stabilizes under an unfair read/write daemon, while preserving silentness.*

4 Conclusion

The results in this paper reaffirm that in the deterministic ID-based shared-variable model, all daemons are equally powerful. An alternative to the shared-variable model is ***link registers*** where each pair of adjacent nodes have a register to pass messages. If the nodes have IDs, then link-registers can be trivially simulated in general memory—a node writes the contents of each link-register next to each neighbor's ID. Thus link-registers and shared-variable are equivalent in ID-based networks. It follows that all daemons are equally powerful in this model too.

An interesting question is of comparing the powers of different daemons in *deterministic anonymous networks*. Link-registers and a distinguished node or root allow one to form a breadth-first-search spanning tree and hence assign IDs; see e.g. [8]. Thus all daemons are equally powerful in this case. However, without a root, the results of [9, 25] and others on rings show that the distributed and central daemon have different powers. That is, there are some problems which have a solution under a central daemon but do not have a solution under a distributed daemon. This difference holds in both the link-register and shared-variable case. One way to proceed might be to determine exactly which problems have solutions. Angluin's well-known arguments [26] about symmetry-breaking provide several limits on what problems can be solved, but are these essentially the only limits?

Acknowledgement. The work was partially supported by NSF Awards CCF 0832582 and DBI-0960586.

References

1. Dijkstra, E.W.: Self stabilizing systems in spite of distributed control. Comm. ACM 17, 643–644 (1974)
2. Herman, T.: Models of Self-Stabilization and Sensor Networks. In: Das, S.R., Das, S.K. (eds.) IWDC 2003. LNCS, vol. 2918, pp. 205–214. Springer, Heidelberg (2003)
3. Dubois, S., Tixeuil, S.: A taxonomy of daemons in self-stabilization. CoRR, abs/1110.0334 (2011)
4. Korman, A., Kutten, S., Masuzawa, T.: Fast and compact self stabilizing verification, computation, and fault detection of an mst. In: Proceedings of the 30th Annual ACM SIGACT-SIGOPS Symposium on Principles of Distributed Computing, PODC 2011, pp. 311–320. ACM, New York (2011)
5. Beauquier, J., Delaët, S., Haddad, S.: A 1-Strong Self-stabilizing Transformer. In: Datta, A.K., Gradinariu, M. (eds.) SSS 2006. LNCS, vol. 4280, pp. 95–109. Springer, Heidelberg (2006)
6. Dolev, S.: Self-Stabilization. MIT Press (2000)
7. Tel, G.: Introduction to Distributed Algorithms. Cambridge University Press (1994)
8. Dolev, S., Israeli, A., Moran, S.: Self-stabilization of dynamic systems assuming only read/write atomicity. Distrib. Comput. 7, 3–16 (1993)
9. Israeli, A., Jalfon, M.: Uniform self-stabilizing ring orientation. Inform. Comput. 104, 175–196 (1993)
10. Dolev, S., Gouda, M.G., Schneider, M.: Memory requirements for silent stabilization. In: PODC 1996 Proceedings of the Fifteenth Annual ACM Symposium on Principles of Distributed Computing, pp. 27–34 (1996)
11. Itkis, G., Lin, C., Simon, J.: Deterministic, Constant Space, Self-Stabilizing Leader Election on Uniform Rings. In: Helary, J.-M., Raynal, M. (eds.) WDAG 1995. LNCS, vol. 972, pp. 288–302. Springer, Heidelberg (1995)

12. Gradinariu, M., Johnen, C.: Self-stabilizing Neighborhood Unique Naming under Unfair Scheduler. In: Sakellariou, R., Keane, J.A., Gurd, J.R., Freeman, L. (eds.) Euro-Par 2001. LNCS, vol. 2150, pp. 458–465. Springer, Heidelberg (2001)

13. Goddard, W., Hedetniemi, S.T., Jacobs, D.P., Srimani, P.K.: Anonymous Daemon Conversion in Self-stabilizing Algorithms by Randomization in Constant Space. In: Rao, S., Chatterjee, M., Jayanti, P., Murthy, C.S.R., Saha, S.K. (eds.) ICDCN 2008. LNCS, vol. 4904, pp. 182–190. Springer, Heidelberg (2008)

14. Beauquier, J., Datta, A.K., Gradinariu, M., Magniette, F.: Self-Stabilizing Local Mutual Exclusion and Daemon Refinement. In: Herlihy, M.P. (ed.) DISC 2000. LNCS, vol. 1914, pp. 223–237. Springer, Heidelberg (2000)

15. Hsu, S.C., Huang, S.T.: A self-stabilizing algorithm for maximal matching. Inform. Process. Lett. 43, 77–81 (1992)

16. Beauquier, J., Gradinariu, M., Johnen, C.: Cross-Over Composition - Enforcement of Fairness under Unfair Adversary. In: Datta, A.K., Herman, T. (eds.) WSS 2001. LNCS, vol. 2194, pp. 19–34. Springer, Heidelberg (2001)

17. Lamport, L.: A new solution of Dijkstra's concurrent programming problem. Communications of the ACM 17, 453–455 (1974)

18. Awerbuch, B., Kutten, S., Mansour, Y., Patt-Shamir, B., Varghese, G.: Time optimal self-stabilizing synchronization. In: STOC 1993 Proceedings of the 25th Annual ACM Symposium on Theory of Computing, pp. 652–661 (1993)

19. Karaata, M.H.: Self-stabilizing strong fairness under weak fairness. IEEE Trans. Parallel Distrib. Systems 12, 337–345 (2001)

20. Gouda, M.G., Haddix, F.: The alternator. In: Proceedings of the Fourth Workshop on Self-Stabilizing Systems (published in association with ICDCS 1999), pp. 48–53. IEEE Computer Society (1999)

21. Mizuno, M., Kakugawa, H.: A Timestamp Based Transformation of Self-Stabilizing Programs for Distributed Computing Environments. In: Babaoğlu, Ö., Marzullo, K. (eds.) WDAG 1996. LNCS, vol. 1151, pp. 304–321. Springer, Heidelberg (1996)

22. Kakugawa, H., Mizuno, M., Nesterenko, M.: Development of self-stabilizing distributed algorithms using transformation: case studies. In: Proceedings of the Third Workshop on Self-Stabilizing Systems, pp. 16–30. Carleton University Press (1997)

23. Nesterenko, M., Arora, A.: Stabilization-preserving atomicity refinement. J. Parallel Distrib. Comput. 62(5), 766–791 (2002)

24. Shukla, S., Rosenkrantz, D., Ravi, S.: Developing self-stabilizing coloring algorithms via systematic randomization. In: Proceedings of the International Workshop on Parallel Processing, pp. 668–673 (1994)

25. Hoepman, J.H.: Uniform Deterministic Self-Stabilizing Ring-Orientation on Odd-Length Rings. In: Tel, G., Vitányi, P.M.B. (eds.) WDAG 1994. LNCS, vol. 857, pp. 265–279. Springer, Heidelberg (1994)

26. Anglۧuin, D.: Global and local properties in networks of processors. In: Proc. 12th Symposium on the Theory of Computing, pp. 82–93 (1980)

Broadcasting in Conflict-Aware Multi-channel Networks

Francisco Claude[1], Reza Dorrigiv[2], Shahin Kamali[1], Alejandro López-Ortiz[1], Paweł Prałat[3], Jazmín Romero[1], Alejandro Salinger[1], and Diego Seco[4]

[1] David R. Cheriton School of Computer Science, University of Waterloo, Canada
[2] Faculty of Computer Science, Dalhousie University, Canada
[3] Department of Mathematics, Ryerson University, Toronto, Canada
[4] Database Laboratory, University of A Coruña, Spain

Abstract. The broadcasting problem asks for the fastest way of transmitting a message to all nodes of a communication network. We consider the problem in conflict-aware multi-channel networks. These networks can be modeled as undirected graphs in which each edge is labeled with a set of available channels to transmit data between its endpoints. Each node can send and receive data through any channel on its incident edges, with the restriction that it cannot successfully receive through a channel when multiple neighbors send data via that channel simultaneously.

We present efficient algorithms as well as hardness results for the broadcasting problem on various network topologies. We propose polynomial time algorithms for optimal broadcasting in grids, and also for trees when there is only one channel on each edge. Nevertheless, we show that the problem is NP-hard for trees in general, as well as for complete graphs. In addition, we consider balanced complete graphs and propose a policy for assigning channels to these graphs. This policy, together with its embedded broadcasting schemes, result in fault-tolerant networks which have optimal broadcasting time.

1 Introduction

Multi-channel networks constitute a class of networks in which communication is achieved via a set of orthogonal *channels*. Two nodes of a multi-channel network can directly communicate if they share at least one common channel. Channels may represent different frequencies in Multi-radio Wireless Networks [9,12], different wavelengths in Free Space Optical Networks (FSON) [1], or different communication buffers in parallel computers [13].

A multi-channel network can be modeled as an undirected graph with multiple labels on edges, where vertices represent nodes in the network and labels represent available channels between connected nodes. Communication is assumed to occur in discrete *rounds* in which a node can transmit data through one of its channels. For a node u and channel c, we say that a *conflict* occurs when two or more neighbors of u send data to u through channel c in the same round, in which case u does not receive data through this channel. This definition of

S.K. Ghosh and T. Tokuyama (Eds.): WALCOM 2013, LNCS 7748, pp. 158–169, 2013.

conflict arises in many practical scenarios; for example, in wireless networks, conflicts represent the interference of radio waves with the same frequency.

Multi-channel networks have been already studied in the context of wireless networks, in which the underlying network is modeled as a geometric graph in Euclidean metric space (e.g., [7,15]). However, geometric graphs are not good representatives of all types of wireless networks. For example, in the case of indoor networks in which walls can block transmissions between pairs of nodes, the underlying network can form any graph topology [14]. There are also several works that provide heuristics for information dissemination in the wireless multi-channel networks, mostly assuming that conflicts do not occur (e.g., [5,6,8]).

In this paper, we present the *conflict-aware multi-channel model*, a comprehensive model that captures several aspects of multi-channel networks that are tied to existing network technologies, in particular conflict awareness and the advantage of simultaneous communication through one channel. Theoretical analysis of this model can provide insights into the capabilities of multi-channel networks for future technology advances, particularly because the model represents a broad spectrum of network technologies such as wireless mesh networks, FSONs, and parallel computers.

The focus of this work is on the *Broadcasting Problem* in multi-channel networks, in which the goal is to transmit one message from a given source node to all other nodes in the minimum number of rounds. In the classical model of broadcasting, each node can send data to at most one of its neighbors via a *telephone call* (hence the model is called telephone model). In contrast, in multi-channel networks, when a node u transmits through one channel c, all the nodes connected to u via channel c will receive the message (if no conflicts occur). Note that the telephone model can be considered as a restricted version of the multi-channel model in which there is a single and unique channel associated with each edge.

Channel Assignment is another problem that has been studied for multi-channel wireless networks [9,10]. We consider the channel assignment problem in complete graphs, in which the goal is to assign channels in a way to perform broadcasting in minimum time. In particular, it is desirable that such channel assignment enables broadcasting of multiple messages in parallel.

Summary of Results. In Section 2, we describe the conflict-aware multi-channel model. In general, it is assumed that there can be any number of channels between a pair of nodes, however in some occasions we consider the case when there is only one channel on each edge of the graph. In Section 3, we show that the broadcasting problem is NP-hard for trees in the general case, while we describe a polynomial time algorithm when there is only one channel on each edge of the tree. We also provide a polynomial time algorithm for optimal broadcasting in grids (in the general case). In Section 4, we show that the broadcasting problem is NP-hard for complete graphs, even if restricted to graphs with only one channel on each edge.

In Section 5, we focus on the special case of complete graphs when there is only one channel on each edge and the channel assignment is balanced, i.e., each node is connected to approximately same number of nodes with each channel. We refer to these graphs as *balanced complete graphs*, and show that broadcasting in these networks requires at least three rounds, when the number of different channels does not grow too fast with the size of the network (which is the case in practical settings). On the positive side, we introduce a channel assignment policy that yields a balanced complete network for which broadcasting can always be completed in two rounds. This channel assignment also enables broadcasting of k messages simultaneously in three rounds, where k is the number of channels in the network.

2   Conflict-Aware Multi-channel Model

A multi-channel network is modeled as an undirected graph $G = (V, E)$ where V is the set of nodes and E the set of edges. Each edge $e \in E$ has a set of labels $C(e) \subseteq \{c_1, c_2, \ldots, c_k\}$ that denotes its set of available channels.

The communication of messages through the network occurs in discrete rounds and is governed by the following assumptions and restrictions. In any given round, a node may be involved in receiving and/or transmitting (sending) messages through the channels on its incident edges. If a node u transmits through a channel c, it cannot transmit through any other channel in the same round, and also cannot receive through channel c. When u sends a message through channel c, the message is simultaneously transmitted through all incident edges of u that have channel c in their set of labels. A key restriction is that a node cannot successfully receive any data through a channel when more than one of its neighbors send data through that channel. More precisely, a node v can only receive a message through channel c in round r if exactly one of the nodes that are adjacent to it with edges labeled with channel c is transmitting through channel c in round r. Otherwise we say there is a conflict at node v on channel c. A node will successfully receive the message if it is transmitted by any of its neighbors through a channel without conflict.

The transmission of a message on any edge completes in one round: if in round r node u transmits a message through channel c, then every node v such that $e = (u, v) \in E$ and $c \in C(e)$ will receive the message during this round, provided that there is no conflict at v on channel c. In this case we say that u *informs* v during round r, and node v is ready to transmit in round $r + 1$ if desired. For any round r during the execution of the broadcast, we say that a node is *active* if it is transmitting the message in round r and *inactive* otherwise.

Given a network represented by a graph G, the broadcasting problem is defined as follows. At the beginning, a single node, called the *source*, has a message. In each round, those vertices that have the message can transmit through one channel to inform some uninformed vertices. The broadcasting completes when all vertices successfully receive the message. The broadcasting problem asks for a scheme that completes this procedure in minimum time. We are interested in centralized broadcasting schemes, i.e., we assume the broadcasting algorithm can be determined in advance and with full knowledge of the network topology.

3   Basic Topologies

3.1   Trees

In this section, we show that the broadcasting problem in the general case is NP-hard even if the network topology is a tree. On the positive side, we show that when there is a single channel on each edge of the tree, there is an algorithm that finds the optimal broadcasting scheme in polynomial time.

Theorem 1. *The broadcasting problem in the conflict-aware multi-channel model is NP-hard for trees.*

Proof. We use a reduction from the set cover problem, which is NP-hard [4]. Recall that an instance of set cover includes a collection of subsets of a universe U, and the goal is to find the minimum number of subsets that cover the universe. Given an instance I of set cover, we create an instance of the broadcasting problem in a tree as follows. We create a tree T with a root node and u children, where $u = |U|$ is the size of the universe. Each child of the root is a leaf of the tree and a member of the universe (hence T is a star). Each subset S in I is assigned a label that represents a channel in the broadcast instance. For each member of S, the label of S is added to the edge that connects the root with that member. For example, if $S = \{x, y\}$, the label of S is added to the edges that connect the root to the leaves x and y. It is not hard to see that there is a set cover of size k if and only if the broadcast finishes in k rounds: assume there is a set cover of size k, then if the root sends the message through the k channels associated with the k subsets (in any order), after k rounds all the nodes of T are informed. This is because there are no conflicts (one channel is used at each round), and all the nodes are covered by k channels. Similarly, if there is a broadcasting scheme that completes in k rounds, the subsets associated with the k channels used by the root cover the universe.   □

The problem becomes easy when there is a single channel on each edge. Consider a tree of n nodes with only one channel on each edge. The optimal broadcasting scheme can be obtained in $O(n \log n)$ time with a simple recursive algorithm. Given a root node v, we compute the cost (number of rounds) of broadcasting from each of v's children recursively, and associate with each outgoing channel of v the cost of the most expensive child connected to v with that channel. We then sort these channels in decreasing order of associated cost and transmit through each one following this order. It is not hard to see that this strategy is optimal. Note as well that there are no conflicts in this topology. A simple implementation of the algorithm runs in $O(n \log n)$ time.

3.2   Grids

Unlike trees, the broadcasting problem can be solved in polynomial time for grids, even if there are multiple channels on edges. In what follows, we describe a scheme for optimal broadcasting in a grid of size $n \times m$.

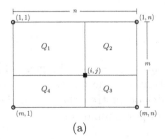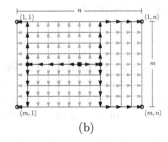

(a)   (b)

Fig. 1. (a) Quadrants defined by the source (i,j) in a grid of size $m \times n$. (b) Example of a broadcast from a source in the center of the grid. The source is a horizontal splitter and there are two splitters in row i, depicted by black discs. Arrows indicate the route of the message to any node; in particular black arrows show the first direction of transmission from each node on the critical path of the scheme. Note that splitters have two black arrows. The broadcasting completes in optimal $(n + m - 2)/2$ rounds.

Consider first the simple case when the source is one of the corner nodes. W.l.o.g., assume the source is on the upper-leftmost node. A simple scheme is to send the message to the nodes in the first row: after receiving the message, each node transmits to its right neighbor through any one of the available channels. This takes $n - 1$ rounds. Then, in parallel, the message is transmitted in each column downwards, again through any available channel. The broadcast finishes in $m + n - 2$ rounds, which matches a trivial lower bound determined by the diameter of the grid. Note that conflicts do not arise in this strategy.

Combinations of small variations of the strategy described above will serve for the general case in which the source is any node (i,j) in the grid. Consider the set of nodes $N = \{(k, \ell)|k = i, \ell \neq j \text{ or } \ell = j, k \neq i\}$, i.e., nodes in the same row or column as the source node, not including the source. Let Q_i denote the i-th quadrant defined by N in G (See Figure 1 (a)). We say that a node $u \in N$ is a *splitter* if it is connected to neighbors in two different quadrants with at least one channel in common. Similarly, we say that the source is a vertical (resp. horizontal) splitter if it is connected to neighbors above and below (resp. to the left and right) with at least one common channel.

Broadcasting schemes may differ depending on the availability of splitters and the relative sizes of the quadrants. If there are no splitters or the sizes of all the quadrants are different, then optimal strategies for broadcasting in grids in the telephone model [3] apply to our model as well. For other cases, we derive optimal strategies by taking advantage of the splitters (See Figure 1 (b) for an example), thus proving the following theorem. The proof requires a tedious case analysis, and a sketch of it appears in the full version of this paper [2].

Theorem 2. *Given an $m \times n$ grid G with k channels and a source node (i,j), where $1 \leq i \leq m$, $1 \leq j \leq n$, an optimal broadcasting scheme can be computed in $O((n + m)k)$ time.*

4   Complete Graphs

In this section we show that the broadcasting problem in multi-channel networks is NP-hard for complete bipartite graphs and complete graphs. Through this section, we assume there is a single channel on each edge of concerned graphs. Using a reduction from the exact cover problem, we show that the broadcasting problem is NP-hard for complete bipartite graphs; then we show a reduction from the broadcasting problem in complete bipartite graphs to the same problem in complete graphs. The proof of the following lemma appears in the full version of this paper.

Lemma 1. *The broadcasting problem is NP-hard for complete bipartite graphs in the conflict-aware multi-channel model (assuming there is a single channel on each edge), even in the special case when there are a total of 2 channels and the source is connected to all its neighbors with the same channel.*

To reduce from broadcasting in complete bipartite graphs to complete graphs, we use *ladder bipartite graphs* which we define as follows:

Definition 1. *A ladder bipartite graph with channels i, j is a balanced complete bipartite graph with n vertices on each side. There is a one-to-one mapping between the vertices of two sides such that the edge connecting a vertex u to its mapped vertex u' has channel j and all the other edges incident to u have channel i.*

The proof of the following lemma appears in the full version of the paper.

Lemma 2. *Assume all vertices on one side of a ladder bipartite graph with channels i, j have received the message. If these vertices need to inform the vertices on the other side in one round, all the vertices should be active in that round, i.e., they need to transmit the message either through channel i or j.*

Theorem 3. *The broadcasting problem in the conflict-aware multi-channel model is NP-hard for complete graphs, when there are at least 8 channels in the network (assuming there is a single channel on each edge).*

Proof. Given an instance (G, r_0) of the broadcasting problem in a complete bipartite graph in which there are two channels and the edges adjacent to the source r_0 are labeled with the same channel, we create an instance of the broadcasting problem in a complete graph in which there are 8 channels. Let L and R denote the two partitions of the vertices of G so that $r_0 \in R$. We create a complete graph H as follows (See Figure 2). We take two copies of L and three copies of $R - \{r_0\}$ (r_0 is the source in the original instance). Call these components L_1, L_2, and R_1, R_2, R_3, respectively, and also add a new vertex r as the new source. The channels of edges connecting vertices in L_1 and L_2 to any of R_1, R_2, R_3 are copied from the original bipartite graph G. Let vertex r be connected to the 5 components via 5 different channels so that the edges connecting r to the vertices in the same component have the same channel.

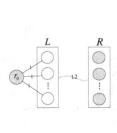

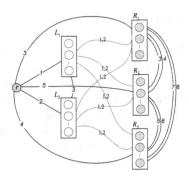

(a) An instance of the problem in a complete bipartite graph G.

(b) The resulting instance in the complete graph H.

Fig. 2. The broadcasting problem in complete bipartite graphs (with one channel on the edges incident to the source) reduces to the broadcasting problem in complete graphs. Here, a number i on the solid edge connecting two components indicate that all edges between the vertices of the two components are labeled with channel i. The channels of the edges between two components connected by curved blue edges are copied from the reduced bipartite graph. Solid and dashed paired lines indicate that the components form a ladder bipartite graph.

Moreover, we assign the channels to the edges connecting vertices in R_1 to vertices in R_2 in a way that these edges form a ladder bipartite graph with channels $3, 4$. Similarly, we set the edges between R_2 and R_3 to form a ladder graph with channels $5, 6$, and between R_1 and R_3 to form a ladder with channels $7, 8$. The edges connecting vertices in L_1 and L_2 get channel 3 and all other edges (the edges inside components) get arbitrary channels. We claim that there is a broadcasting scheme for the instance (G, r_0) that completes in 2 rounds if and only if there is a broadcasting scheme for (H, r) that also takes 2 rounds.

Assume there is a broadcasting scheme for (G, r_0) that completes in 2 rounds. In the first round r_0 informs the vertices of L via its single channel, so in the new instance r can inform the vertices of L_1 via the single channel that connects them (channel 1 in Figure 2). In the second round of the broadcast in (G, r_0), a subset of L informs all vertices of R. In the new instance the same subset can inform all vertices of R_1, R_2, R_3 (via the same edges used in the first instance), while r informs L_2 via the unique connecting channel (channel 2 in Figure 2). Hence, the broadcast completes in 2 rounds.

Now assume that there is a broadcasting scheme for (H, r) that completes in 2 rounds. First, we show that r cannot inform any of R_1, R_2, R_3 in the first round. For the sake of contradiction, suppose r informs R_1 in the first round (the same reasoning holds for R_2 and R_3); in the second round r cannot inform both L_1 and L_2. Thus, at least one vertex in R_1 should use channels 1 or 2 to inform some vertices of L_1 and L_2. Since the edges between R_1 and R_2 form a ladder bipartite graph and at least one vertex of R_1 is busy informing vertices of L_1 and L_2, by Lemma 2, R_1 cannot inform all vertices of R_2. Thus, some vertices of

R_2 are to be informed by the source. Similarly, some vertices of R_3 are also left for the source to inform them. However, the source is connected to R_2 and R_3 with two different channels, thus it cannot inform both in a single round. Hence, the broadcast cannot be completed in 2 rounds and we get a contradiction. As a result, we may assume that in the first round r informs either L_1 or L_2.

Assume r informs L_1 in the first round (the same reasoning holds for L_2). Since in the second round r can inform at most one of the R_i's, the other two should be informed via L_1, which implies a subset of vertices in L_1 can inform all vertices in two R_i's. The same subset can be used for the instance (G, r_0) to inform all the vertices on the right in the second round. Therefore, there is a broadcasting scheme for (G, r_0) that completes in 2 rounds.   □

5   Balanced Complete Graphs

As the broadcasting problem is NP-hard for complete graphs, we consider a particular case of complete graphs in which there is a single channel on each edge, and every node is connected to $\frac{n-1}{k}$ nodes through edges with the same channel. Thus, all the nodes use k different channels. We refer to this subset of complete graphs as *balanced complete graphs*. Since this would restrict us from considering networks where n is not congruent to one modulo k, we relax the condition slightly in order to include almost balanced assignments. For a given $\epsilon \geq 0$, we require that for every node v and every channel i, the number of nodes connected to v using channel i is at least $(1 - \epsilon)(n - 1)/k$ and at most $(1 + \epsilon)(n - 1)/k$. We call this family of graphs ϵ-*balanced complete graphs*.

In this setting, k corresponds to a trivial upper bound on the broadcast time. It suffices that the source transmits once through each channel and, since the graph is complete, the broadcasting is done. If we ignore all possible conflicts, it is easy to obtain a simple lower bound on the transmission time. Consider a graph where at any round a node can transmit to at most $(1+\epsilon)(n-1)/k$ nodes without conflicts. It is clear then that after the first round, we have at most $(1 + \epsilon)(n - 1)/k + 1$ nodes informed. The general formula for an upper bound for the number of nodes that have been informed after i rounds is $((1 + \epsilon)(n - 1)/k + 1)^i$, and thus we get a lower bound for the total number of rounds to inform all nodes.

Lemma 3. *Let $\epsilon \geq 0$. For ϵ-balanced complete graphs, at least $\lceil \log n / \log((1 + \epsilon)(n - 1)/k + 1) \rceil$ rounds are required to complete a broadcast.*

When $k = n - 1$ and $\epsilon < 1$ (i.e., each node is connected to exactly one node using each channel) a simple greedy algorithm finds the optimal broadcasting scheme and it takes $\lceil \log_2 n \rceil$ rounds. This is because there are no conflicts when receiving the message, since all channels are different. The solution matches the lower bound in Lemma 3. This example shows that there are some cases where the broadcast time is not as bad as the trivial upper bound of k. When aiming at practical applications, a more interesting scenario is one in which the number of channels is relatively small compared to the number of nodes. Note that for

$k \leq (1+\epsilon)(n-1)/(\sqrt{n}-1) = O(\sqrt{n})$ the lower bound in Lemma 3 asserts that the broadcast requires at least 2 rounds. Therefore, it would be desirable to have the property that there exists a constant $C > 0$ such that for every ϵ-balanced complete graph G with at most $C\sqrt{n}$ channels, a broadcast can always be completed in 2 rounds. Unfortunately, we can show that this is not true by constructing a counterexample using a random assignment of channels.

For given natural numbers n and k, let $G(n,k)$ be a complete graph with node set $[n] = \{1, 2, \ldots, n\}$ in which two nodes are connected via channel $c \in [k]$ with probability $1/k$, independently for each such pair. As is typical in random graph theory, we shall consider only asymptotic properties of $G(n,k)$ as $n \to \infty$, where k may and usually does depend on n. We say that an event in a probability space holds *asymptotically almost surely (a.a.s.)* if its probability tends to one as n goes to infinity. The following theorem implies that there are ϵ-balanced complete graphs for which the broadcast requires 3 rounds.

Theorem 4. *Let $\epsilon > 0$, $c_0 = 1 - 1/e$, $f = f(n)$ be any function tending to infinity together with n, $k' = k'(M) = \log_{1/c_0} n - 3 \log_{1/c_0} \log n - M$, $k'' = \log_{1/c_0} n + f$, and $k''' = \sqrt{n/(2 \log n)}$. Then, there exists a sufficiently large constant M such that the following holds a.a.s.:*
- $G(n,k)$ is an ϵ-balanced complete graph for any k such that $2 \leq k \leq k'''$,
- Broadcasting in $G(n,k)$ requires at least 3 rounds for any k such that $3 \leq k \leq k'(M)$,
- Broadcasting in $G(n,k)$ requires 2 rounds for any k such that $k'' \leq k \leq k'''$.

The proof appears in the full version of the paper. Since there are ϵ-balanced complete graphs with bounded number of channels ($k = O(1)$) in which broadcasting requires at least 3 rounds, it is interesting to design ϵ-balanced complete graphs (or even, balanced complete graphs) that can be broadcasted in 2 rounds. Since the topology is fixed (a complete graph), such design is equivalent to a promising channel assignment. Our channel assignment algorithm relies on the following known result for edge coloring.

Lemma 4. *[11, problem 16.5, p. 133] The minimum number of colors required for an edge coloring of a complete graph K_n is $n-1$ if n is even, and n otherwise.*

A constructive proof of this lemma leads to the following edge-coloring algorithm. When n is odd, assign the color $((i+j) \mod n) + 1$ to each edge $e = (v_i, v_j)$ with $v_i, v_j \in V = \{v_1, v_2, \ldots, v_n\}$ (an edge-coloring for K_3 is shown in Figure 3(b)). We say that a node v_i *uses a color* c if there is an edge (v_i, v_j), $i \neq j$ colored with c. For even values of n, the graph K_{n-1} is colored using the above method, any edge $e = (v_i, v_n)$ incident to the remaining vertex v_n is colored with the color not used by v_i (See Figure 3(a)). Note that *coloring* an edge e and *assigning* a channel to e are assumed to be equivalent terms.

Relying on this result, we obtain the following theorem (our construction algorithm follows immediately from its constructive proof).

Theorem 5. *Given an odd number of channels k and a positive integer t, it is possible to construct a balanced complete graph with $kt + 1$ nodes (i.e., K_{kt+1}).*

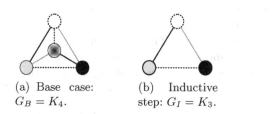
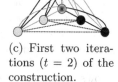

(a) Base case: $G_B = K_4$.   (b) Inductive step: $G_I = K_3$.   (c) First two iterations ($t = 2$) of the construction.

Fig. 3. Construction example using K_4 as base case and K_3 in the inductive step

Proof. We use induction on t to prove a stronger statement as follows. For given values of k and t, there is a complete graph with $kt + 1$ nodes that satisfies the following properties: (i) the vertices of the graph can be classified in k classes with t vertices in each class (and one root vertex in no class), (ii) vertices in the same class are all connected with one channel, and are connected to the root with the same channel, and (iii) for each pair of classes, all edges connecting vertices in the two classes are connected with the same channel. It is not hard to see that proving this statement proves the theorem.

Let $G_B = K_{k+1}$ be the base case. As we define k to be odd, G_B is a complete graph with an even number of nodes. Hence, we can assign k different channels to G_B in such a way that no two edges adjacent to the same node use the same channel (by Lemma 4). Note that each node in G_B uses a different channel to connect with the other k nodes. Define the last node added by the coloring given by Lemma 4 as the root. We assign each non-root node to a class defined by the channel that connects it to the root. In Figure 3(a), the root is the center node, and we name each non-root node with one of the 3 channels (black, gray, and dashed). For the inductive step, assume G_t is a complete graph with $kt + 1$ nodes satisfying the desired properties. We add k new nodes to G_t to form G_{t+1}. For this sake, we connect all vertices of G_t to the vertices of a complete graph $G_I = K_k$. Thus G_{t+1} is a complete graph with $kt + 1 + k = k(t+1) + 1$ vertices. Since k is odd, we can assign k different channels to G_I in such a way that no two edges adjacent to the same node use the same channel (by Lemma 4). By construction, each node in G_I uses $k - 1$ different channels. We assign each node to the class corresponding to the channel it does not use. Consequently, G_{t+1} satisfies (i).

Let $class(c)$ be the set of nodes in G_{t+1} that belong to the class corresponding to channel c. We assign channel c to each edge (u, v) such that $u, v \in class(c)$, and also to each edge $(u, root), \forall u \in class(c)$. Thus G_{t+1} satisfies (ii), and all the nodes in the same class are interconnected and connected with the root using the channel that defines the class. The remaining step is to assign channels to edges with end-points in different classes. Consider two classes c_1 and c_2. By property (iii) all edges in G_t connecting nodes in these classes are labeled with the same channel. We assign this channel to all edges (u, v) such that $u \in class(c_1)$ and $v \in class(c_2)$, with $u \in G_I$ and $v \in G_t$. This step is repeated for all pairs of classes. Finally, since the color assignment for G_B given by Lemma 4 builds on the assignment for G_I, for any pair of classes, edges connecting vertices in these classes have the same colors in both G_B and G_I. Thus for all pairs of classes c_1 and c_2, the edge (u, v) with $u, v \in G_I$ and $u \in class(c_1)$ and $v \in class(c_2)$ has the same color of the edges in G_t connecting vertices in $class(c_1)$ to vertices in $class(c_2)$. Hence G_{t+1} satisfies (iii), which completes the proof.   □

Figure 3 shows an example of the construction algorithm with $k = 3$ channels (thus, a balanced complete graph with $3t+1$ nodes). K_4 with 3 different channels is used as the base case in the inductive construction. The graph used in the inductive steps is a K_3 designed using a channel assignment with 3 different channels. The algorithm iteratively adds K_3 at each step. Figure 3(c) shows how the construction algorithm connects G_I and G_B to obtain the final graph.

Theorem 6. *Let G be a complete graph with k channels and at least $k^2 - 2k + 1$ nodes constructed according to the inductive algorithm described in Theorem 5. Then, a broadcast in G from any node can be completed in 2 rounds.*

The proof of this theorem appears in the full version of the paper. In fact, the claimed broadcasting scheme follows directly from the construction in the proof of Theorem 5. Notice that the broadcasting scheme together with the channel assignment constitute a fault-tolerant system. The network may be much larger than $k^2 - 2k + 1$ nodes, and this broadcasting scheme will still work when some of the nodes fail. More precisely, if the root and $k - 2$ nodes in each class do not fail, a message can still be broadcasted to all functioning nodes in 2 rounds.

The described channel assignment is also efficient when several messages need to be broadcasted from different sources at the same time. Specifically, up to k messages can be broadcasted simultaneously, and all the broadcasts complete in 3 rounds. The fault-tolerance property that holds for the broadcast of one message holds as well for this scheme. We formalize this in the following theorem, the proof appears in the full version of the paper.

Theorem 7. *Let G be a complete graph with k channels and at least $k^2 - 2k + 1$ nodes constructed according to the inductive algorithm described in Theorem 5. Then, broadcasting k messages from any k different nodes in G can be completed in 3 rounds.*

6   Conclusions

We studied the broadcasting problem in conflict-aware multi-channel networks, and presented positive and negative results for various network topologies. These include polynomial time algorithms that give optimal broadcasting schemes for grids, and also for trees when there is a single channel on each edge. We proved that the problem is NP-hard for trees in general case, and also for complete graphs even in the restricted case with only one channel on each edge. We studied the balanced complete graphs as a subclass of complete graphs in which each node is connected to roughly the same number of nodes with each channel. In this setting, we proposed a channel assignment that results in broadcasting schemes that complete in two rounds, which is optimal for non-trivial networks. Besides, we proved that broadcasting in some balanced complete graphs requires at least three rounds, thus justifying the significance of our construction. The construction results in fault-tolerant networks that enable efficient broadcasting of multiple messages at the same time.

References

1. Bloom, S., Korevaar, E., Schuster, J., Willebrand, H.: Understanding the performance of free-space optics. Journal of Optical Networking 2(6), 178–200 (2003)
2. Claude, F., Dorrigiv, R., Kamali, S., López-Ortiz, A., Prałat, P., Romero, J., Salinger, A., Seco, D.: Broadcasting in conflict-aware multi-channel networks. Tech. Rep. CS-2012-22, School of Computer Science, University of Waterloo
3. Farley, A.M., Hedetniemi, S.T.: Broadcasting in grid graphs. In: Proc. 9th S-E Conf. Combinatorics, Graph Theory, and Computing, pp. 275–288 (1978)
4. Garey, M.R., Johnson, D.S.: Computers and Intractability; A Guide to the Theory of NP-Completeness. W. H. Freeman & Co., New York (1990)
5. Kondareddy, Y., Agrawal, P.: Selective broadcasting in multi-hop cognitive radio networks. In: IEEE Sarnoff Symposium, pp. 1–5 (2008)
6. Li, L., Qin, B., Zhang, C., Li, H.: Efficient Broadcasting in Multi-radio Multi-channel and Multi-hop Wireless Networks Based on Self-pruning. In: Perrott, R., Chapman, B.M., Subhlok, J., de Mello, R.F., Yang, L.T. (eds.) HPCC 2007. LNCS, vol. 4782, pp. 484–495. Springer, Heidelberg (2007)
7. Mahjourian, R., Chen, F., Tiwari, R., Thai, M., Zhai, H., Fang, Y.: An approximation algorithm for conflict-aware broadcast scheduling in wireless ad hoc networks. In: Proceedings of the 9th ACM International Symposium on Mobile Ad Hoc Networking and Computing, MobiHoc 2008, pp. 331–340 (2008)
8. Qadir, J., Chou, C.T., Misra, A., Lim, J.G.: Localized minimum-latency broadcasting in multi-radio multi-rate wireless mesh networks. In: WOWMOM 2008, pp. 1–12 (2008)
9. Ramachandran, K.N., Belding, E.M., Almeroth, K.C., Buddhikot, M.M.: Interference-aware channel assignment in multi-radio wireless mesh networks. In: INFOCOM 2006, pp. 1–12 (2006)
10. Raniwala, A., Gopalan, K., Cker Chiueh, T.: Centralized channel assignment and routing algorithms for multi-channel wireless mesh networks. Mobile Computing and Communications Review 8(2), 50–65 (2004)
11. Soifer, A.: The Mathematical Coloring Book. Springer (2009)
12. Subramanian, A., Buddhikot, M., Miller, S.: Interference aware routing in multi-radio wireless mesh networks. In: WiMesh 2006, pp. 55–63 (2006)
13. Wilkinson, B., Allen, M.: Parallel Programming: Techniques and Applications Using Networked Workstations and Parallel Computers. Prentice Hall (2005)
14. Zagalj, M., Hubaux, J.P., Enz, C.C.: Minimum-energy broadcast in all-wireless networks: NP-completeness and distribution issues. In: MOBICOM 2002, pp. 172–182 (2002)
15. Zhang, X., Shin, K.G.: Chorus: collision resolution for efficient wireless broadcast. In: Proceedings of the 29th Conference on Information Communications, INFOCOM 2010, pp. 1747–1755 (2010)

Shared-Memory Parallel Frontier-Based Search

Shogo Takeuchi[1], Jun Kawahara[1], Akihiro Kishimoto[2], and Shin-ichi Minato[3,1]

[1] JST ERATO Minato Discrete Structure Manipulation System Project, Japan
[2] Department of Mathematical and Computing Sciences, Graduate School of
Information Science and Engineering, Tokyo Institute of Technology, Japan
[3] Graduate School of Information Science and Technology, Hokkaido University

Abstract. Knuth's Simpath algorithm is an efficient algorithm enumerating all paths between two locations. This paper presents three approaches to parallelizing frontier-based search in Simpath in shared-memory environments: *node-based*, *range-based* and *edge-based* approaches. Our results on solving grid graphs show that the lock-free edge-based approach performs best and achieves seven-fold speedup with 32 CPU cores, while the others suffer from severe synchronization overhead due to locks, resulting in performance saturation with more than 12 cores.

Keywords: graph algorithm, enumeration, Simpath, and parallelization.

1 Introduction

Enumerating all solutions efficiently has been a subject of algorithm research for decades. In particular, computing all the paths between two locations has been a fundamental research topic due to many real-world applications such as network reliability analysis [4], solving and generating puzzle instances [16] and finding configurations minimizing the loss of energy in the electric power network [5].

Given two vertices s and t in graph G, Knuth's Simpath algorithm presented in the latest volume of "The Art of Computer Programming" (exercise 225 in Section 7.1.4, [8]) computes all the loop-free paths from s to t with a compact representation of Zero-suppressed Binary Decision Diagram (ZDD) [10], a variant of Binary Decision Diagram (BDD) [2].

Simpath performs breadth-first search called *frontier-based search (FBS)* to build a binary decision graph by marking edge e_i in G as selected or unselected and by checking whether selecting/not selecting e_i leads to a dead-end or an actual s-t path. When FBS generates two nodes n_1 and n_2 with the identical set of *frontier vertices* S used to enumerate all possible connections among S and t, n_1 and n_2 are merged into one node to avoid duplicate search effort. FBS continues this procedure until considering all the combinations of edges. Simpath then reduces the binary decision graph to a ZDD. To our best knowledge, Simpath is so far the most efficient algorithm that is difficult to achieve further performance improvement. However, enumerating paths is still a computationally intensive, difficult task, because its computational complexity is #P-complete.

Parallel computing is one way for achieving speedups and has become important due to the wider availability of multi-core CPUs. Moreover, since the speed

S.K. Ghosh and T. Tokuyama (Eds.): WALCOM 2013, LNCS 7748, pp. 170–181, 2013.
© Springer-Verlag Berlin Heidelberg 2013

of the individual CPU core has been less rapidly improved recently, parallelizing algorithms will become the only way to obtain benefits from the hardware soon.

Efficient parallelization of Simpath is a non-trivial issue. For example, serial FBS uses the hash table to check if two nodes are merged. In shared-memory parallel FBS, if duplicate nodes are allocated to various threads, the hash table must be shared among the threads for duplicate detection. That is, parallel Simpath may incur the synchronization overhead (idle time) caused by mutual exclusion (mutex) lock on shared data, which never arises in serial FBS. Not merging duplicates is not a choice of parallel Simpath, since it would result in an exponential increase of searching extra nodes explored only by parallel search.

This paper presents the first attempt to parallelize FBS of Simpath. The advantage of our work is that the speed of FBS is improved for free, once parallel FBS is implemented and when new techniques for increasing the number of cores are developed from the hardware perspective. We develop three shared-memory parallel algorithms: *node-based*, *range-based* and *edge-based* approaches. These approaches guarantee that two nodes with the same set of frontier vertices are always detected as a duplicate. The node-based approach uses a shared task queue among threads. However, non-negligible overhead caused by mutex lock operations is incurred for managing the task queue. Although the range-based approach alleviates the overhead of the node-based approach by exploiting locality of task allocation, it still suffers from the overhead regarding locks. In contrast, the edge-based approach uses a different work distribution strategy based on the levelized structure of the binary decision graph. This approach can access the hash tables and task queues with no locks. Although one drawback is that the edge-based approach requires larger hash tables, they fit into memory of modern PCs in our experiments. Although Simpath has a procedure of reducing nodes to build a ZDD, parallelizing this procedure currently remains future work.

We ran experiments to measure the performance of the above algorithms using up to 32 CPU cores on grid graphs, a representative domain used to investigate ideas for many real-world applications including geographical information processing and network reliability analysis. Our results show that the edge-based approach performed best and yielded about seven-fold speedups on 32 cores.

2   Sequential Simpath

This section first deals with a naive approach to path enumeration and briefly introduces ZDD. It then describes the sequential Simpath algorithm.

2.1   Naive Approach and ZDD

Let $\{e_1, e_2, \cdots, e_m\}$ be a set of edges and $\{s, v_1, v_2, \cdots, v_\ell, t\}$ be a set of vertices of graph G. Vertices s and t are respectively the source and destination. Edges $e_1, \cdots, e_m$ are ordered in a breadth-first manner starting at s. The task of Simpath is to calculate all the paths from s to t without forming any cycles.

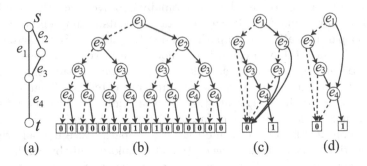

Fig. 1. Binary decision tree, Simpath's DAG and ZDD

A naive approach to solve this problem is first to assign either 0 or 1 to each variable e_i which indicates that edge e_i is respectively unselected or selected and then to check if the set of selected edges constructs an s-t path. This can be seen as a process of building a binary decision tree that represents a Boolean function. Figure 1 (b) illustrates an example of the binary decision tree representing all the s-t paths of a graph shown in Figure 1 (a). The outgoing dotted and solid lines (called *0-arc* and *1-arc*, respectively[1]) from circle e_i (called a *node*) indicate respectively values 0 and 1 are set to e_i. A Boolean value inside a square (called a *terminal node*) indicates whether or not an s-t path is formed with a full assignment to $e_1, e_2, \cdots, e_m$. For example, since selecting edges e_1 and e_4 forms an s-t path, the outcome for assignment $\{e_1 = 1, e_2 = 0, e_3 = 0, e_4 = 1\}$ is 1.

The binary decision tree requires $2^{m+1} - 1$ nodes to represent all the paths for the graph with m edges. In contrast, Simpath leverages ZDD that compresses the set of paths as a directed acyclic graph (DAG) by removing all nodes without which the equivalent set of paths can be represented. In building a ZDD, if node n whose 1-arc directly points the terminal node with the value of 0 (called the *0-terminal node*), n is removed from the current DAG and the subgraph pointed by n's 0-arc is directly connected to n's parents p with the arc that used to connect n and p. A unique form of ZDD is obtained until no node can be removed with this reduction rule (see [10] for details). Figure 1 (d) illustrates the ZDD with an equivalent representation to Figure 1 (b). In case of $e_1 = e_2 = 0$, no s-t path can be generated irrespective of the value assignment of e_3 and e_4. The nodes with assigning values to e_3 and e_4 do not therefore exist in the ZDD.

2.2   The Simpath Algorithm

As in [4,13], instead of first building a binary decision tree and then transforming it to its corresponding ZDD, Simpath directly constructs a DAG that is later efficiently reduced to the ZDD for the sake of time and space efficiency.

---

[1] Although they are usually called "edges" in the BDD research community, we call them arcs to avoid the confusion with edges in the graph.

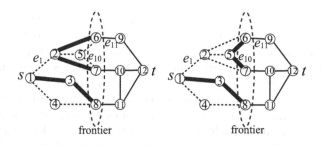

Fig. 2. An Example of nodes considered to be identical

Algorithm 1. Frontier-Based Search

1: $N_1 \leftarrow \{n_{\text{root}}\}$ // n_{root} is the root node labeled by e_1.
2: $N_i \leftarrow \emptyset$ for $i = 2, \ldots, m + 1$.
3: **for** $i = 1$ to m **do**
4:     **for each** $n \in N_i$ **do**
5:         **for each** $x \in \{0, 1\}$ **do** // Assign $e_i = x$.
6:             $n' \leftarrow$ CheckTerminal(n, i, x) // Returns 0-terminal, 1-terminal or nil.
7:             **if** $n' = $ **nil then** // n' is not 0/1-terminal.
8:                 Create a new node n' and set it to n'
9:                 **if** there exists $n'' \in N_i$ s.t. n'' is equivalent to n' **then**
10:                     $n' \leftarrow n''$
11:                 **else**
12:                     $N_{i+1} \leftarrow N_{i+1} \cup \{n'\}$
13:                 **end if**
14:             **end if**
15:             Create x-arc to connect n and n'.
16:         **end for**
17:     **end for**
18: **end for**

For example, Simpath directly builds the DAG shown in Figure 1 (c) for the graph in Figure 1 (a) and then transforms it to ZDD in Figure 1 (d). An explanation of the algorithm reducing the constructed DAG to ZDD is omitted here (see [8, pp 216–218]), since parallel reduction is currently beyond the scope of the paper.

To build a DAG, Simpath performs breadth-first search in a top-down manner from s, which we call *frontier-based search (FBS)*. During the breadth-first search, FBS prunes out combinations of edge selections that generate no s-t path. For example, if node n with the partial assignment of $\{e_1 = val_1, \cdots, e_k = val_k\}$ forms either a cyclic path or a spanning branch connected to s or t, no s-t path can be formed irrespective of the remaining value assignment. Therefore, n is connected to the 0-terminal node with the arc of value val_k.

The heart of the enhancement to FBS is to merge nodes that always return the identical binary outcomes with assignments for the remaining variables. Figure 2 illustrates two nodes that are merged into one. Assume that the binary values are already assigned to variables $e_1, e_2, \cdots, e_{10}$ and that FBS is about to assign either 0 or 1 to variable e_{11}. In both diagrams vertices 1, 6, 7 and 8 are the tip vertices of two partial paths. This indicates that vertices 6, 7 and 8 called *frontier vertices* must be passed through in building an *s-t* path with the remaining assignment. That is, irrespective of $e_1, e_2, \cdots, e_{10}$, if a partial assignment A for the remaining variables contributes to generating an *s-t* path in the left diagram, it also holds for the right diagram. Analogously, if A forms no *s-t* path in the left diagram, it does not yield a path in the right diagram either. As a result, FBS safely regards these nodes to be the "same" to process the unprocessed edges. The "mate" structure is used to efficiently detect frontier vertices and cyclic paths (see [8] for details). The hash table is used to check whether two nodes are equivalent or not in terms of their frontier vertices. The hash keys are computed based on the mate information of the nodes.

Algorithm 1 shows the pseudo-code of FBS. CheckTerminal checks if a node n with an assignment of $e_i = x$ generates the terminal node (i.e. an *s-t* path or a dead-end). Duplicate node detection is performed in lines 9-13.

3   Parallel Frontier-Based Algorithms

This section presents our new shared-memory parallel frontier-based algorithms.

3.1   Node-Based Approach

Let a node of variable e_i be a node with *level i* and N_i be a set of nodes with level i. The node-based approach (NBA) starts processing the nodes in N_{i+1} after all the nodes in N_i are expanded and all the possible nodes generated from the nodes in N_i are saved in N_{i+1}. This indicates that the synchronization overhead is incurred whenever NBA switches to processing nodes in the next level.

Assume that NBA is about to process N_i to generate the successor nodes that are saved in N_{i+1}. NBA uses two shared first-in-first-out (FIFO) task queues to represent N_i and N_{i+1}, respectively, and one shared hash table for duplicate node detection. In this paper we assume that the hash table is implemented as chaining. Each thread dequeues one node n from N_i with a mutex lock operation. The lock is released when the thread obtains n from N_i. The thread then generates two nodes n_1 and n_2 by respectively assigning variable e_i to 0 and 1 and checks the shared hash table to check whether n_1 and n_2 are duplicates or not. If n_1 (or n_2) is a new node, it is saved in the shared hash table with a lock operation[2]. The thread next stores n_1 (or n_2) in N_{i+1} with a lock operation.

NBA is the simplest strategy that tries to evenly distribute work to threads. However, since N_i and N_{i+1} are accessed very frequently by all the threads, the synchronization overhead caused by the contention for N_i and N_{i+1} limits the performance improvement especially with a large number of threads.

---

[2] If read operations to the hash table are lock-free, some other threads may have saved n_1 or n_2 in the hash table after the first duplicate check. Therefore, the duplicate check procedure must be performed again.

3.2   Range-Based Approach

The range-based approach (RBA) is an improvement to NBA. When threads deal with nodes in N_i, RBA alleviates the synchronization overhead by eliminating the mutex lock operations on N_i in NBA. More specifically, when RBA starts to proceed to level i, N_i is partitioned into disjoint sets so that each thread can locally deal with the nodes with no locks. The FIFO queue for N_{i+1} is shared among the threads as in NBA. The lock operation is therefore required to consistently manage N_{i+1}, which still incurs the synchronization overhead. Additionally, locks are required for accesses to the shared hash table.

3.3   Edge-Based Approach

The edge-based approach (EBA) is a completely lock-free approach by modifying the way of managing nodes. Let k be the number of threads and j be i mod k ($1 \le i \le m$) where m is the number of edges. In EBA, thread j must hold N_i. Each of N_i is represented as a FIFO queue. When thread j dequeues node n in N_i and generates n_1 and n_2 by assigning variable e_i to 0 and 1, respectively, n_1 and n_2 are moved to thread l where l is $(i+1)$ mod k. The duplication check of n_1 and n_2 is performed by thread l by using thread l's local hash table that requires no lock operations to read/write information there.

At a first sight, locks are apparently required in case of enqueueing from and dequeueing to N_i a node, because N_i is accessed by two threads. However, at the price of occasionally increasing the synchronization overhead, N_i can be represented as a lock-free structure by leveraging the property that there is only one writer thread and one reader thread to access N_i. More specifically, assume that the FIFO queue is implemented as an array and two indexes are prepared: one pointing a node to dequeue and the other indicating the place to enqueue a node. Then, the writer increments the index of N_i after it stores a node in N_i. When the writer's index is not incremented yet but the node is already in N_i, the reader might consider that no node exists in N_i. In this case, the reader tries to find work in another N_n or waits until the writer increments its index for N_i.

In a way, EBA behaves like a depth-first search except the fact that when successor nodes are generated, they must be moved to a different thread. When each thread has enough amount of work, EBA starts initiating parallelism.

EBA assumes that m is much larger than k. This assumption holds for many practical domains including [4,16,5]. In our test domain explained in Section 4, m ranges between 456 and 480 and k is at most 32.

Unlike NBA and RBA, EBA processes nodes with different levels in parallel. This indicates that EBA must preserve nodes previously stored in the hash table during FBS. In contrast, as described in [8], if NBA and RBA start processing N_i, the information on N_{i-1} can be safely discarded from the hash table by outputting N_{i-1} to a file that is later used to reduce the nodes to build a ZDD. This indicates that the amount of memory required for EBA is larger than the others. However, Knuth's node reduction approach using the file is a sequential algorithm. If the node reduction approach is parallelized, the nodes removed from memory must be most likely to be preserved in memory, which implies that both NBA and RBA would require the same amount of hash tables as EBA.

Because the order of node expansion in each level in EBA is identical to that in sequential FBS, both algorithms construct the same DAG. In contrast, NBA and RBA might construct a different DAG with the same size. However, the node reduction algorithm reduces these different DAGs to the same ZDD.

4   Experiments

4.1   Experimental Setting

We performed experiments on a machine that consists of eight quad-core 3.1 GHz AMD Opteron 8393 SE processors (i.e, 32 CPU cores in total) and 512 GB DDR2 memory shared among cores. Each implementation used at most 10 GB memory that fits into the memory size of modern PCs. All the algorithms were implemented in C++ with Boost C++ (version 1.46.1) and Standard Template Libraries and were compiled with the version 4.1.2 of the GNU C/C++ compiler. To avoid high overhead for locks, we used not only the atomic function offered by GNU C++ but also a highly efficient spin lock implementation in the Open Shogi Library[3], which uses the "xchgl" assembly operation.

Due to inconsistent executions in parallel algorithms, we ran each method three times in solving problems and calculated the average runtime.

4.2   Experimental Results Using a Complete Grid Graph

We prepared the 16×16 complete grid graph consisting of 256 vertices and 480 edges to perform thorough empirical analysis of each algorithm. Since there are many paths leading to the same vertex, grid graphs are a difficult domain to enumerate paths, often resulting in an astronomically large number of paths. Knuth also used this domain to test Simpath.

The original search space of this problem is 3.12×10^{144} and the number of solution is 2,266,745,568,862,672,746,374,567,396,713,098,934,866,324,885,408, 319,028. Such solutions are represented as a ZDD with 464,004,180 nodes and sequential FBS' DAG contains 883,640,712 nodes.

Figure 3 shows the execution time and speedups of NBA, RBA and EBA when the number of CPU cores (shown as "#Threads") is varied. The horizontal solid line indicates sequential baseline FBS.

Although all the approaches with more than four cores solved the problem more quickly than sequential FBS, our results show that EBA clearly outperformed NBA and RBA. RBA performed better than NBA. However, if the number of cores is increased to more than 12, the speedups of both NBA and RBA saturated and resulted in only 1.59 and 2.27 fold with 32 cores, respectively. In contrast, EBA scaled well up to 24 cores. Although the speedup improvement

---

[3] http://gps.tanaka.ecc.u-tokyo.ac.jp/gpsshogi/pukiwiki.php?OpenShogiLib

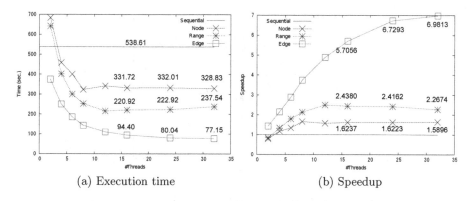

(a) Execution time   (b) Speedup

Fig. 3. Performance graphs on solving the 16×16 complete grid graph. We measured the cases of using 2, 4, 6, 8, 12, 16, 24 and 32 cores.

between 24 and 32 cores was small, which might imply the performance saturation with additional cores, EBA achieved a 6.98-fold speedup with 32 cores, indicating the importance of the lock-free approach. With 32 cores, the execution time was reduced from 539 seconds to 77 seconds.

In EBA, since each thread used about 300 MB memory for its own hash table to detect duplicate nodes, it used about a total of 10 GB memory with 32 cores. In contrast, the other approaches used only 300 MB memory for the hash table shared among cores. Although the additional memory consumption is a drawback of EBA, the size of 10 GB memory fits into the physical memory of recent standard PCs. Additionally, as discussed in Section 3, when parallel algorithms to reduce nodes are developed in the future, this memory overhead might also become a price to pay for NBA and RBA.

There are several reasons parallel efficiency was degraded in each approach. The *start-up overhead* indicates the overhead incurred to start up a parallel algorithm such as thread creation and allocation of the additional structures used only by the parallel algorithm. The *synchronization overhead* refers to the overhead of idle time starving for work and accessing shared data by using locks. The *termination overhead* is the overhead incurred when the threads are terminated and their additional data structures are de-allocated.

We easily measured the start-up and termination overheads. However, we could not measure the synchronization overhead because lock acquisitions/releases occurred very frequently and the duration regarding each lock operation was very short. We therefore approximated the synchronization overhead as follows:

$$Synchronization\ overhead = \frac{(T_{par} - T_{start} - T_{term}) - \frac{T_{seq}}{k}}{T_{par}} \times 100\ (\%)$$

where T_{par} and T_{seq} are respectively the execution time of parallel and sequential algorithms, T_{start} and T_{term} are respectively the runtimes related to the start-up and termination overheads, and k is the number of cores.

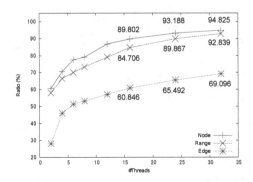

Fig. 4. Synchronization overhead on the complete grid graph

Figure 4 shows the synchronization overhead (SO). As the number of cores increases, SO of all the approaches increased. In particular, although RBA slightly reduced SO, both NBA and RBA had significantly high SO. Even in RBA, 92.8 % of the execution time with 32 cores was related to SO, which was a major reason why it achieved only a 2.27-fold speedup. In contrast, EBA's lock-free approach reduced SO to 69 %, resulting in a better speedup value than NBA and RBA. However, because EBA still suffered from non-negligible SO, reducing this overhead remains a challenge to maximize the efficiency of parallel algorithms.

Compared to T_{par}, $T_{start} + T_{term}$ for NBA and RBA was very small (the ratio of 0.05 % to T_{par} with 32 cores). However, this number was larger for EBA (the ratio of 8.82 % to 32 cores). This might be related to the fact that EBA had to allocate/de-allocate larger hash tables than NBA and RBA.

Load balance refers to how evenly the work is distributed among the threads and is defined as the ratio of the maximal number of nodes allocated to a core to the average number of nodes allocated to each core. With 32 cores, the load balance of all approaches ranged 1.02–1.05, which was reasonably effective.

4.3   Experimental Results Using Incomplete Grid Graphs

We prepared ten 16×16 grid graphs with some edges removed as a more practical test suite (called *incomplete grid graphs*) to calculate the average runtime per problem. In constructing each grid graph, we randomly removed 24 edges from the complete grid graph. Each algorithm performed FBS to enumerate all the paths from the top-left corner to the bottom-right corner in the graphs.

Figure 5 shows the execution time and speedups of all the approaches. Again, EBA significantly outperformed NBA and RBA. However, EBA's speedup value was lower than that of the complete grid graph. EBA suffered from performance degradation when the number of cores was increased from 24 to 32. This is not surprising, not only because sequential FBS solved each incomplete grid graph more quickly compared to the case of the complete grid graph, but also because the number of nodes in each level varied significantly in incomplete grid graphs (see Figure 6), which made EBA harder to evenly distribute work.

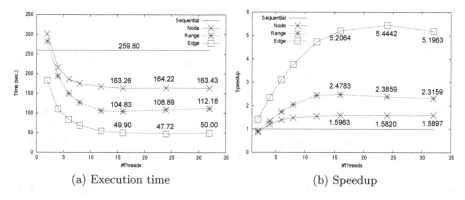

(a) Execution time   (b) Speedup

Fig. 5. Performance graphs on 16×16 incomplete grid graphs

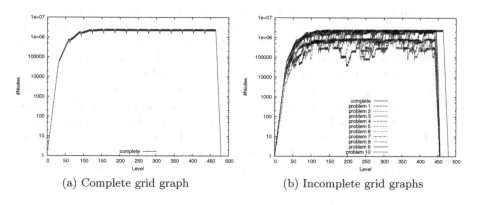

(a) Complete grid graph   (b) Incomplete grid graphs

Fig. 6. The number of nodes generated at each level

5   Related Work

No prior work exists on parallelizing Simpath. We therefore review the related literature on parallelizing other algorithms.

Kumar et al. presented two parallel A* search algorithms [9]. In their centralized strategy, the open list, which corresponds to the task queue of N_i in FBS, is shared among processors. In this sense, their approach is similar to NBA. However, while NBA's task queue is FIFO, their open list is a priority queue that incurs a higher operational overhead. Additionally, while their parallel A* must place generated successors back to the shared open list, NBA places them to N_{i+1}, which is different than N_i. Their parallel A* would therefore suffer from severer synchronization overhead than NBA. In contrast, in their distributed strategy, the open list is partitioned over processors as in RBA. While this approach exchanges nodes in the local open lists among processors to achieve load balancing, RBA does so with no communications. The most essential difference between our work and Kumar et al.'s is that their search space is not a DAG but a tree. That is, because of no requirement for detecting duplicates, their closed list can be easily implemented. In contrast, our approaches must consider effective duplication detection techniques using the hash tables.

Despite a number of existing parallel BDD construction algorithms, many of them intend to work well on specific hardware such as vector machines where data structures must be vectorized [11] and distributed-memory machines where identical BDD nodes must be detected efficiently in the presence of much higher communication overhead than shared-memory machines (e.g., [1,14]). Chen and Banerjee's shared-memory parallel algorithms use depth-first search with very different work distribution schemes that are not easily applicable to FBS [3].

In Ranjan et al.'s breadth-first BDD construction on a PC cluster, each PC manages several levels of BDD nodes to use a vast amount of memory [12], while EBA splits nodes per level. However, their algorithm was not parallelized.

Yang and O'Hallaron's parallel BDD construction algorithm [15] performs node expansions in both breadth-first and depth-first manners. Both EBA and their approach manage nodes in a breadth-first manner but they are often expanded in a depth-first manner. However, while their approach controls the amount of depth-first node expansions based on the threshold, EBA automatically controls the right balance between depth-first and breadth-first search.

Karp and Zhang presented a random work allocation strategy in which generated successors are sent to randomly selected processors [6,7]. While this approach achieves load balancing, their search space is not a DAG but a tree. Their algorithms must be combined with an effective duplication detection scheme, which remains a challenge as future work. In contrast, EBA not only balances the workload effectively but also efficiently detects duplicates with no locks.

6   Conclusions and Future Work

We have developed three shared-memory parallel FBS algorithms: NBA, RBA, and EBA. By eliminating the overhead on locks, EBA outperformed RBA and NBA in solving grid graphs which are an abstract domain for real-world geographical information processing and network reliability analysis. As a result, EBA yielded a 7-fold speedup using 32 cores in the best case. Considering the fact that this paper is the first attempt to parallelize FBS, this is an encouraging result. Additionally, once the presented approaches are implemented, the speed of FBS will be automatically improved with advances in multi-core CPUs.

There are a number of possible extensions to our work. First of all, we are currently trying to parallelize the node reduction algorithm to develop a completely parallel Simpath algorithm in shared-memory environments. Next, investigating ideas to improve the performance of presented approaches is of importance. Particularly, reducing the synchronization overhead is a key feature to obtain better speedups. Finally, developing parallel Simpath in distributed-memory environments will be challenging since it is notoriously hard to achieve satisfactory speedups there due to the communication overhead.

References

1. Bianchi, F., Corno, F., Rebaudengo, M., Reorda, M.S., Ansaloni, R.: Boolean Function Manipulation on a Parallel System Using BDDs. In: Hertzberger, B., Sloot, P.M.A. (eds.) HPCN-Europe 1997. LNCS, vol. 1225, pp. 916–928. Springer, Heidelberg (1997)
2. Bryant, R.E.: Graph-based algorithms for boolean function manipulation. IEEE Transactions on Computers C-35(8), 677–691 (1986)
3. Chen, J.-S., Banerjee, P.: Parallel constuction algorithms for BDDs. In: Proceedings of the 1999 IEEE International Symposium on Circuits and Systems, ISCAS 1999, vol. 1, pp. 318–322 (1999)
4. Hardy, G., Lucet, C., Limnios, N.: K-terminal network reliability measures with binary decision diagrams. IEEE Transactions on Reliability 56(3), 506–515 (2007)
5. Inoue, T., Takano, K., Watanabe, T., Kawahara, J., Yoshinaka, R., Kishimoto, A., Tsuda, K., Minato, S., Hayashi, Y.: Finding all configurations satisfying operational constraints in delivery networks by ZDDs. In: Proceedings of the Institute of Electrical Engineers of Japan National Conference (2012) (in Japanese)
6. Karp, R., Zhang, Y.: A randomized parallel branch-and-bound procedure. In: Proceedings of the 20th ACM Symposium on Theory of Computing, STOC, pp. 290–300 (1988)
7. Karp, R., Zhang, Y.: Randomized parallel algorithms for backtrack search and branch-and-bound computation. Journal of the Association for Computing Machinery 40(3), 765–789 (1993)
8. Knuth, D.E.: The Art of Computer Programming, 1st edn. Combinatorial Algorithms, Part 1, vol. 4A. Addison-Wesley Professional (March 2011)
9. Kumar, V., Ramesh, K., Rao, V.N.: Parallel best-first search of state-space graphs: A summary of results. In: Proceedings of the 10th National Conference Artificial Intelligence, AAAI, pp. 122–127. Press (1988)
10. Minato, S.: Zero-suppressed BDDs for set manipulation in combinatorial problems. In: Proceedings of the 30th ACM/IEEE Design Automation Conference, pp. 272–277 (1993)
11. Ochi, H., Yasuoka, K., Yajima, S.: Breadth-first manipulation of SBDD of boolean functions for vector processing. In: Procedings of the 28th ACM/IEEE Design Automation Conference, pp. 413–416 (1991)
12. Ranjan, R.K., Sanghavi, J.V., Brayton, R.K., Sangiovanni-Vincentelli, A.: Binary decision diagrams on network of workstations. In: IEEE International Conference on Computer Design: VLSI in Computers and Processors, ICCD 1996, pp. 358–364 (1996)
13. Sekine, K., Imai, H., Tani, S.: Computing the Tutte Polynomial of a Graph of Moderate Size. In: Staples, J., Katoh, N., Eades, P., Moffat, A. (eds.) ISAAC 1995. LNCS, vol. 1004, pp. 224–233. Springer, Heidelberg (1995)
14. Stornetta, T., Brewer, F.: Implementation of an efficient parallel BDD package. In: Proceedings of the 33rd Annual Design Automation Conference, DAC 1996, pp. 641–644. ACM, New York (1996)
15. Yang, B., O'Hallaron, D.R.: Parallel breadth-first BDD construction. In: Proceedings of the 9th ACM SIGPLAN Symposium on Principles and Practice of Parallel Programming, pp. 145–156. ACM Press (1997)
16. Yoshinaka, R., Saitoh, T., Kawahara, J., Tsuruma, K., Iwashita, H., Minato, S.: Finding all solutions and instances of numberlink and slitherlink by ZDDs. Algorithms 5(2), 176–213 (2012)

Smoothed Analysis of Belief Propagation for Minimum-Cost Flow and Matching[*]

Tobias Brunsch[1], Kamiel Cornelissen[2], Bodo Manthey[2], and Heiko Röglin[1]

[1] University of Bonn
brunsch@cs.uni-bonn.de, heiko@roeglin.org
[2] University of Twente
{k.cornelissen,b.manthey}@utwente.nl

Abstract. Belief propagation (BP) is a message-passing heuristic for statistical inference in graphical models such as Bayesian networks and Markov random fields. BP is used to compute marginal distributions or maximum likelihood assignments and has applications in many areas, including machine learning, image processing, and computer vision. However, the theoretical understanding of the performance of BP is unsatisfactory. Recently, BP has been applied to combinatorial optimization problems. It has been proved that BP can be used to compute maximum-weight matchings and minimum-cost flows for instances with a unique optimum. The number of iterations needed for this is pseudo-polynomial and hence BP is not efficient in general.

We study belief propagation in the framework of smoothed analysis and prove that with high probability the number of iterations needed to compute maximum-weight matchings and minimum-cost flows is bounded by a polynomial if the weights/costs of the edges are randomly perturbed. To prove our upper bounds, we use an isolation lemma by Beier and Vöcking (SIAM J. Comput., 2006) for matching and generalize an isolation lemma for min-cost flow by Gamarnik, Shah, and Wei (Oper. Res., 2012). We also prove almost matching lower tail bounds for the number of iterations that BP needs to converge.

1 Belief Propagation

Belief propagation (BP) is a message-passing algorithm that is used for solving probabilistic inference problems on graphical models. It has been introduced by Pearl in 1988 [8]. Typical graphical models to which BP is applied are Bayesian networks and Markov random fields. There are two variants of the BP algorithm. The sum-product variant is used to compute marginal probabilities. The max-product or min-sum variant is used to compute maximum a posteriori (MAP) probability estimates.

Recently, BP has experienced great popularity. It has been applied in a large number of fields, such as machine learning, image processing, computer vision, and statistics. For an introduction to BP and several applications, we refer to

[*] This research was supported by ERC Starting Grant 306465 (BeyondWorstCase) and NWO grant 613.001.023 (Smoothed Analysis of Belief Propagation). A full version of this paper is available at http://arxiv.org/abs/1211.3299.

S.K. Ghosh and T. Tokuyama (Eds.): WALCOM 2013, LNCS 7748, pp. 182–193, 2013.
© Springer-Verlag Berlin Heidelberg 2013

Yedidia et al. [14]. There are basically two main reasons for the popularity of BP. First of all, it is generally applicable and easy to implement because of its simple and iterative message-passing nature. In addition, it performs well in practice in numerous applications.

If the graphical model is tree-structured, BP computes exact marginals/MAP estimates. In case the graphical model contains cycles, convergence and correctness of BP have been shown only for specific classes of graphical models. To improve the general understanding of BP and to gain new insights about it, the performance of BP as either a heuristic or an exact algorithm for several combinatorial optimization problems has been studied. Amongst others it has been applied to the maximum-weight matching (MWM) problem, the minimum spanning tree problem [3], the minimum-cost flow (MCF) problem, and the maximum-weight independent set problem [11].

Bayati et al. [4] have shown that max-product BP correctly computes the MWM in bipartite graphs in pseudo-polynomial time if it is unique. Gamarnik et al. [6] have shown that max-product BP computes the MCF in pseudo-polynomial time if it is unique.

1.1 Belief Propagation for Matching and Flow Problems

Bayati et al. [4] have shown that the max-product BP algorithm correctly computes the maximum-weight matching in bipartite graphs if it is unique. Convergence of BP takes pseudo-polynomial time and depends linearly on the weight of the heaviest edge and on $1/\delta$, where δ is the difference in weight between the best and second-best matching. In Section 2 we describe BP for MWM in detail.

Belief propagation has also been applied to finding maximum-weight perfect matchings in arbitrary graphs and to finding maximum-weight perfect b-matchings [2,10], where a perfect b-matching is a set of edges such that every vertex is incident to exactly b edges in the set. For arbitrary graphs the BP algorithm for MWM does not necessarily converge [10]. However, Bayati et al. [2] and Sanghavi et al. [10] have shown that the BP algorithm converges to the optimal matching if the relaxation of the corresponding linear program has an optimal solution that is unique and integral. The number of iterations needed until convergence depends again linearly on the reciprocal of the parameter δ. Bayati et al. [2] have also shown that the same result holds for the problem of finding maximum-weight b-matchings that do not need to be perfect.

It turns out that BP can, to some extent, solve the relaxation of the corresponding linear program for matching, even if it has a non-integral optimal solution. Bayati et al. [2] have shown that it is possible to solve the LP relaxation by considering so-called *graph covers*, in which they compute a bipartite matching. In case of an optimum that is unique and integral, the optimal solution in the graph cover corresponds to the optimal solution. In case of a unique but fractional optimal solution, the average of the estimates of two consecutive iterations (both of which are perfect matchings in the graph cover) yield a value of 0, 1/2, or 1 for any edge, which then equals its value in the optimal solution of the relaxed LP. Sanghavi et al. [10] have shown that BP remains uninformative for some edges (and outputs "?" for those), but computes the correct values for all edges that have a fixed integral value in all optimal solutions.

Gamarnik et al. [6] have shown that BP can be used to find a minimum-cost flow, provided that the instance has a unique optimal solution. The number of iterations until convergence is pseudo-polynomial and depends again linearly on the reciprocal of the difference in cost between the best and second-best integer flow. In addition, they have proved a discrete isolation lemma [6, Theorem 8.1] that shows that the edge costs can be slightly randomly perturbed to ensure that, with probability at least 1/2, the perturbed MCF instance has a unique optimal solution. Using this result, they have constructed an FPRAS for MCF using BP.

1.2 Smoothed Analysis

Smoothed analysis has been introduced by Spielman and Teng [12] in order to explain the performance of the simplex method for linear programming. It is a hybrid of worst-case and average-case analysis and an alternative to both: An adversary specifies an instance, and this instance is then slightly randomly perturbed. The perturbation can, for instance, model noise from measurement. Since its invention in 2001, smoothed analysis has been applied in a variety of contexts. We refer to two recent surveys [7, 13] for a broader picture.

We apply smoothed analysis to BP for min-cost flow and maximum-weight matching. To do this, we consider the following general probabilistic model.

- The adversary specifies the graph $G = (V, E)$ and, in case of min-cost flow, the integer capacities of the edges and the integer budgets (both are not required to be polynomially bounded). Additionally the adversary specifies a probability density function $f_e : [0, 1] \to [0, \phi]$ for every edge e.
- The costs (for min-cost flow) or weights (for matching) of the edges are then drawn independently according to their respective density function.

The parameter ϕ controls the adversary's power: If $\phi = 1$, then we have the average case. The larger ϕ, the more powerful the adversary. The role of ϕ is the same as the role of $1/\sigma$ in the classic model of smoothed analysis, where instances are perturbed by independent Gaussian noise with standard deviation σ. In that model the maximum density ϕ is proportional to $1/\sigma$.

1.3 Our Results

We prove upper and lower tail bounds for the number of iterations that BP needs to solve maximum-weight matching problems and min-cost flow problems. Our bounds match up to a small polynomial factor.

In Sections 3 and 4 we prove that the probability that BP needs more than t iterations is bounded by $O(n^2 m\phi/t)$ for the min-cost flow problem and $O(nm\phi/t)$ for various matching problems, where n and m are the number of nodes and edges of the input graph, respectively. The upper bound for matching problems holds for the variants of BP for the maximum-weight matching problem in bipartite graphs [4] as well as for the maximum-weight (perfect) b-matching problem in general graphs [2, 10]. For the latter it is required that the polytope corresponding to the relaxation of the matching LP is integral. If this is not the case, we can still solve the relaxation of the matching LP with a slightly modified BP algorithm [2] using graph covers (see the comments at the end of Section 4.1). To prove the upper tail bound for BP for MCF we use a continuous isolation lemma that is similar to the discrete isolation lemma by Gamarnik et al. [6, Theorem 8.1]. We need the continuous version since we do not only want to have a unique optimal solution, but we also need to quantify the gap between the best and the second-best solution.

These upper tail bounds are not strong enough to yield any bound on the expected number of iterations. Indeed, in Section 5 we show that this expectation is not finite by providing a lower tail bound of $\Omega(n\phi/t)$ for the probability that t iterations do not suffice to find a maximum-weight matching in bipartite graphs. This lower bound even holds in the average case, i.e., if $\phi = 1$, and it carries over to the variants of BP for the min-cost flow problem and the minimum/maximum-weight (perfect) b-matching problem in general graphs mentioned above [2, 4, 6, 10]. The lower bound matches the upper bound up to a factor of $O(m)$ for matching and up to a factor of $O(nm)$ for min-cost flow. The smoothed lower bound even holds for complete (i.e., non-adversarial) bipartite graphs.

Finally, let us remark that, for the min-cost flow problem, we bound only the number of iterations that BP needs until convergence. The messages might be super-polynomially long. For all matching problems, however, the size of each message is polynomial in the input size and linear in the number of iterations.

2 Definitions and Problem Statement

2.1 Maximum-Weight Matching and Minimum-Cost Flow

First we define the maximum-weight matching problem on bipartite graphs. Consider an undirected weighted bipartite graph $G = (U \cup V, E)$ with $U = \{u_1, \ldots, u_n\}$, $V = \{v_1, \ldots, v_n\}$, and $E \subseteq \{(u_i, v_j) = e_{ij}, 1 \leq i, j \leq n\}$. Each edge e_{ij} has weight $w_{ij} \in \mathbb{R}^+$. A collection of edges $M \subseteq E$ is called a matching if each node of G is incident to at most one edge in M. We define the weight of a matching M by $w(M) = \sum_{e_{ij} \in M} w_{ij}$. The maximum-weight matching $M^\star$ of G is defined as $M^\star = \operatorname{argmax}\{w(M) \mid M \text{ is a matching of } G\}$.

A b-matching $M \subseteq E$ in an arbitrary graph $G = (V, E)$ is a set of edges such that every node from V is incident to at most b edges from M. A b-matching is called perfect if every node from V is incident to exactly b edges from M. Also for these problems we assume that each edge $e \in E$ has a certain weight w_e and we define the weight of a b-matching M accordingly.

In the min-cost flow problem (MCF), the goal is to find a cheapest flow that satisfies all capacity and budget constraints. We are given a graph $G = (V, E)$ with $V = \{v_1, \ldots, v_n\}$. In principle we allow multiple edges between a pair of nodes, but for ease of notation we consider simple directed graphs. Each node v

has a budget $b_v \in \mathbb{Z}$. Each directed edge $e = e_{ij}$ from v_i to v_j has capacity $u_e \in \mathbb{N}_0$ and cost $c_e \in \mathbb{R}^+$. For each node $v \in V$, we define E_v as the set of edges incident to v. For each edge $e \in E_v$ we define $\Delta(v, e) = 1$ if e is an out-going edge of v and $\Delta(v, e) = -1$ if e is an in-going edge of v. In the MCF one needs to assign a flow f_e to each edge e such that the total cost $\sum_{e \in E} c_e f_e$ is minimized and the flow constraints $0 \leq f_e \leq u_e$ for all $e \in E$, and budget constraints $\sum_{e \in E_v} \Delta(v, e) f_e = b_v$ for all $v \in V$ are satisfied. We refer to Ahuja et al. [1] for more details.

Let us remark that we could have allowed also rational values for the budgets and capacities. As our results do not depend on these values, they are not affected by scaling all capacities and budgets by the smallest common denominator.

Note that finding a perfect minimum-weight matching in a bipartite graph $G = (U \cup V, E)$ is a special case of the min-cost flow problem [1].

2.2   Belief Propagation

For convenience, we describe the BP algorithm used by Bayati et al. [4]. For the details of the other versions of BP for the (perfect) maximum-weight b-matching problem and the min-cost flow problem we refer to the original works [2,6,10]. When necessary, we discuss the differences between the different versions of BP in Sections 4 and 5.

The BP algorithm used by Bayati et al. [4] is an iterative message-passing algorithm for computing maximum-weight matchings (MWM). Bayati et al. define their algorithm for complete bipartite graphs $G = (U \cup V, E)$ with $|U| = |V| = n$. In each iteration t, each node u_i sends a message vector $\vec{M}_{ij}^t = [\vec{m}_{ij}^t(1), \vec{m}_{ij}^t(2), \ldots, \vec{m}_{ij}^t(n)]$ to each of its neighbors v_j. The messages can be interpreted as how 'likely' the sending node thinks it is that the receiving node should be matched to a particular node in the MWM. The greater the value of the message $\vec{m}_{ij}^t(r)$, the more likely it is according to node u_i in iteration t that node v_j should be matched to node u_r. Similarly, each node v_j sends a message vector $\overleftarrow{M}_{ji}^t$ to each of its neighbors u_i. The messages are initialized as

$$\vec{m}_{ij}^0(r) = \begin{cases} w_{ij} & \text{if } r = i, \\ 0 & \text{otherwise} \end{cases} \quad \text{and} \quad \overleftarrow{m}_{ji}^0(r) = \begin{cases} w_{ij} & \text{if } r = j, \\ 0 & \text{otherwise.} \end{cases}$$

The messages in iterations $t \geq 1$ are computed from the messages in the previous iteration as follows:

$$\vec{m}_{ij}^t(r) = \begin{cases} w_{ij} + \sum_{k \neq j} \overleftarrow{m}_{ki}^{t-1}(j) & \text{if } r = i, \\ \max_{q \neq j} \left[w_{iq} + \sum_{k \neq j} \overleftarrow{m}_{ki}^{t-1}(q) \right] & \text{otherwise,} \end{cases} \quad \text{and}$$

$$\overleftarrow{m}_{ji}^t(r) = \begin{cases} w_{ij} + \sum_{k \neq i} \vec{m}_{kj}^{t-1}(i) & \text{if } r = j, \\ \max_{q \neq i} \left[w_{qj} + \sum_{k \neq i} \vec{m}_{kj}^{t-1}(q) \right] & \text{otherwise.} \end{cases}$$

The beliefs of nodes u_i and v_j in iteration t are defined as $b_{u_i}^t(r) = w_{ir} + \sum_k \overleftarrow{m}_{ki}^t(r)$ and $b_{v_j}^t(r) = w_{rj} + \sum_k \vec{m}_{kj}^t(r)$. The beliefs can be interpreted as the 'likelihood' that a node should be matched to a particular neighbor. The greater the value of $b_{u_i}^t(j)$, the more likely it is that node u_i should be matched to node v_j. We denote the estimated MWM in iteration t by $\tilde{M}^t$. The estimated matching $\tilde{M}^t$ matches each node u_i to node v_j, where $j = \text{argmax}_{1 \leq r \leq n} \{b_{u_i}^t(r)\}$. Note that $\tilde{M}^t$ does not always define a matching, since multiple nodes may be matched to the same node. However, Bayati et al. [4] have shown that if the MWM is unique, then for t large enough, $\tilde{M}^t$ is a matching and equal to the MWM.

3 Isolation Lemma

3.1 Maximum-Weight Matchings

Beier and Vöcking [5] have considered a general scenario in which an arbitrary set $S \subseteq \{0,1\}^m$ of feasible solutions is given and to every $x = (x_1, \ldots, x_m) \in S$ a weight $w \cdot x = w_1 x_1 + \ldots + w_m x_m$ is assigned by a linear objective function. As in our model they assume that every coefficient w_i is drawn independently according to an adversarial density function $f_i : [0,1] \to [0,\phi]$ and they define δ as the difference in weight between the best and the second-best feasible solution from S, i.e., $\delta = w \cdot x^\star - w \cdot \hat{x}$ where $x^\star = \text{argmax}_{x \in S} w \cdot x$ and $\hat{x} = \text{argmax}_{x \in S \setminus \{x^\star\}} w \cdot x$. They prove a strong isolation lemma that, regardless of the adversarial choices of S and the density functions f_i, the probability of the event $\delta \leq \varepsilon$ is bounded from above by $2\varepsilon \phi m$ for any $\varepsilon \geq 0$.

If we choose S as the set of incidence vectors of all matchings or (perfect) b-matchings in a given graph, Beier and Vöcking's results yield for every $\varepsilon \geq 0$ an upper bound on the probability that the difference in weight δ between the best and second-best matching or the best and second-best (perfect) b-matching is at most ε. Combined with the results in Section 1 on the number of iterations needed by BP in terms of δ, this can immediately be used to obtain an upper tail bound on the number of iterations of the BP algorithm for these problems.

3.2 Min-Cost Flows

The situation for the min-cost flow problem is significantly more difficult because the set S of feasible integer flows cannot naturally be expressed with binary variables. If one introduces a variable for each edge corresponding to the flow on that edge, then $S \subseteq \{0,1,2,\ldots,u_{\max}\}^m$ where $u_{\max} = \max_{e \in E} u_e$. Röglin and Vöcking [9] have extended the isolation lemma to the setting of integer, instead of binary, vectors. However, their result is not strong enough for our purposes as it bounds the probability of the event $\delta \leq \varepsilon$ by $\varepsilon \phi m (u_{\max} + 1)^2$ from above for any $\varepsilon \geq 0$. As this bound depends on $u_{\max}$ it would only lead to a pseudo-polynomial upper tail bound on the number of iterations of the BP algorithm when combined with the results of [6]. Our goal is, however, to obtain a polynomial tail bound that does not depend on the capacities. In the remainder of this section, we prove that the isolation lemma for integer programs [9] can be significantly strengthened when structural properties of the min-cost flow problem are exploited.

In the following we consider the residual network for a flow f [1]. For each edge e_{ij} in the original network that has less flow than its capacity u_{ij}, we include an edge e_{ij} with capacity $u_{ij} - f_{ij}$ in the residual network. Similarly, for each edge e_{ij} that has flow greater than zero, we include the backwards edge e_{ji} with capacity f_{ij} in the residual network.

As all capacities and budgets are integers, there is always a min-cost flow that is integral. An additional property of our probabilistic model is that with probability one there do not exist two different integer flows with exactly the same costs. This follows directly from the fact that all costs are continuous random variables. Hence, without loss of generality we restrict our presentation in the following to the situation that the min-cost flow is unique.

In fact, Gamarnik et al. [6] have not used δ, the difference in cost between the best and second-best integer flow, to bound the number of iterations needed for BP to find the unique optimal solution of MCF, but they have used another quantity Δ. They have defined Δ as the length of the cheapest cycle in the residual network of the min-cost flow $f^\star$. Note that Δ is always non-negative. Otherwise, we could send one unit of flow along a cheapest cycle. This would result in a feasible integral flow with lower cost. With the same argument we can argue that Δ must be at least as large as δ because sending one unit of flow along a cheapest cycle results in a feasible integral flow different from $f^\star$ whose costs exceed the costs of $f^\star$ by exactly Δ. Hence any lower bound for δ is also a lower bound for Δ and so it suffices for our purposes to bound the probability of the event $\delta \leq \varepsilon$ from above.

The isolation lemma we prove is based on ideas that Gamarnik et al. [6, Theorem 8.1] have developed to prove that the optimal solution of a min-cost flow problem is unique with high probability if the costs are randomly drawn integers from a sufficiently large set. We provide a continuous counterpart of this lemma, where we bound the probability that the second-best integer flow is close in cost to the optimal integer flow.

Lemma 1. *The probability that the cost of the optimal and the second-best integer flow differs by at most $\varepsilon \geq 0$ is bounded from above by $2\varepsilon\phi m$.*

The isolation lemma (Lemma 1) together with the discussion about the relation between δ, the difference in cost between the best and second-best integer flow, and Δ, the length of the cheapest cycle in the residual network of the min-cost flow $f^\star$, immediately imply the following result.

Corollary 2. *For any $\varepsilon > 0$, we have $\mathbb{P}(\Delta \leq \varepsilon) \leq 2\varepsilon\phi m$.*

4   Upper Tail Bounds

4.1   Maximum-Weight Matching

We first consider the BP algorithm of Bayati et al. [4], which computes maximum-weight matchings in complete bipartite graphs G in $O(nw^\star/\delta)$ iterations on all instances with a unique optimum. Here $w^\star$ denotes the weight of the heaviest edge and δ denotes the difference in weight between the best and the second-best matching. Even though it is assumed that G is a complete bipartite graph, this is not strictly necessary. If a non-complete graph is given, missing edges can just be interpreted as edges of weight 0.

With Beier and Vöcking's isolation lemma (see Section 3) we obtain the following tail bound for the number of iterations needed until convergence when computing maximum-weight perfect matchings in bipartite graphs using BP.

Theorem 3. *Let τ be the number of iterations until Bayati et al.'s BP [4] for maximum-weight perfect bipartite matching converges. Then $\mathbb{P}(\tau \geq t) = O(nm\phi/t)$.*

This tail bound is not strong enough to yield any bound on the expected running-time of BP for bipartite matchings. But it is strong enough to show that BP has smoothed polynomial running-time with respect to the relaxed definition adapted from average-case complexity [5], where it is required that the expectation of the running-time to some power $\alpha > 0$ is at most linear. However, a bound on the expected number of iterations is impossible, and the tail bound proved above is tight up to a factor of $O(m)$ (see Section 5).

As discussed in Section 1, BP has also been applied to finding maximum-weight (perfect) b-matchings in arbitrary graphs [2, 10]. The result is basically that BP converges to the optimal matching if the optimal solution of the relaxation of the corresponding linear program is unique and integral. The number of iterations needed until convergence depends again on "how unique" the optimal solution is. For Bayati et al.'s variant [2], the number of iterations until convergence depends on $1/\delta$, where δ is again the difference in weight between the best and the second-best matching. For Sanghavi et al.'s variant [10], the number of iterations until convergence depends on $1/c$, where c is the smallest rate by which the objective value will decrease if we move away from the optimal solution.

However, the technical problem in transferring the upper bound for bipartite graphs to arbitrary graphs is that the adversary can achieve that, with high probability or even with a probability of 1 (for larger ϕ), the optimal solution of the LP relaxation is not integral. Already in the average-case, i.e., for $\phi = 1$, where the adversary has no power at all, the optimal solution of the LP relaxation has some fractional variables with high probability.

Still, we can transfer the results for bipartite matching to both algorithms for arbitrary matching if we restrict the input graphs to be bipartite, since in this case the constraint matrix of the associated LP is totally unimodular.

Theorem 4. *Let τ be the number of iterations until Bayati et al.'s [2] or Sanghavi et al.'s [10] BP for general matching, restricted to bipartite graphs as input, converges. Then $\mathbb{P}(\tau \geq t) = O(nm\phi/t)$.*

Bayati et al. [2] and Sanghavi et al. [10] have also shown how to compute b-matchings with BP. If b is even, then the unique optimum to the LP relaxation is integral. Thus, we circumvent the problem that the optimal solution might be fractional. Hence, following the same reasoning as above, the probability that BP for b-matching for even b runs for more than t iterations until convergence is also bounded by $O(mn\phi/t)$.

Furthermore, Bayati et al. [2, Section 4] have shown how to compute the optimal solution of the relaxation of the matching LP with *graph covers*. They obtain the same $O(n/\delta)$ bound for the number of iterations until convergence as for ordinary matching. However, since we are no longer talking about integer solutions, we cannot directly apply the isolation lemma of Beier and Vöcking [5]. To see that δ is still unlikely to be small in the same way (with a slightly worse constant), we can apply the isolation lemma of Röglin and Vöcking [9] since the matching polytope is half-integral. Thus, if we scale the right-hand side with a factor of 2, then we obtain a 0/1/2 integer program. Because of this, we obtain the same $O(mn\phi/t)$ tail bound for the probability that the number of iterations until convergence exceeds t.

4.2   Min-Cost Flow

The bound for the probability that Δ is small (Corollary 2) together with the pseudo-polynomial bound of Gamarnik [6] yield a tail bound for the number of iterations that BP needs until convergence.

Theorem 5. *Let τ be the number of iterations until BP for min-cost flow [6] converges. Then $\mathbb{P}(\tau \geq t) = O(n^2 m\phi/t)$.*

5   Lower Tail Bounds

We show that the expected number of iterations necessary for convergence of BP for maximum-weight matching (MWM) is unbounded. To do this, we prove a lower tail bound on the number of iterations that matches the upper tail bound from Section 4 with respect to t. The lower bound holds even for a two by two complete bipartite graph with edge weights drawn independently and uniformly from the interval $[0, 1]$. In the following analysis, we consider the BP variant introduced by Bayati et al [4]. Our results can be extended to other versions of BP for matching and min-cost flow [2,6,10] in a straightforward way. We discuss these extensions at the end of this section.

5.1   Computation Tree

For proving the lower bounds, we need the notion of a *computation tree*, which we define analogously to Bayati et al. [4].

Let $G = (U \cup V, E)$ be a bipartite graph with $U = \{u_1, \ldots, u_n\}$ and $V = \{v_1, \ldots, v_n\}$. We denote the level-k *computation tree* with the root labeled $x \in U \cup V$ by $T^k(x)$. The tree $T^k(x)$ is a weighted rooted tree of height $k + 1$. The root node in $T^0(x)$ has label x, its degree is the degree of x in G, and its children are labeled with the adjacent nodes of x in G. $T^{k+1}(x)$ is obtained recursively from $T^k(x)$ by attaching children to every leaf node in $T^k(x)$. Each child of a former leaf node labeled y is assigned one vertex adjacent to y in G as a label, but the label of the former leaf node's parent is not used. (Thus, the number of children is the degree of y minus 1.) Edges between nodes with label u_i and label v_j in the computation tree have a weight of w_{ij}.

We call a collection Λ of edges in the computation tree $T^k(x)$ a T-matching if no two edges of Λ are adjacent in $T^k(x)$ and each non-leaf node of $T^k(x)$ is the endpoint of exactly one edge from Λ. Leaves can be the endpoint of either one or zero edges from Λ. Let $t^k(u_i; r)$ be the weight of a maximum weight T-matching in $T^k(u_i)$ that uses the edge (u_i, v_r) at the root.

5.2   Average-Case Analysis

Consider the undirected weighted complete bipartite graph $K_{2,2} = (U \cup V, E)$, where $U = \{u_1, u_2\}$, $V = \{v_1, v_2\}$, and $(u_i, v_j) \in E$ for $1 \leq i, j \leq 2$. Each edge $(u_i, v_j) = e_{ij}$ has weight w_{ij} drawn independently and uniformly from $[0, 1]$. We define the event E_ε for $0 < \varepsilon \leq \frac{1}{8}$ as the event that $w_{11} \in \left[\frac{7}{8}, 1\right]$, $w_{12} \in \left(\frac{1}{2}, \frac{5}{8}\right]$, $w_{21} \in \left(\frac{5}{8}, \frac{3}{4}\right]$, and $w_{22} \in [w_{12} + w_{21} - w_{11} - \varepsilon, w_{12} + w_{21} - w_{11})$. Consider the two possible matchings $M_1 = \{e_{11}, e_{22}\}$ and $M_2 = \{e_{12}, e_{21}\}$. If event E_ε occurs, then the weight of M_2 is greater than the weight of M_1 and the weight differs by at most ε. In addition, w_{11} is greater than w_{12} and the weight differs by at least $1/4$. See Figure 1 for a graphical illustration.

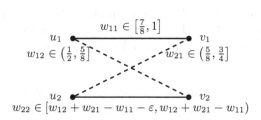

Fig. 1. If E_ε occurs, then the weight of the dashed matching $M_2 = \{e_{12}, e_{21}\}$ is greater than the weight of the solid matching $M_1 = \{e_{11}, e_{22}\}$ and the weight difference is at most ε. In addition w_{11} is greater than w_{12} and the weight difference is at least $\frac{1}{4}$.

Lemma 6. *The probability of event E_ε is $\varepsilon/8^3$.*

Lemma 7. *If event E_ε occurs, then the belief of node u_1 of $K_{2,2}$ at the end of the $4k$-th iteration is incorrect for all integers $k \leq \frac{1}{8\varepsilon} - 1$.*

By Lemma 6 and Lemma 7, we have a lower tail bound for the number of iterations that BP for MWM needs to converge for $K_{2,2}$.

Theorem 8. *The probability that BP for MWM needs at least t iterations to converge for $K_{2,2}$ with edge weights drawn independently and uniformly from $[0,1]$ is at least $\frac{1}{ct}$ for some constant $c > 0$.*

By using copies of $K_{2,2}$ we can extend the result of Theorem 8 to larger graphs.

Corollary 9. *There exist bipartite graphs on $n \geq 4$ nodes, where n is a multiple of 4, with edge weights drawn independently and uniformly from $[0,1]$, for which the probability that BP for MWM needs at least t iterations to converge is $\Omega\left(\frac{n}{t}\right)$ for $t \geq n/c'$ for some constant $c' > 0$.*

5.3   Smoothed Analysis

In this section, we consider complete bipartite graphs $K_{n,n}$ in the smoothed setting. We denote by $X \sim U[a,b]$ that the random variable X is uniformly distributed on interval $[a,b]$. In the following we assume that $\phi \geq 26$ and $n \geq 2$ and even. Similarly to the average case, we define the event E_ε^ϕ for $K_{2,2}$ and for $0 < \varepsilon \leq 1/\phi$ as the event that $w_{11} \in \left[1 - \frac{1}{\phi}, 1\right]$, $w_{12} \in \left(\frac{23}{26}, \frac{23}{26} + \frac{1}{\phi}\right]$, $w_{21} \in \left(\frac{23}{26}, \frac{23}{26} + \frac{1}{\phi}\right]$, and $w_{22} \in [w_{12} + w_{21} - w_{11} - \varepsilon, w_{12} + w_{21} - w_{11})$. Consider the two possible matchings $M_1 = \{e_{11}, e_{22}\}$ and $M_2 = \{e_{12}, e_{21}\}$. If event E_ε^ϕ occurs, then the weight of M_2 is greater than the weight of M_1 and the weight difference is at most ε. In addition w_{11} is greater than w_{12} and the weight difference is at least $\frac{3}{26} - \frac{2}{\phi}$. On this $K_{2,2}$, BP needs at least t rounds with a probability of $\Omega(\phi/t)$.

By taking $n/2$ copies of this $K_{2,2}$ and connecting all nodes in different parts of the bipartite graph by edges whose weights are drawn independently according to $U[0, \frac{1}{\phi}]$, we obtain a $K_{n,n}$ on which BP requires at least t rounds with a probability of $\Omega(\phi n/t)$.

Theorem 10. *There exist probability distributions on $[0,1]$ for the weights of the edges, whose densities are bounded by ϕ, such that the probability that BP for MWM needs at least t iterations to converge for $K_{n,n}$ is $\Omega(n\phi/t)$ for $t \geq n\phi/c$ for some constant $c > 0$.*

5.4   Other Versions of BP

The results of this section also hold for other versions of belief propagation for minimum/maximum-weight (perfect) b-matching and min-cost flow [2,6,10] applied to the matching problem on bipartite graphs. The difference in the number of iterations until convergence differs no more than a constant factor. We omit the technical details but provide some comments on how the proofs need to be adjusted.

Some of the versions of BP consider minimum-weight perfect matching [2] or min-cost flow [6] instead of maximum-weight perfect matching. For these versions we get the same results if we have edge weights $\tilde{w}_e = 1 - w_e$ for all edges e.

For some of the versions of BP [6, 10] the root of the computation tree is an edge instead of a node . If we choose the root of this tree suitably, then we have that the difference in weight between the two matchings M_1 and M_2 of at most ε not only has to 'compensate' the weight difference $\Delta w(e_1, e_2)$ between an edge e_1 in M_1 and an edge e_2 in M_2, but the entire weight w_e of an edge e in M_1 or M_2. However, the probability distributions for the edge weights in Section 5 are chosen such that $\Delta w(e_1, e_2)$ and w_e do not differ more than a constant factor.

References

1. Ahuja, R.K., Magnanti, T.L., Orlin, J.B.: Network flows: theory, algorithms, and applications. Prentice-Hall (1993)
2. Bayati, M., Borgs, C., Chayes, J., Zecchina, R.: Belief-propagation for weighted b-matching on arbitrary graphs and its relation to linear programs with integer solutions. SIAM Journal on Discrete Mathematics 25(2), 989–1011 (2011)
3. Bayati, M., Braunstein, A., Zecchina, R.: A rigorous analysis of the cavity equations for the minimum spanning tree. Journal of Mathematical Physics 49(12), 125206 (2008)
4. Bayati, M., Shah, D., Sharma, M.: Max-product for maximum weight matching: Convergence, correctness, and LP duality. IEEE Transactions on Information Theory 54(3), 1241–1251 (2008)
5. Beier, R., Vöcking, B.: Typical properties of winners and losers in discrete optimization. SIAM Journal in Computing 35(4), 855–881 (2006)
6. Gamarnik, D., Shah, D., Wei, Y.: Belief propagation for min-cost network flow: Convergence & correctness. Operations Research 60(2), 410–428 (2012)
7. Manthey, B., Röglin, H.: Smoothed analysis: Analysis of algorithms beyond worst case. IT – Information Technology 53(6), 280–286 (2011)
8. Pearl, J.: Probabilistic Reasoning in Intelligent Systems: Networks of Plausible Inference. Morgan Kaufmann (1988)
9. Röglin, H., Vöcking, B.: Smoothed analysis of integer programming. Mathematical Programming 110(1), 21–56 (2007)
10. Sanghavi, S., Malioutov, D.M., Willsky, A.S.: Belief propagation and LP relaxation for weighted matching in general graphs. IEEE Transactions on Information Theory 57(4), 2203–2212 (2011)
11. Sanghavi, S., Shah, D., Willsky, A.S.: Message passing for maximum weight independent set. IEEE Transactions on Information Theory 55(11), 4822–4834 (2009)
12. Spielman, D.A., Teng, S.-H.: Smoothed analysis of algorithms: Why the simplex algorithm usually takes polynomial time. Journal of the ACM 51(3), 385–463 (2004)
13. Spielman, D.A., Teng, S.-H.: Smoothed analysis: An attempt to explain the behavior of algorithms in practice. Communications of the ACM 52(10), 76–84 (2009)
14. Yedidia, J.S., Freeman, W.T., Weiss, Y.: Understanding belief propagation and its generalizations. In: Lakemeyer, G., Nebel, B. (eds.) Exploring Artificial Intelligence in the New Millennium, ch. 8, pp. 239–269. Morgan Kaufmann (2003)

Triangle-Partitioning Edges of Planar Graphs, Toroidal Graphs and k-Planar Graphs

Jiawei Gao[1,*], Ton Kloks[**], and Sheung-Hung Poon[2,***]

[1] Software School, Fudan University, 220 Handan Rd., Shanghai, China
gaojw76@gmail.com
[2] Dept. of Computer Science & Inst. of Information Systems and Applications
National Tsing Hua University, No. 101, Sec. 2, Kuang Fu Rd., Hsinchu, Taiwan
spoon@cs.nthu.edu.tw

Abstract. We consider the question whether the edges of a graph can be partitioned into a set of triangles. We propose a linear-time algorithm to partition the edges of a planar graph into triangles. We also obtain a polynomial-time algorithm for toroidal graphs. On the other hand, we show that it is NP-complete for k-planar graphs, where $k \geq 8$.

1 Introduction

To partition the edges into a minimal number of cliques, and to cover them with a minimal number of cliques, are problems that receive a lot of attention lately. Both problems are NP-complete. Covering the edges with a minimal number of cliques remains NP-complete for planar graphs and for graphs with maximal degree at most six [5,12]. The problem is polynomial for graphs with maximal degree at most five. Also for linegraphs the problem can be solved in polynomial time [19,20]. Approximating the minimum number of cliques to cover the edges within a constant factor smaller than two remains NP-complete [14]. Approximations with polynomial factors are obtained eg in [1]. A result of Gyárfás implies that the problem of covering the edges with k cliques can be reduced to a kernel with $O(2^k)$ vertices [9,10]. Recently, it was shown that, under assumption of the exponential time hypothesis, there is no polynomial algorithm which reduces the problem to a kernel of size $O(2^{o(k)})$ [6]. This contrasts the problem of partitioning the edges into cliques. A result of De Bruijn and Erdös implies that there are only two ways to partition the edges of a clique [4]. Actually, the only clique K_n that has a nontrivial partition of the edges into triangles is K_7. Mujuni and Rosamond exploit this fact to derive a fixed-parameter algorithm

---

* Supported in part by grants from the NSFC (no. 60973026), the Shanghai Leading Academic Discipline Project (no. B114), and the Shanghai Committee of Science and Technology (nos. 08DZ2271800 and 09DZ2272800) in China.
** Supported in part by grants from the NSC (nos. 100-2218-E-007-007 and 101-2218-E-007-001) in Taiwan.
*** Supported in part by grants from the NSC (nos. 100-2218-E-007-007, 101-2218-E-007-001 and 100-2628-E-007-020-MY3) in Taiwan.

S.K. Ghosh and T. Tokuyama (Eds.): WALCOM 2013, LNCS 7748, pp. 194–205, 2013.

which reduces the problem to partition the edges into cliques to a kernel of size $O(k^2)$. Fleischer and Wu obtain linear kernels for K_4-free graphs and for planar graphs [7]. Partitioning the edges of a graph into $|E|/3$ triangles remains NP-complete for K_4-free graphs [11,21].

In this paper, we show that the problem can be solved in linear time for planar graphs. Considering that the triangle covering problem for planar graphs is NP-complete [5], this is a very surprising result. Moreover, we also show that the problem for toroidal graphs can be solved in polynomial time. However, we find that the problem is NP-complete for k-planar graphs.

2   Partitioning Planar Graphs

We say that a graph has a partition if its edges can be partitioned into triangles. We assume that the planar graph G is given together with a plane embedding, *i.e.*, G is a plane graph. We remark here that if the original planar graph G has a partition, our following algorithm will always find a solution in any given embedding of G on the plane. In this section, our goal is to show the following theorem.

Theorem 1. *There is a linear-time algorithm to check if the edges of a planar graph can be partitioned into triangles, and find a partition if it exists.*

First, we need the following definitions.

Definition 1. *Let G be a plane graph. Let T be a triangle in G. The triangle divides the plane into one closed region and one open region. We refer to the closed region and the open region as the* inside *and* outside *of T, respectively. If both regions contains vertices of the graph $G - T$ then T is* separating.

Definition 2. *Let T be a separating triangle in G and let $v \in T$. The* inside degree *of v is the number of edges that contain v and of which the other endpoint is inside T. A separating triangle is* even *if the inside degrees of its three vertices are all even.*

Definition 3. *A separating triangle is* outermost *if it has no vertices inside any other separating triangle.*

2.1   The Dual Graph

Let H be the dual of plane graph G. First the following observation is obvious.

Lemma 1. *When G has a partition then every edge is contained in a triangle. Furthermore, if an edge of G is contained in only one triangle, say with edges e_1, e_2 and e_3, then G has a partition if and only if $G - \{e_1, e_2, e_3\}$ has a partition.*

Disconnnected components of a graph can be edge-partitioned separately. A connected graph can be divided into several biconnected components, each of which again can be handled separately. Thus we may assume from now on that G is biconnected. Furthermore, since an edge contained in two triangles has both its endpoints being of degree at least three, then by Lemma 1, we can see that an incident edge of a vertex of degree two belongs to at most one triangle in G. Thus in the following, we may also assume that G has no vertices of degree two.

Lemma 2. *If G has a partition, then H is bipartite.*

Proof. Consider a partition of G. Let C be a cycle in H. There is a one-to-one correspondence between the edges of G and H. The edges of C correspond with a cut set in G. Every triangle of the triangle partition has either all vertices on one side of the cut or it has two vertices on one side, and one vertex on the other side. Thus the cut has an even number of edges. This proves that every cycle of H is even, and so H is a bipartite multigraph. It is easy to check that H has no loops or multiple edges, since G has a partition and no vertex of degree two. □

2.2   Triangle Partitioning Algorithm

A graph with some odd-degree vertex does not have a triangle partition. By Lemma 2, we also see that a plane graph with non-bipartite dual does not have a triangle partition. Thus from now on, we assume that the given plane graph G satisfying the following conditions:

1. G is biconnected,
2. the dual H of G is bipartite,
3. every vertex in G has even degree at least four, and
4. every edge of G is in at least two triangles.

We consider two cases, namely, the graph G has separating triangles or not in the following subsections.

Graphs without Separating Triangles. First, we consider the case where G has no separating triangle. That is, every triangle is a face. Since every edge is in two faces, it follows that every edge is in exactly two facial triangles and that every face is a triangle.

Lemma 3. *Assume that G has no separating triangle. Then G has a partition if and only if every vertex of one color class H_1 of the dual H has degree three.*

Proof. Assume that all vertices in one color class H_1 of H have degree three. Since H is bipartite, H_1 forms a vertex cover of H. All dual faces of vertices in H_1 are triangles and so each vertex of H represents one triangle of G. Therefore, the color class H_1 of H forms a partition of the edges of G into triangles.

Assume that G has a partition. The dual vertices of the triangles in the partition are degree-three vertices in H. Between any two of them, the distance is even, and so they form a color class of H. □

Graphs with Separating Triangles. Consider a partition $\mathcal{P}$ of the edges of G into triangles. We distinguish three types of separating triangles.

Definition 4. *A separating triangle* $S = \{x, y, z\}$ *is one of the following three types.*

Type 1: *Either* $S \in \mathcal{P}$ *or the three edges* $\{x, y\}$, $\{x, z\}$ *and* $\{y, z\}$ *are in triangles of* $\mathcal{P}$ *with the third vertex inside* S.
Type 2: *The three edges* $\{x, y\}$, $\{x, z\}$ *and* $\{y, z\}$ *are in triangles of* $\mathcal{P}$ *of which the third vertex is outside* S.
Type 3: *Some of the edges of* $\{x, y\}$, $\{x, z\}$ *and* $\{y, z\}$ *are in triangles of* $\mathcal{P}$ *with the third vertex inside* S *and some of them are in triangles with the third vertex outside* S.

The following lemma shows that a separating triangle of Type 3 cannot be a single triangle in any partition.

Lemma 4. *If a separating triangle* S *is even, then it is of Type 1 or 2 in any partition. If* S *is not even, then it is of Type 3 in any partition.*

Proof. Let $\mathcal{P}$ be a partition and let $S = \{x, y, z\}$ be a separating triangle. Consider the graph G' induced by the vertices inside S, including S. First assume that $S \in \mathcal{P}$. Then S is even, otherwise there is no partition of the edges in G'.

Assume that $S \notin \mathcal{P}$. Assume that $\{x, y\}$ is in a triangle of $\mathcal{P}$ with the third vertex outside S. Assume that the other two edges of S are in triangles of $\mathcal{P}$ with the third vertex inside S. Thus S is of Type 3. Remove the edge $\{x, y\}$ from the graph G'. There is a partition of the edges of $G' - \{x, y\}$ which implies that the degree of x and y is even. Then S is not even in G, a contradiction.

The other cases are similar. This proves that S is even if and only if S is of Type 1 or Type 2 with respect to $\mathcal{P}$.   □

We suppose that all even separating triangles have been identified. (In later subsection 2.3, we in fact design a linear-time algorithm to compute them.) Our main algorithm traverses G starting from its outer boundary, and search for all outermost even separating triangles. Our search stops at those outermost even separating triangles when they are reached. Thus the interior of any outermost even separating triangle is considered as being removed since our algorithm does not go into it at the current step. The interior subgraph of each outermost, even, separating, triangle will be dealt with in a later recursive step.

Removing the interiors of even, separating triangles, turns these outmost even separating triangles into triangular faces. Let's denote this new graph as G'. Then G' has no more even separating triangles. We call a face of G', which corresponds to an outermost even separating triangle in G, a *region*.

Remark 1. Lemma 4 generalizes to the regions and faces of G'. Any even region or face is one two types.

Type 1: The region or face is a triangle in the partition $\mathcal{P}$ of G'.

Type 2: All the edges of the boundary are in triangles of $\mathcal{P}$ with the third vertex outside the region.

Notice that even, separating triangles in G that are of Type 1 correspond to regions of G' that are Type 1. Similarly, even, separating triangles in G of Type 2 correspond to regions in G' of Type 2. Of course, faces of G' that are not triangles are automatically Type 2. In the next lemma, we show that the two color classes of the dual of G' correspond with the two types.

Lemma 5. *Assume that G' has a partition and let H' be its dual. Let H_1 and H_2 be the two color classes of H'. Then all the vertices of H_1 are of one of the two Types 1 or 2 and all the vertices of H_2 are of the opposite type.*

Proof. Notice that the vertices of H' along any path alternate between the two types and, since H' is connected, this proves the theorem.   □

We use Lemma 5 to partition G', and we can see that there are at most two ways to partition G'. Then we proceed to process a recursive step for the subgraph inside an outmost even separating triangle S of G. If triangle S is labeled Type 1 in G', then the related subgraph inside S to be processed in this recursive step includes S; if triangle S is labeled Type 2, the interior subgraph of S to be processed does not include S. Since all recursive steps are processed on separate subgraphs of G, it is clear that the whole recursive procedure runs in linear time. Thus the last remaining task for us is to find all even separating triangles of G in linear time, which will be done in next subsection.

2.3   Finding Even Separating Triangles

Definition 5. *Let $G = (V, E)$ be a plane graph. A level decomposition partitions the vertices into levels $L_1, L_2, \ldots$ defined as follows.*

(a) *L_1 is the outerface of G and,*
(b) *for $i > 1$, L_i is the outerface of*

$$G - \bigcup_{j=1}^{i-1} L_j.$$

Given a plane graph G, a level decomposition can be obtained in linear time [16] (see also [2,17]). Notice that any consecutive sequence of k levels induces a k-outerplanar graph. It is well-known that k-outerplanar graphs have treewidth at most $3k + 1$ (see eg [3]) and therefore they have $O(k^3 n)$ triangles. Each level induces an outerplanar graph. A graph is outerplanar if and only if each of its biconnected components is formed by a set of cycles connecting together as a tree structure such that neighboring cycles in the tree structure have one edge in common and each edge is contained in at most two of these cycles (see eg [13]). Next, we show how to find all even separating triangles of G using the level decomposition in linear time.

Lemma 6. *All even separating triangles of a plane graph G can be found in linear time.*

Proof. Consider a level decomposition. The outerface L_1 is outerplanar. Each biconnected component of L_1 induces a tree of cycles. Assume there is a cycle in this tree of cycles which is a triangle T. Assume that $L_1 \neq T$ and that the inside of T is nonempty. Then T is a separating triangle. Using the clockwise orientation of each neighborhood we can determine if it is even.

Consider a cycle C of L_1. Assume that the inside of C is nonempty. Then it contains a component of L_2. First create a list of triangles that have at least one vertex of C and at least one vertex of the part of L_2 which is inside C. Since this graph has treewidth at most 7 we can make a list of these triangles in linear time. Check which triangles are even and have a nonempty interior using the clockwise orientation of the neighborhoods.

When all even, separating triangles are determined that contain at least one vertex of L_1, then the vertices of L_1 are deleted, and the algorithm continues with the remaining graph in a similar manner as described above.

Using some suitable data structures this algorithm can be implemented to run in linear time. This proves the lemma.   □

With the triangle partition algorithm in Section 2.2 and Lemma 6, we thus have a linear-time algorithm to partition the edges of plane graph G into triangles. This completes the proof of Theorem 1.

3   Partitioning Toroidal Graphs

A graph is *toroidal* if it can be embedded on the torus. Toroidal graphs [22] generalize planar graphs in many ways dramatically. For example, cliques with up to seven vertices are toroidal. By the graph minor theorem toroidal graphs are characterized by a finite collection of forbidden minors or topological obstructions. By Kuratowski's or Wagner's theorem, for planar graphs this obstruction set has only two elements. For toroidal graphs these obstructions are still not completely known. One has identified 16,629 forbidden minors and 239,322 forbidden topological obstruction [8].

It is convenient to consider drawings of toroidal maps using a rectangular or a square piece of paper. Opposite edges of the paper are point-by-point identified (in the same direction); an edge of the graph which runs out on the right edge of the square, comes back in on the left edge of the square, and similarly edges wrap around on the top- and bottom-edge of the square. As an example one may have a look at the embedding of K_7 on a torus, ie, a representation of Császár's, or Szilassi's polyhedron. Let G be a toroidal embedding of a graph. We distinguish the following types of cycles in G.

Contractible Cycles. These are the boundaries of areas homeomorphic to open discs, or faces.

Noncontractible and Nonseparating Cycles. Consider the drawing of the graph on a square piece of paper. These cycles consist of a path connecting the top- and bottom-edge (with identified edges), or the left- and right-edge (with identified edges). The removal of these cycles reduces the graph to a planar graph, drawn on a cylinder.

Noncontractable, Separating Cycles. These are the cycles whose removal separate the graph into an inside component and an outside component, just as in the planar case.

Lemma 7. *Let G be a toroidal embedding of a graph. Assume that G has a partition $\mathcal{P}$. Assume that all triangles are contractible. Then the dual is bipartite. Furthermore, the triangles of $\mathcal{P}$ consist of one color class of the dual.*

Proof. All triangles of G are faces. Since every edge is in exactly one triangle, any path in the dual alternates between faces that are in $\mathcal{P}$ and faces that are not in $\mathcal{P}$. Thus all cycles of the dual are even. Furthermore, every path between two faces of $\mathcal{P}$ has even length, so $\mathcal{P}$ consists of exactly the faces of one color class of the dual.   □

Consider representation of G on a rectangular planar region. Consider a left to right ordering of the nonseparating triangles that wrap around the top- and bottom-end of the region. Let T_1 and T_2 be two triangles with T_1 left of T_2 in this order. Possibly T_1 and T_2 have some vertices in common, but we assume that $T_1 \neq T_2$. The *piece* $G(T_1, T_2)$ consists of the vertices and edges that are in the region between T_1 and T_2.

Definition 6. *Consider a piece $G(T_1, T_2)$. A bridge is either an edge or a path of length two between two vertices, on in $T_1 \setminus T_2$ and the other in $T_2 \setminus T_1$, which wraps around the right- and left edge of the plane region.*

Consider two vertices $x \in T_1 \setminus T_2$ and $y \in T_2 \setminus T_1$. Assume that x and y are adjacent such that the edge $\{x, y\}$ is embedded in $G(T_1, T_2)$. A bridge between x and y of length two, together with the edge $\{x, y\}$ creates a nonseparating triangle. Similarly, a path of length two from x to y embedded in the piece $G(T_1, T_2)$ together with a bridge which is an edge, is a nonseparating triangle.

Theorem 2. *There is a polynomial-time algorithm to check if the edges of a toroidal graph can be partitioned into triangles.*

Proof (Sketch). Separating triangles are treated in exactly the same manner as in the planar case or, alternatively, via dynamic programming using Tarjan's decomposition tree [23]. Tarjan describes an $O(nm)$ algorithm to find a binary decomposition tree which decomposes a graph using clique separators. Using dynamic programming on this decomposition tree we can obtain, for each even separating triangle T, a table with boolean entries which tells us whether the graph G_T induced by the triangle and the inside has a partition $\mathcal{P}$ with

(a) $T \in \mathcal{P}$, and
(b) all the edges of T are in triangles of $\mathcal{P}$ with some vertex outside T.

The algorithm determines the feasible partitions of pieces $G(T_1, T_2)$ by dynamic programming. For each T_i it has a boolean value which indicates if there is a partition with T_i as a triangle, and for each edge in T_i whether there is a partition with the edge in a triangle with a third vertex inside the piece or outside the piece.

Consider a piece $G(T_1, T_2)$. The table also needs to keep track of the triangles in partitions that use some vertex of T_1 and some vertex of T_2 are a bridge. For any two vertices $x \in T_1 \setminus T_2$ and $y \in T_2 \setminus T_1$, for which there is a triangle which uses a bridge, there are at most n such triangles. Furthermore, at most one of them can be an element of a partition. The algorithm builds a table which lists all partitions of the edges of the piece into triangles. For each pair of vertices x and y an entry of the table contains the information whether a triangle is used in the partition that uses a bridge of length one or two from x to y. Triangles that use bridges cut the piece into parts.

The pieces are processed as follows. The triangles that use a bridge cut the piece into smaller strips. Each strip is bounded on the top and bottom by paths of length one or two. The table for the piece is computed using dynamic programming on the strips. For each edge in the border of the strip, the information is kept whether the border is a triangle in the partition, or which edges are in triangles with the third vertex inside or outside the piece.

By dynamic programming the algorithm computes a table for all pieces in order of increasing size. To write down the dynamic programming algorithm is a standard technique. Its details is omitted due to lack of space.   □

4   NP-Completeness for k-Planar Graphs

A graph is k-planar if it has an embedding in the plane such that every edge crosses at most k other edges. Note that 0-planar graphs are simply planar graphs. In this section, we show that the partition problem for k-planar graphs is NP-complete for all $k \geq 8$.

Theorem 3. *The triangle partition problem for k-planar graphs is NP-complete, where $k = 8$.*

First we show that the problem is in NP. Suppose we are given a triangle partition of the edges of the given graph G. We can easily verify that whether the given partition is a triangle partition of the edges of G.

Next we show that the triangle partition problem is NP-hard for k-planar graphs. We reduce the planar one-in-three 3SAT problem [15] to this problem. We reduce from the 3SAT problem. The input instance for the planar one-in-three 3SAT problem is a set $\{x_1, x_2, \ldots, x_n\}$ of n variables, and a Boolean formula $F = c_1 \wedge c_2 \wedge \ldots \wedge c_m$ of m clauses, where each clause consists of exactly three literals, such that the variable clause graph of the input instance is planar. The planar one-in-three 3SAT problem asks for whether there exists a truth assignment to the variables so that each clause in given formula F has exactly one true literal and exactly two false literals. In the following, we will describe the construction of variable gadgets, literal gadgets, and clause gadgets, respectively, for our polynomial-time reduction. In the construction, we repeatedly use the following construction unit, called an ω-tube. An ω-tube of length ℓ and of width ω is a graph consisting of an integer grid of vertices $\{(x, y)\}$ with $0 \leq x < \ell$ and $0 \leq y < \omega$ for some positive integers ℓ and ω. The edge set of the ω-tube is formed by performing the following steps: (Note that the plus operations relating to y indices here are all modulo ω.)

1. Connect an edge between (x, y) and $(x, y + 1)$ for $0 \leq x < \ell$ and $0 \leq y < \omega$;
2. Connect an edge between (x, y) and $(x+1, y)$ for $0 \leq x < \ell-1$ and $0 \leq y < \omega$; and
3. Connect an edge between (x, y) and $(x + 1, y + 1)$ for $0 \leq x < \ell - 1$ and $0 \leq y < \omega$.

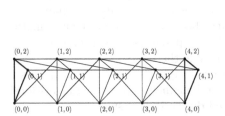

Fig. 1. A variable gadget, which is a 3-tube

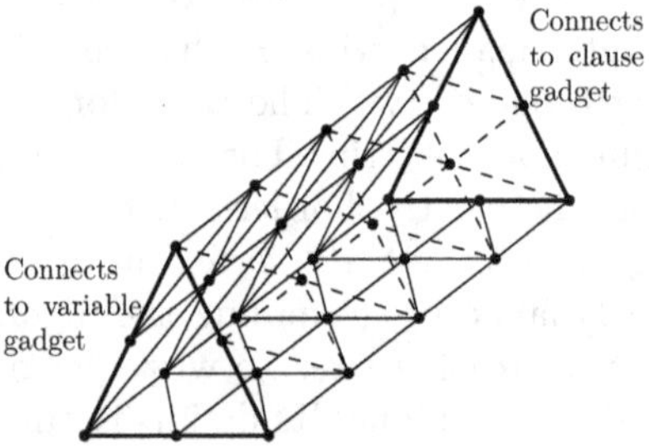

Fig. 2. A literal gadget, which is a 6-tube

See Figure 1 for an example of a 3-tube. The polygons $\{(0,0), (0,1), \ldots, (0, \omega - 1)\}$ and $\{(\ell - 1, 0), (\ell - 1, 1), \ldots, (\ell - 1, \omega - 1)\}$ at the both ends of the ω-tube are called the *end polygons* of the tube.

4.1 Variable Gadget

A variable gadget is a 3-tube of length N. See Figure 1 as an example. First we show in the following lemma that a variable gadget has exactly two partitions.

Lemma 8. *A variable gadget has exactly two partitions.*

Proof. A 3-tube has only 3 types of triangles:

1. α-triangle: $\{(x, y), (x, y + 1), (x + 1, y + 1)\}$,
2. β-triangle: $\{(x, y), (x + 1, y), (x + 1, y + 1)\}$, and
3. γ-triangle: $\{(x, 0), (x, 1), (x, 2)\}$.

A triangle partition of a variable gadget must contain at least one α-triangle, or one β-triangle, because γ-triangles do not contain any $((x, y), (x + 1, y + 1))$ edge. If a partition contains an α-triangle, then the partition nearly only contains α-triangles, except for one end γ-triangle $\{(0,0),(0,1),(0,2)\}$. If a partition contains a β-triangle, then the partition nearly only contains β-triangles, except for one end γ-triangle $\{(N-1,0),(N-1,1),(N-1,2)\}$.   □

The former partition corresponds to the true value for the variable, and thus is called the *true partition*; the latter partition corresponds to the false value for the variable, and thus is called the *false partition*.

Later on, if variable v appears as a literal v in clause c, we plan to delete an α-triangle in the variable gadget of v and a hexagonal hole is thus formed. Such a hole will identify with an end-hexagon of a literal gadget later on. If variable v appears as a literal $\bar{v}$ in clause c, we plan to delete a β-triangle to create a hexagonal hole for connecting a literal gadget later on.

4.2   Literal Gadget

A literal gadget is a subgraph that connects a variable gadget to a clause gadget. We form a literal gadget as a 6-tube of length M. See Figure 2 for an example. It serves the function of propagating a partition from one of its ends to the other.

Of the two end-hexagons of the literal gadget, one end merges with a hexagonal hole of a variable gadget as mentioned previously, the other will connect to a clause gadget later on. In one partition of the literal gadget, it contains all α-triangles of the literal gadget; however, in such a so-called partition, the edges of the clause-end hexagon has not been included in any triangle of this partition. This corresponds to the false value of this literal, and thus such a partition is called the *false partition* of the literal gadget. In the other partition of the literal gadget, it contains all β-triangles of the literal gadget; however, in such a so-call partition, the edges of the variable-end hexagon has not been included in any triangle of this partition. This corresponds to the true value of this literal, and thus such a partition is called the *true partition* of the literal gadget.

4.3   Clause Gadget

A clause gadget for a clause c is formed by simply identifying the clause-end hexagons of the three corresponding literal gadgets. See Figure 4.3 for an example. Because the clause end-hexagons of the three literal gadgets of clause gadget are identified as one hexagon H, the edges of H lies in the triangle partition of exactly one literal gadget among the three literal gadgets for clause c, but not in partitions of the other two literal gadgets. This means that one literal gadget has the true partition and the other two have false partitions. That is to say, exactly one of the three literals is true and the other two literals are false. Moreover, if variable v has the true value, the variable gadget is partitioned as the true partition. For clauses with literal v, the literal gadget is partitioned as the true partition, and the hexagon of the clause gadget is partitioned in the partition of the literal gadget of v whereas the variable-end hexagon of this literal gadget is partitioned in the partition of the variable gadget of v. For clauses with literal

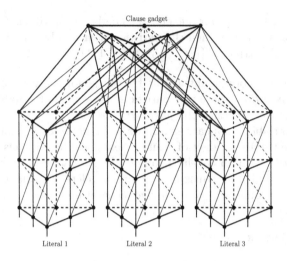

Fig. 3. The structure of a clause gadget

$\bar{v}$, the literal gadget is partitioned as the false partition, and the clause gadget hexagon is not partitioned by current literal gadget whereas the variable-end hexagon of this literal gadget is partitioned in the partition of this literal gadget. If variable v have the false value, the variable gadget is partitioned as the false partition. For clauses with literal $\bar{v}$, the literal gadget is partitioned as the true partition and the hexagon of the clause gadget is partitioned in the partition of the literal gadget of $\bar{v}$ whereas the variable-end hexagon of this literal gadget is partitioned in the partition of the variable gadget of v. For clauses with v, the literal gadget is partitioned as the false partition, and the clause gadget hexagon is not partitioned by current literal gadget whereas the variable-end hexagon of this literal gadget is partitioned in the partition of this literal gadget. Hence, the whole constructed graph has a triangle partition if and only if there is a truth assignment for all variables so that for any clause in formula F, exactly one literal is true and the other two literals are false. Therefore, the planar 1-in-3 3SAT problem has a solution if and only if the constructed graph has a triangle partition. This completes our reduction proof. However, we still need to obtain the constant k for the k-planarity of the constructed graph, which is the following lemma. Its proof is omitted due to lack of space.

Lemma 9. *The graph constructed above is 8-planar.*

References

1. Ausiello, G., Creszensi, P., Gambosi, G., Kann, V., Marchetti-Spaccamela, A., Protasi, M.: Complexity and Approximation: Combinatorial optimization problems and their approximability properties. Springer (1999)
2. Baker, B.: Approximation algorithms for NP-complete problems. Journal of the ACM 41, 153–180 (1994)

3. Bodlaender, H.: A partial k-arboretum of graphs of bounded treewidth. Theoretical Computer Science 209, 1–45 (1998)
4. de Bruijn, N., Erdös, P.: On a combinatorial problem. Indagationes Mathematicae 10, 421–423 (1948)
5. Chang, M.-S., Müller, H.: On the Tree-Degree of Graphs. In: Brandstädt, A., Le, V.B. (eds.) WG 2001. LNCS, vol. 2204, pp. 44–54. Springer, Heidelberg (2001)
6. Cygan, M., Pilipczuk, M., Pilipczuk, M.: Known algorithms for edge clique cover are probably optimal. Manuscript ArXiV: 1203.1754v1 (2012)
7. Fleischer, R., Wu, X.: Edge Clique Partition of K_4-Free and Planar Graphs. In: Akiyama, J., Bo, J., Kano, M., Tan, X. (eds.) CGGA 2010. LNCS, vol. 7033, pp. 84–95. Springer, Heidelberg (2011)
8. Gagarin, A., Myrvold, W., Chambers, J.: The obstructions for toroidal graphs with no $K_{3,3}$'s. Discrete Mathematics 309, 513–520 (2009)
9. Gramm, J., Guo, J., Hüffner, F., Niedermeier, R.: Data reduction, exact, and heuristic algorithms for clique cover. In: Proceedings ALENEX, pp. 86–94. SIAM (2006)
10. Gyárfás, A.: A simple lowerbound on edge clique covering by cliques. Discrete Mathematics 85, 103–104 (1990)
11. Holyer, I.: The NP-completeness of some edge-partition problems. SIAM J. Comput. 10, 713–717 (1981)
12. Hoover, D.: Complexity of graph covering problems for graphs of low degree. JCMCC 11, 187–208 (1992)
13. Kloks, T.: Treewidth. LNCS, vol. 842. Springer, Heidelberg (1994)
14. Kou, L., Stockmeyer, L., Wong, C.: Covering edges by cliques with regard to keyword conflicts and intersection graphs. Comm. ACM 21, 135–139 (1978)
15. Laroche, P.: Planar 1-in-3 satisfiability is NP-complete. Comptes Rendus de l'Acade'mie des Sciences, Se'rie 1, Mathe'matique 316(4), 389–392 (1993)
16. Lipton, R., Tarjan, R.: A separator theorem for planar graphs. SIAM Journal on Applied Mathematics 36, 177–189 (1979)
17. Lipton, R., Tarjan, R.: Applications of a planar separator theorem. SIAM Journal on Computing 9, 615–627 (1980)
18. Mujuni, E., Rosamond, F.: Parameterized complexity of the clique partition problem. In: Proceedings CATS, vol. 77, pp. 75–78. Australian Computer Society (2008)
19. Orlin, J.: Contentment in graph theory: covering graphs with cliques. Proceedings of the Nederlandse Academie van Wetenschappen, Amsterdam. Series A 80, 406–424 (1977)
20. Pullman, N.: Clique covering of graphs IV. Algorithms. SIAM Journal on Computing 13, 57–75 (1984)
21. Shaohan, M., Wallis, W., Lin, W.: The complexity of the clique partition number problem. In: Nineteenth Southeastern Conference on Combinatorics, Graph Theory and Computing, Congr. Numer., vol. 67, pp. 59–66 (1988)
22. Surhone, L., Tennoe, M., Henssonow, S. (eds.): Toroidal graph. Betascript Publishing (2010)
23. Tarjan, R.: Decomposition by clique separators. Discrete Mathematics 55, 221–232 (1985)
24. Valiant, L.: The complexity of computing the permanent. Theoretical Computer Science 8, 189–201 (1979)

Alliances and Bisection Width for Planar Graphs

Martin Olsen[1] and Morten Revsbæk[2]

[1] AU Herning
Aarhus University, Denmark
martino@hih.au.dk
[2] MADALGO*, Department of Computer Science
Aarhus University, Denmark
mrevs@madalgo.au.dk

Abstract. An alliance in a graph is a set of vertices (allies) such that each vertex in the alliance has at least as many allies (counting the vertex itself) as non-allies in its neighborhood of the graph. We show that any planar graph with minimum degree at least 4 can be split into two alliances in polynomial time. We base this on a proof of an upper bound of n on the bisection width for 4-connected planar graphs with an odd number of vertices. This improves a recently published $n + 1$ upper bound on the bisection width of planar graphs without separating triangles and supports the folklore conjecture that a general upper bound of n exists for the bisection width of planar graphs.

1 Introduction

An *alliance* is a set of vertices (allies) such that any vertex in the alliance has at least as many allies (including the vertex itself) as non-allies in its neighborhood of the graph. The alliance is said to be *strong* if this holds even without including the vertex itself among the allies. Alliances of vertices in graphs were introduced by Kristiansen et al. [11] to model among other things alliances of individuals or nations but appear many places in the literature under different names: Flake et al. [8] refer to a strong alliance as a *community* and base their work on the assumption that web pages related to each other form communities in the web graph. Gerber and Kobler [10] look at what they refer to as the *Satisfactory Graph Partition Problem* where the objective is to partition a graph into two strong alliances. A partition of a graph into strong alliances can also be viewed as a so called *Nash stable partition* of an *Additive Hedonic Game* [13]. As mentioned above, alliances have been used to model scenarios that might be planar of nature, so in this paper we focus on the problem of partitioning a *planar* graph into two alliances. In Section 2 we show how to compute such a partition in polynomial time for any planar graph with minimum degree at least 4. To prove this, we need an upper bound of n on the bisection width of 4-connected planar graphs with an odd number of vertices. We prove this upper bound in Section 3.

---

* Center for Massive Data Algorithms, a center of the Danish National Research Foundation.

S.K. Ghosh and T. Tokuyama (Eds.): WALCOM 2013, LNCS 7748, pp. 206–216, 2013.
© Springer-Verlag Berlin Heidelberg 2013

This tight upper bound is an improvement over the recently published [7] $n + 1$ upper bound for planar graphs without separating triangles, and it supports the folklore conjecture [7], that a general upper bound of n exists for the bisection width of planar graphs.

1.1   Preliminaries

Consider the connected graph G with vertex set V and edge set E where $|V| = n$ and $|E| = m$. The degree $d(v)$ of a vertex v in G is the number of edges incident to v in G. Similarly, for a subset $X \subseteq V$ we define the degree $d_X(v)$ of a vertex v in the subgraph of G induced by $X \cup \{v\}$ as $d_X(v) = |\{u \in X : \{v, u\} \in E\}|$. We denote the minimum degree of the vertices in G as δ. A graph G is k-connected when at least k vertices are required to be removed in order to disconnect G. A *clique* is a fully connected graph and a *maximal planar graph* is a planar graph with the property that the addition of any new edge destroys planarity. An *alliance* in G is a non empty set $A \subseteq V$ such that $\forall u \in A : d_A(u) + 1 \geq d_{V-A}(u)$. Throughout this paper when considering a planar graph, we will implicitly consider an embedding of the graph. A *separating triangle* in a planar graph is a triangle where both the interior and the exterior are non-empty. This definition can be tightened giving the notion of a *strong alliance* which is a non empty set $A \subseteq V$ such that $\forall u \in A : d_A(u) \geq d_{V-A}(u)$. A *partition* of G is a collection of non-empty disjoint subsets $V_1 \ldots V_l$ of V such that $\bigcup_{i=1}^{l} V_i = V$. For a partition of G into two subsets V_1 and V_2 we will denote the set of edges crossing this partition as $e(V_1, V_2) = \{\{u, v\} \in E : u \in V_1 \wedge v \in V_2\}$. A *bisection* of G is a partition of G into V_1 and V_2 such that $||V_1| - |V_2|| \leq 1$ and the *bisection width* of G is defined as the minimum $|e(V_1, V_2)|$ over all bisections.

1.2   Related Work

The problem of partitioning a graph into two strong alliances is NP-hard if we put no restrictions on the graph [2]. There are however classes of graphs for which we can decide whether a partition into strong alliances exists and compute it in polynomial time. Examples of such classes are graphs with maximum degree at most 4 and graphs with girth at least 5 and minimum degree at least 3 [2,3].

For a general graph G, the computational complexity of partitioning G into two alliances is an open problem [4]. Fricke et al. [9] show that any graph G contains an efficiently computable alliance with no more than $\lceil \frac{n}{2} \rceil$ vertices, while the problem of deciding whether an alliance with less than k members exists in G is NP-complete if k is part of the input. This even holds if G is planar [6].

Fan et al. [7] prove an upper bound of $n + 1$ for the bisection width for planar graphs without a separating triangle and an upper bound on $n - 2$ for the bisection width for any triangle-free planar graph. The latter upper bound has subsequently been strengthened by Li et al. [12].

2   Alliances in Planar Graphs

In this section we show that for planar graphs with minimum degree at least 4 there exists a partition of the vertices into two alliances and that this partition can be computed in polynomial time. This is trivially also true for planar graphs with minimum degree 1 (let one alliance consist of a single vertex with degree 1), while for planar graphs with minimum degree 2 and 3 it is easy to find examples which show that not all such graphs can be partitioned into alliances. See Figure 1.

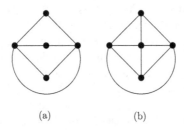

(a)   (b)

Fig. 1. Examples of planar graphs that can not be partitioned into alliances

In Lemma 1 we characterize a group of general graph partitions that can be refined into an alliance partition using a polynomial time algorithm. In the proof of Theorem 1 we then show how to construct such a partition for planar graphs with minimum degree 4 in polynomial time. Lemma 1 is a precise formulation of the well known principle [4,9] that a partition into two sets of vertices forms a good starting point for obtaining a partition into alliances if the number of crossing edges is relatively small compared to the cardinality of the smallest set of vertices.

Lemma 1. *A graph G can be partitioned into two alliances if there exists a partition V_1, V_2 of G such that*

$$|e(V_1, V_2)| - 2\min(|V_1|, |V_2|) < \delta - 2 \ . \tag{1}$$

The alliances can be computed in polynomial time if V_1 and V_2 can be obtained in polynomial time.

Proof. Let V_1, V_2 be a partition of G satisfying (1). We now run the following simple algorithm:

1. Let $A_1 = V_1$ and $A_2 = V_2$.
2. If A_1 and A_2 both are alliances or if one of them is empty we stop. Otherwise we go to step 3.
3. Assume that A_1 is not an alliance (otherwise we process A_2 similarly). There must be a $u \in A_1$ with $d_{A_1}(u) + 1 < d_{A_2}(u)$. We now move u from A_1 to A_2 and go to step 2.

The number of crossing edges $|e(A_1, A_2)|$ decreases with 2 or more every time step 3 is executed so the algorithm must stop after no more than $\frac{m}{2}$ steps. Assume that the algorithm stops because A_1 is empty and let u be the last vertex to leave A_1. We now consider the point in time where $A_1 = \{u\}$:

$$d_V(u) = |e(A_1, A_2)| \le |e(V_1, V_2)| - 2(\min(|V_1|, |V_2|) - 1) \ .$$

We obtain a contradiction since (1) implies that the right hand side is less than δ. We conclude that the algorithm can not stop emptying A_1 or A_2. It has to stop with A_1 and A_2 being alliances.   □

Theorem 1. *Any planar graph with $\delta \ge 4$ can be partitioned into two alliances in polynomial time.*

Proof. We start by expanding the graph by adding edges until it is a maximal planar graph which can be done in polynomial time. We now consider two cases:

The expanded graph has a separating triangle: A separating triangle has vertices both inside and outside of the triangle. Let V_1 be the vertices on the side of the triangle containing the fewest vertices and let $V_2 = V \setminus V_1$. There can be no more than one vertex in V_1 having edges to all three vertices in the separating triangle so $|e(V_1, V_2)| \le 2|V_1| + 1$. This inequality also holds in the original graph so we can now use Lemma 1. The detection and processing of the separating triangle case is easily done in polynomial time.

The expanded graph does not have a separating triangle: In this case the graph is 4-connected since all maximal planar graphs without a separating triangle are 4-connected [5] and thus contains a hamiltonian cycle computable in linear time [1]. Fan et al. [7] show how to efficiently compute a bisection V_1, V_2 of V with $|e(V_1, V_2)| \le n + 1$ for such a graph. This makes it possible for us to apply Lemma 1 in the case where n *is even* but for n *odd* an upper bound on n for the bisection width is needed to make inequality (1) hold. In Section 3 we prove Theorem 2 stating the existence of an efficiently computable bisection V_1, V_2 with $|e(V_1, V_2)| \le n$ for any 4-connected planar graph $G(V, E)$ with an odd number of vertices. We now use Lemma 1 in the case where n is odd.   □

As mentioned above, Fricke et al. [9] have shown that any graph contains an alliance with no more than $\lceil \frac{n}{2} \rceil$ members. We can now improve this upper bound for planar graphs with $\delta \ge 4$:

Corollary 1. *Any planar graph with $\delta \ge 4$ contains an alliance with no more than $\lfloor \frac{n}{2} \rfloor$ members.*

3   An Upper Bound for the Bisection Width

We now show that a bisection V_1, V_2 with $|e(V_1, V_2)| \le n$ can be computed in polynomial time for any 4-connected planar graph with an odd number of vertices. Some of the techniques used are similar to the techniques used by Fan et al. [7] but we also use other techniques and the analysis is considerably more complicated compared to the analysis of Fan et al.. Since the bisection width never increases when removing edges from a graph, it is sufficient to only consider maximal 4-connected planar graphs with an odd number of vertices.

Lemma 2. *A maximal 4-connected planar graph with an odd number of vertices has a vertex u with $d(u) \geq 5$ such that $G - u$ is Hamiltonian. The vertex u and the hamiltonian cycle of $G - u$ can be found in polynomial time.*

Proof. Consider a maximal 4-connected planar graph G with an odd number of vertices. There is at least one node u in G with $d(u) \geq 5$ since otherwise we would have $\sum_{v \in V} d(v) = 2m = 2(3n - 6) \leq 4n$ that could only happen if $n \leq 5$ which would contradict 4-connectedness. The graph G is 4-connected so the graph $G - u$ has a Hamiltonian cycle computable in polynomial time as showed by Thomas and Yu [14].   □

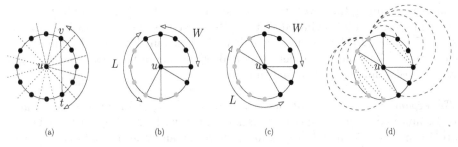

(a)   (b)   (c)   (d)

Fig. 2. Illustrations of a configuration. Figure (a) shows the hamiltonian cycle with its k hamiltonian bisections (the dotted lines) and the cycle length of edge $\{v, t\}$. Figure (b) shows a single hamiltonian bisection where the vertices are colored according to which side of the bisection they belong to. Also, it shows L and W for the configuration. Figure (c) shows a compacted neighbor configuration with L and W. Figure (d) shows a heavy compacted neighbor configuration where the dotted edges are the inner edges of the configuration and the dashed edges are the outer edges of the configuration.

Let G be a maximal 4-connected graph with an odd number of vertices and let u be a vertex in G with $d(u) \geq 5$ and C a Hamiltonian cycle in $G - u$. We will say that the tuple (G, u, C) represents a *configuration* of G. For any such configuration, there are essentially $k = \lfloor \frac{n}{2} \rfloor$ different ways to split C into two connected and equally sized parts. From these parts, we construct a *hamiltonian bisection* V_1, V_2 of G by adding u to the part where it has the most neighbors i.e. the part that minimizes $|e(V_1, V_2)|$ (ties are broken arbitrarily). Refer to Figure 2(a) and 2(b). In the following we let $T(G, u, C)$ denote the sum of $|e(V_1, V_2)|$ over the k possible hamiltonian bisections of (G, u, C). We will sometimes omit the arguments if they are clear from the context. The *cycle length* of an *edge* $\{v, t\}$ in $G - u$ is the minimum distance between v and t in the graph induced by the cycle. The contribution to $T(G, u, C)$ of an edge in $G - u$ is precisely the cycle length of the edge. Refer to Figure 2(a). We let L denote the length of the longest path along C starting and ending at a neighbor from u but visiting no other neighbors of u and let W denote the length of the second longest such path. Refer to Figure 2(b).

We refer to the configuration (G, u, C) as a *compacted neighbor configuration* if the neighbors of u can be divided into two subsets N_1 and N_2 of size $\left\lfloor \frac{d(u)}{2} \right\rfloor$ and $\left\lceil \frac{d(u)}{2} \right\rceil$ respectively such that each subset occupies a connected subpath of the hamiltonian cycle C. Refer to Figure 2(c). The *inner edges* are the edges on the same side of C as u. The inner edges that are not incident to u are naturally grouped into (at most) two groups in a compacted neighbor configuration. A compacted neighbor configuration is called *heavy* if the edges from both these groups have cycle lengths $2, 3, 4, \ldots, k, k - 1, k - 2, k - 3, \ldots$ (for both groups we start the sequence from the left) and if the set of *outer edges* has two edges of length 2, two edges of length 3, $\ldots$, two edges of length $k - 1$ and one edge of length k. Refer to Figure 2(d).

In what follows, we will show that $T(G, u, C) < k(n+1)$ for any configuration (G, u, C) of a maximal 4-connected planar graph with an odd number of vertices. Since $T(G, u, C)$ is the sum of bisection sizes for the k hamiltonian bisections this implies that there exists at least one hamiltonian bisection V_1, V_2 such that $|e(V_1, V_2)| \le n$ which then gives us the upper bound on the bisection width. To prove $T(G, u, C) < k(n + 1)$ we will first show that the heavy compacted neighbor configurations can be considered as a set of worst case configurations such that for any configuration (G, u, C) there exists a heavy compacted neighbor configuration (G', u', C') where $T(G, u, C) \le T(G', u', C')$. We then exploit that the heavy compacted neighbor configurations are reasonably simple such that $T(G', u', C') < k(n + 1)$ can be shown for this set of configurations.

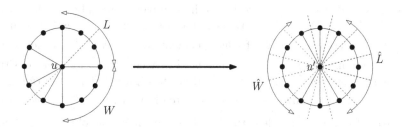

Fig. 3. In the configuration (G, u, C) to the left, the hamiltonian bisections where edges incident to u contribute with $\lfloor d(u)/2 \rfloor$ are shown with dotted lines. Similarly, in the configuration $(\hat{G}, u', C)$ to the right, the hamitonian bisections where edges incident to u' contribute with $\lfloor d(u')/2 \rfloor$ are shown with dotted lines.

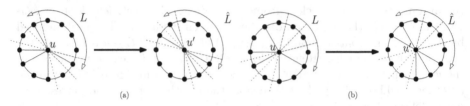

Fig. 4. In (a) we show those hamiltonian bisections which fully contain the vertices on the cycle path corresponding to L in (G, u, C) (to the left) and in $(\hat{G}, u', C)$ (to the right). In (b) we show those hamiltonian bisections which does not fully contain the vertices in (G, u, C) (to the left) and in $(\hat{G}, u', C)$ (to the right).

Lemma 3. *If (G, u, C) is a configuration then it is possible to construct a heavy compacted neighbor configuration (G', u', C) where G and G' have the same number of vertices and $d(u) = d(u')$ such that $T(G, u, C) \leq T(G', u', C)$.*

Proof. Let (G, u, C) represent an arbitrary configuration. We now remove those edges in G that are not on C and not incident to u. We then replace u (and the edges incident to u) with a vertex u' with $d(u) = d(u')$ such that the resulting configuration $(\hat{G}, u', C)$ is a compacted neighbor configuration. Finally, we put in edges to create the graph G' such that (G', u', C) is a heavy compacted neighbor configuration. Below, we first argue that the contribution to T of edges incident to u in G is not higher than the contribution to T of edges incident to u' in G'. Secondly, we argue that the contribution to T of edges in $G - u$ is not higher than the contribution to T of edges in $G' - u'$.

Edges incident to u': We separate our analysis into a case analysis based on the value of L in G. The values of L and W in $\hat{G}$ are denoted by $\hat{L}$ and $\hat{W}$ respectively.

Case 1: $L \leq k$: We consider the following subcases:

- If $2L + d(u) - 2 \leq 2k$ we build the compacted neighbor configuration $(\hat{G}, u', C)$ such that $\hat{L} - \hat{W}$ is minimized (0 or 1). Refer to Figure 3. The contribution to T of edges incident to u' is $k \left\lfloor \frac{d(u)}{2} \right\rfloor$ which is the maximum obtainable value since u' always chooses to join the partition which contributes the least to T. Refer to Figure 3. The condition $2L + d(u) - 2 \leq 2k$ makes it possible for us to obtain the $k \left\lfloor \frac{d(u)}{2} \right\rfloor$ contribution to T from edges incident to u' and at the same time obtain $\hat{L} \geq L$ and $\hat{W} \geq W$ that is important when we consider the contribution from the other edges.
- If $2L + d(u) - 2 > 2k$ we build the compacted neighbor configuration $(\hat{G}, u', C)$ such that $L = \hat{L}$ and such that the nodes forming the long paths along C with no neighbors of u of u' respectively are the same. Refer to Figure 4. For each of the k hamiltonian bisections in $(\hat{G}, u', C)$

we now show that the number of crossing edges incident to u' has not decreased compared to the corresponding (same partition of C) hamiltonian bisection in (G, u, C).

- We first consider a bisection V_1, V_2 where the vertices on the path along C of length $L = \hat{L}$ is fully contained within either V_1 or V_2 – say V_1. In this case, u' must choose to join V_2. The number of neighbors of u' in V_1 is at least as high as the number of neighbors of u in V_1 in G so the number of crossing edges for such a bisection has not decreased. Refer to Figure 4(a).

- We now consider a bisection where the vertices on the path along C of length $L = \hat{L}$ are not fully contained within either side of the bisection. When u' has chosen a side of the bisection u' has only crossing edges to members of either N_1 or N_2 (the two groups of neighbors of u'). If u' has $\left\lfloor \frac{d(u')}{2} \right\rfloor$ crossing edges the case is clear. Otherwise, the number crossing edges has not dropped since every node on the other side of the cut and not on the long path is a neighbor to u'. Refer to Figure 4(b).

Case 2: $L > k$: We build the compacted neighbor configuration $(\hat{G}, u', C)$ with $L = \hat{L}$. Consider a bisection V_1, V_2 of $(\hat{G}, u', C)$. When u' chooses side of the bisection u' can not have crossing edges to both N_1 and N_2. If there are no crossing edges the same would be the case for the corresponding bisection of the original configuration. Refer to Figure 5(a). If there are crossing edges then the number of neighbors on the other side can not have decreased. Refer to Figure 5(b)

Edges not incident to u': Since C is in both G and G' the edges on C obviously contribute with the same to T. We now consider the edges in G not incident to u and not on C and the edges of G' not incident to u' and not on C. Fan et al. [7] show how to eliminate any triangle of such edges and obtain a new set of edges with higher cycle lengths by replacing some of the edges and Fan et al. also argue that repeated elimination of triangles will produce a heavy configuration – we refer to [7] for more details. The fact that $\hat{L} \geq L$ and $\hat{W} \geq W$ makes it possible to use this technique and obtain a one-to-one correspondence between the two sets of edges considered such that any edge in the G-set is matched with an edge in the G'-set with the same cycle length or a bigger cycle length. The contribution to T of these edges can consequently not decrease during the transformation. $\square$

Lemma 4. *Let (G, u, C) be a heavy compacted neighbor configuration with $d(u)$ even. The contribution to $T(G, u, C)$ of the edges incident to u is*

$$\frac{d(u)^2}{4} + W \frac{d(u)}{2} - \frac{d(u)}{2} .$$

Proof. We group the edges incident to u into pairs such that a pair of edges cuts C into two pieces with the same number of neighbors of u. For a given

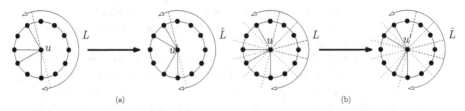

Fig. 5. In (a) we illustrate the case where there are no crossing edges for a hamiltonian bisection in G and the corresponding bisection of $\hat{G}$. In (b) we show the bisections where there are crossing edges in which case the number of crossing edges can not have decreased in $\hat{G}$.

hamiltonian bisection the contribution to $T(G, u, C)$ of a pair is 1 if the endpoints of the edges are separated and 0 otherwise. There are $W + \frac{d(u)}{2} - 1$ bisections that separate each pair so it is now easy to compute the contribution to $T(G, u, C)$ of the edges incident to u:

$$\frac{d(u)}{2}\left(W + \frac{d(u)}{2} - 1\right) .$$

□

Lemma 5. *If (G, u, C) is a heavy compacted neighbor configuration then we have the following:*

$$T(G, u, C) < k(n + 1) .$$

Proof. We divide the proof into three cases.

Assume that $L \geq k - 1$ and that $d(u)$ is even: We compute T in the following way:

$$T = 2k + \left(k + 2\sum_{i=2}^{k-1} i\right) + \left(k + 2\sum_{i=2}^{k-1} i - \sum_{i=1}^{d(u)-3}(W + i)\right) + \left(\frac{d(u)^2}{4} + W\frac{d(u)}{2} - \frac{d(u)}{2}\right) .$$

The first term is the sum of cycle lengths from the edges on the cycle, the second term is the sum of cycle lengths for the outer edges, the third term is the sum of cycle lengths for the inner edges not incident to u, and the fourth term is the contribution from edges incident to u given by Lemma 4. We now use $\sum_{i=2}^{k-1} i = \left(\frac{(k-1)k}{2} - 1\right)$ and $n = 2k + 1$:

$$T - k(n + 1) = \left(\frac{d(u)^2}{4} + W\frac{d(u)}{2} - \frac{d(u)}{2}\right) - \sum_{i=1}^{d(u)-3}(W + i) - 4 .$$

We now work on a part of this sum multiplied by 4 in order to exclusively have integers in the computation:

$$4\left(\left(\frac{d(u)^2}{4} + W\frac{d(u)}{2} - \frac{d(u)}{2}\right) - \sum_{i=1}^{d(u)-3}(W + i)\right)$$

$$= (d(u) - 2)d(u) + 2Wd(u) - 4(d(u) - 3)W - 2(d(u) - 3)(d(u) - 2)$$
$$= -d(u)^2 + 8d(u) - 12 + 12W - 2Wd(u) = -(d(u) - 6)(d(u) - 2) + 12W - 2Wd(u)$$
$$= -d(u)^2 + 8d(u) - 12 + 12W - 2Wd(u) = -(d(u) - 6)(d(u) - 2 + 2W) ,$$

and finally we get

$$T - k(n+1) = -\frac{(d(u) - 6)(d(u) - 2 + 2W)}{4} - 4 < 0 , \tag{2}$$

where we have used that the degree of u is at least 6.

Now assume that $L \leq k - 2$ and that $d(u)$ is even: In this case we get

$$T = \sum_{i=2}^{L} i + \sum_{i=2}^{W} i + \frac{d(u)^2}{4} + W\frac{d(u)}{2} - \frac{d(u)}{2} + 2k + k^2 - 2$$

$$= \frac{L(L+1)}{2} - 1 + \frac{W(W+1)}{2} - 1 + \frac{d(u)^2}{4} + W\frac{d(u)}{2} - \frac{d(u)}{2} + 2k + k^2 - 2$$

implying

$$4T - 4k(2k+2) = 2L(L+1) - 8 + 2W(W+1) + d(u)^2 + 2Wd(u) - 2d(u) + 8k + 4k^2 - 8 - 4k(2k+2)$$

$$= 2L(L+1) + 2W(W+1) + d(u)(d(u) + 2W - 2) - 4k^2 - 16 .$$

We now use $W + L - 2 + d(u) = 2k$:

$$4T - 4k(2k+2) = 2L(L+1) + 2W(W+1) + (2k+2-W-L)(2k+W-L) - 4k^2 - 16$$

$$= 3L^2 + W^2 - 4kL + 4W + 4k - 16. \tag{3}$$

We now use $L \geq W$ in (3):

$$4T - 4k(2k+2) \leq 4L^2 + (4 - 4k)L + 4k - 16$$

$$= 4((L - 1)(L - k + 2) - 2) .$$

implying

$$T - k(n+1) \leq (L - 1)(L - k + 2) - 2 < 0 \text{ for } L \in \{1, 2, \ldots, k - 2\} .$$

Now assume that $d(u)$ is odd: We remove the edge of u from the group with $\lceil \frac{d(u)}{2} \rceil$ edges that is closest to the path along the cycle corresponding to W. It is not hard to see that the contribution to T of the edges of u is unchanged after the removal of this edge. For $d(u) > 5$ there is consequently a heavy compacted neighbor configuration considered above with a higher value of T compared to the original graph. If $d(u) = 5$ we can use (2) with $d(u) = 4$ and W replaced by $W + 1$ if we subtract $W + 1$ (when the edge of u is removed as described above we *insert* an edge with cycle length $W + 1$ and obtain a heavy compacted neighbor configuration with $d(u) = 4$):

$$T - k(n+1) = -\frac{(4 - 6)(4 - 2 + 2(W + 1))}{4} - 4 - (W + 1) = -3 < 0 .$$

□

Theorem 2. *A bisection V_1, V_2 exists with $|e(V_1, V_2)| \leq n$ for any 4-connected planar graph $G(V, E)$ with an odd number of vertices and such a bisection can be obtained in polynomial time.*

Proof. Let $G(V, E)$ be a 4-connected planar graph with an odd number of vertices. As noted earlier, we can assume that G is a maximal planar graph without loss of generality. Lemma 2 assures that we can efficiently obtain a configuration (G, u, c). We now examine all the k hamiltonian bisections of the configuration. By using Lemma 3 and Lemma 5 we know that at least one of the hamiltonian bisections satisfies $|e(V_1, V_2)| \leq n$.   □

References

1. Asano, T., Kikuchi, S., Saito, N.: A linear algorithm for finding hamiltonian cycles in 4-connected maximal planar graphs. Discrete Applied Mathematics 7(1), 1–15 (1984)
2. Bazgan, C., Tuza, Z., Vanderpooten, D.: The satisfactory partition problem. Discrete Appl. Math. 154, 1236–1245 (2006)
3. Bazgan, C., Tuza, Z., Vanderpooten, D.: Efficient algorithms for decomposing graphs under degree constraints. Discrete Appl. Math. 155(8), 979–988 (2007)
4. Bazgan, C., Tuza, Z., Vanderpooten, D.: Satisfactory graph partition, variants, and generalizations. European Journal of Operational Research 206(2), 271–280 (2010)
5. Chen, C.: Any maximal planar graph with only one separating triangle is hamiltonian. J. Comb. Optim. 7(1), 79–86 (2003)
6. Enciso, R.I.: Alliances in graphs: parameterized algorithms and on partitioning series-parallel graphs. PhD thesis, University of Central Florida (2009)
7. Fan, G., Xu, B., Yu, X., Zhou, C.: Upper bounds on minimum balanced bipartitions. Discrete Mathematics 312(6), 1077–1083 (2012)
8. Flake, G., Lawrence, S., Lee Giles, C.: Efficient identification of web communities. In: Proc. 6th ACM SIGKDD International Conference on Knowledge Discovery and Data Mining, pp. 150–160. ACM Press (2000)
9. Fricke, G.H., Lawson, L.M., Haynes, T.W., Hedetniemi, S.M., Hedetniemi, S.T.: A note on defensive alliances in graphs. Bulletin ICA 38, 37–41 (2003)
10. Gerber, M.U., Kobler, D.: Classes of graphs that can be partitioned to satisfy all their vertices. Australasian Journal of Combinatorics 29, 201–214 (2004)
11. Kristiansen, P., Hedetniemi, S.M., Hedetniemi, S.T.: Alliances in graphs. Journal of Combinatorial Mathematics and Combinatorial Computing 48, 157–177 (2004)
12. Li, H., Liang, Y., Liu, M., Xu, B.: On minimum balanced bipartitions of triangle-free graphs. Journal of Combinatorial Optimization, 1–10 (2012)
13. Sung, S.C., Dimitrov, D.: Computational complexity in additive hedonic games. European Journal of Operational Research 203, 635–639 (2010)
14. Thomas, R., Yu, X.X.: 4-connected projective-planar graphs are hamiltonian. Journal of Combinatorial Theory, Series B 62(1), 114–132 (1994)

The Cyclical Scheduling Problem

Binay Bhattacharya[1], Soudipta Chakraborty[2],
Ehsan Iranmanesh[1], and Ramesh Krishnamurti[1]

[1] Simon Fraser University, Canada
{binay,iranmanesh,ramesh}@sfu.ca
[2] Indian Institute of Technology, Kharagpur, India
soudipta.c@gmail.com

Abstract. We consider the $(n-2, n)$ cyclical scheduling problem which assigns a shift of $n-2$ consecutive periods among a total of n periods to workers. We solve this problem by solving a series of b-matching problems on a cycle of n vertices. Each vertex has a capacity, and edges have costs associated with them. The objective is to maximize the total cost of the matching. The best known algorithm for this problem uses network flow, which runs in $O(n^2 \log n)$ on a cycle. We provide an $O(n \log n)$ algorithm for this problem. Using this, we provide an $O(n \log n \log nb_{max})$ algorithm for the $(n-2, n)$ cyclical scheduling problem, where b_{max} is the maximum capacity on a vertex.

1 Introduction

The cyclical scheduling problem is used to schedule shifts for workers in a factory. Given a set of n work periods, each worker is assigned a shift where he works for $n-2$ consecutive periods, and takes off the remaining 2 periods. Thus, for $n = 7$, a typical shift may be to work from Monday to Friday and take off Saturday and Sunday. Another shift may be to work from Friday to Tuesday and take off Wednesday and Thursday (there are 7 such shifts in a week). Each shift may also have a cost associated with it. In addition, the factory requires that a given number of workers be available each period (this requirement may vary from period to period). The objective is to assign a shift to each worker such that the daily requirement is fulfilled and the total cost of the shifts is minimized. We solve the cyclical scheduling problem by solving a series of b-matching problems.

Tibrewala et. al (1972) [7] provide an integer programming formulation for the problem when all shift costs are equal. They also provide a simple algorithm to solve the problem optimally. Bartholdi and Ratliff (1978) [3] solve the cyclical scheduling problem by considering the complementary problem where given the number of workers w in the factory, they maximize the number of workers who are off for two consecutive periods. There is an upper bound on the number of workers who are off during each period. The number of workers they start out with, w, is adequate if the objective function value is at least as much as w. A binary search procedure is used to find the minimum value for w such that w is adequate.

S.K. Ghosh and T. Tokuyama (Eds.): WALCOM 2013, LNCS 7748, pp. 217–232, 2013.

Bartholdi et. al (1980) [4] extend this approach to the general (k, n) cyclic scheduling problem, where there is a total of n periods, and each worker is assigned a shift of k consecutive periods to work (and is off for the remaining $(n-k)$ periods). There is a cost associated with each shift, and the objective is to minimize the total cost of shifts assigned to workers. They provide a parametric solution to this general problem by solving a series of network flow problems, one for each guess for the workforce size. In addition, they also provide a solution using linear programming and roundoff. Using the above notation, we address the $(n-2, n)$ cyclic scheduling problem in this paper. Alfares (1998) [1] provides computational results using linear programming for the $(5, 7)$ cyclic scheduling problem. Alfares (2003) [2] proposes an integer programming formulation for the cyclic scheduling problem where each worker gets 3 days off each week, with additional constraints, such as having at least 2 of the 3 off days consecutive. In a more recent paper, Hochbaum and Levin (2006) [6] provide a quadratic algorithm when all shift costs are equal. Hochbaum and Levin (2006) [5] generalize the cyclical scheduling problem where each worker is assigned multiple shifts, where each shift is a sequence of consecutive work periods followed by a sequence of off periods. They show that this problem is NP-hard when there are two or more such shifts, and propose approximation algorithms for the problem.

2   The b-Matching Formulation

We provide the formulation for the cyclical scheduling problem with n periods, with each shift comprising $n - 2$ consecutive work periods and 2 off periods. Shift s_i assigns off periods i and $(i + 1) \bmod n$ and the remaining as working periods. x_i denotes the number of workers assigned to shift s_i and c_i denotes its cost. b_i denotes the number of workers required to work during period i. The total cost of the assignment is given by $\sum_{i=0}^{n-1} c_i x_i$. The left hand side of each constraint is the number of workers working on shift s_i, and is given by $x_{(i+2) \bmod n} + x_{(i+3) \bmod n} + \cdots + x_{(i-1) \bmod n}$. This should be no less than the requirement b_i. Without loss of generality, we assume all index arithmetic is done   mod n in the formulations for Problem I and Problem II that follow:

Problem I

$$
\begin{aligned}
&\text{Minimize}\quad S = \sum_{i=0}^{n-1} c_i x_i \\
&\quad\text{s.t.} \\
&\qquad x_{i+2} + x_{i+3} + \cdots + x_{i-1} \geq b_i \qquad \forall i, 0 \leq i \leq n - 1 \\
&\qquad x_i \geq 0 \text{ and integer} \qquad\qquad\quad \forall i, 0 \leq i \leq n - 1
\end{aligned}
$$

The above can be reformulated as a b-matching problem. If w denotes the total number of workers, the number of workers working on shift s_i during period i is given by $w - x_{i-1} - x_i$. Letting $d_i = w - b_i$, each of the above inequalities can be rewritten as $x_{i-1} + x_i \leq d_i$, where d_i denotes the number of workers that are off during period i. We thus obtain the formulation for Problem II below, given by Bartholdi and Ratliff (1978) [3]:

$$
\textbf{Problem II} \quad
\begin{bmatrix}
\text{Maximize} \ \ S = \sum_{i=0}^{n-1} c_i x_i \\
\text{s.t.} \\
\ \ x_{i-1} + x_i \leq d_i \qquad \forall i, 0 \leq i \leq n-1 \\
\ \ x_i \geq 0 \text{ and integer} \qquad \forall i, 0 \leq i \leq n-1
\end{bmatrix}
$$

The above formulation for Problem II can be visualised as a graph $G = (V, E)$ with vertex set $V = \{v_0, v_1, \ldots, v_{n-1}\}$ and edge set $E = \{e_0, e_1, \ldots, e_{n-1}\}$, where $e_0 = (v_0, v_1)$, $e_1 = (v_1, v_2)$, $\ldots$, $e_{n-1} = (v_{n-1}, v_0)$, We associate capacities $d_0, d_1, \ldots, d_{n-1}$ with the vertices $v_0, v_1, \ldots, v_{n-1}$ respectively.

We describe in this paper an algorithm based on augmenting paths to solve the above problem. We start with an estimate for w, the total number of workers. If $\bar{w} = \sum_{i=0}^{n-1} x_i < w$, where $\bar{w}$ is the optimal solution to Problem II, and w is our estimate of the number of workers, then we revise this estimate upward. As observed by Bartholdi and Ratliff (1978) [3], a binary search may be performed to find the smallest value of w for which $\bar{w} \geq w$. By solving Problem II for this value of w, we obtain the optimal assignment of shifts to workers.

If w^* is the optimal value of w returned by the binary search procedure, then $b_{max} \leq w^* \leq min(\sum b_i, nb_{max})$, where $b_{max} = max_i\{b_i\}$. Bartholdi and Ratliff (1978) [3] also prove the following inequalities for a particular value $w = w'$ and its corresponding solution $\{x_i\}$:

$$
\sum_{i=0}^{n-1} x_i < w' \quad \text{iff} \quad w' < w^* \tag{1}
$$

$$
\sum_{i=0}^{n-1} x_i \geq w' \quad \text{iff} \quad w^* \leq w' \tag{2}
$$

They then provide the following simple algorithm to find w^*.

Step 1. Restrict w^* within the interval $b_{max} \leq w^* \leq min(\sum b_i, nb_{max})$ where w^* is an integer.

Step 2. Perform binary search through this interval to locate w^*. At each iteration w is fixed at a value w', and the corresponding version of Problem II is solved; then the optimal objective function value $\sum x_i$ is compared to w' and checked using inequalities (1) and (2) to further restrict the location of w^* to $w' < w^*$ or $w^* \leq w'$.

For the rest of the paper, we focus on an efficient solution to Problem II, the b-matching problem. We provide an $O(n \log n)$ algorithm to solve the b-matching problem. The solution to the b-matching problem may be used in conjunction with the binary search procedure above to obtain a solution to the cyclical scheduling problem (Problem I) in $O(n(\log n)(\log nb_{max}))$ time.

To describe the algorithm for the b-matching problem, we first define saturated vertices, unsaturated vertices, alternating paths, and augmenting paths. All the above definitions are with respect to a feasible solution $x = \{x_0, x_1, \ldots, x_{n-1}\}$.

3   Definitions

Without loss of generality, all index arithmetic in the definitions below are done mod n.

Definition 1. *A vertex v_i is* **saturated** *if $x_{i-1} + x_i = d_i$. It is* **unsaturated** *otherwise.*

Definition 2. *A* **path** *is a sequence of vertices and edges v_i, $e_i = (v_i, v_{i+1})$, v_{i+1}, ..., $e_{j-1} = (v_{j-1}, v_j)$, v_j, where no vertex or edge is repeated. The length of a path is the number of edges in it. Alternatively, the above path can also be described as the reversed sequence of vertices and edges v_j, $e_{j-1} = (v_{j-1}, v_j)$, ..., v_{i+1}, $e_i = (v_i, v_{i+1})$, v_i.*

Definition 3. Alternating Path of Type I: *The path from v_i to v_j (from v_j to v_i) is an alternating path of Type I if the feasible solution x assigns a strictly positive value to edges that occur at even positions, namely edges e_{i+1}, e_{i+3}, etc (e_{j-2}, e_{j-4}, etc).*

Definition 4. Alternating Path of Type II: *The path from v_i to v_j (from v_j to v_i) is an alternating path of Type II if the feasible solution x assigns a strictly positive value to edges that occur at odd positions, namely edges e_i, e_{i+2}, etc (e_{j-1}, e_{j-3}, etc).*

Definition 5. Augmenting Path of Type I: *An alternating path of Type I from v_i to v_j (from v_j to v_i) is augmenting if it is of odd length, vertices v_i and v_j are unsaturated, intermediate vertices are saturated, and $\sum_{e \text{ at odd position}} c_e > \sum_{e \text{ at even position}} c_e$. An alternating path of Type I is augmenting if it is of even length, vertex v_i is unsaturated, vertices v_{i+1}, v_{i+2}, ..., v_{j-1} are saturated, and $\sum_{e \text{ at odd position}} c_e > \sum_{e \text{ at even position}} c_e$.*

Definition 6. Augmenting Path of Type II: *An alternating path of Type II from v_i to v_j (from v_j to v_i) is augmenting if it is of odd length, all intermediate vertices v_{i+1}, v_{i+2}, ..., v_{j-1} are saturated, and $\sum_{e \text{ at even position}} c_e > \sum_{e \text{ at odd position}} c_e$. An alternating path of Type II from v_i to v_j is augmenting if it is of even length, all intermediate vertices v_{i+1}, v_{i+2}, ..., v_{j-1} are saturated, vertex v_j is unsaturated, and $\sum_{e \text{ at even position}} c_e > \sum_{e \text{ at odd position}} c_e$.*

We note that an even length augmenting path of Type I from vertex v_i to v_j is also an even length augmenting path of Type II from vertex v_j to vertex v_i. Also, an even length alternating path of Type I (Type II) may be augmenting if it is considered from v_i to v_j, but may not be augmenting if it is considered from v_j to v_i.

4   A Simple Augmentation Algorithm

In this section, we outline a simple augmentation algorithm to find the optimal solution to the b-matching problem. The correctness of the algorithm may be shown by showing that the non-existence of an augmenting path (of either type) with respect to a solution implies that the solution is optimal. We start with a description of this simple algorithm that terminates when there are no augmenting paths. The description below does not specify how an augmenting path is obtained, or what type of augmenting path is obtained.

Algorithm 1. Algorithm Simple Augmentation

1: Initialize $x_i = 0$ for $i = 0, 1, \ldots, n - 1$.
2: **while** there is an augmenting path **do**
3:    Obtain an augmenting path p from v_i to v_j. For odd (even) length augmenting path of Type I, let $\delta = \min\{x_e | e$ is even edge, $d_i - x_i, d_j - x_j\}$ ($\delta = \min\{x_e | e$ is even edge, $d_i - x_i\}$). Augment solution by adding δ to each odd edge, and subtracting δ from each even edge. For odd (even) length augmenting path of Type II, let $\delta = \min\{x_e | e$ is odd edge$\}$ ($\delta = \min\{x_e | e$ is odd edge, $d_j - x_j\}$). Augment solution by subtracting δ from each odd edge, and adding δ to each even edge.
4: **end while**

We now prove the correctness of the algorithm by showing that if there are no augmenting paths with respect to a feasible solution x, then cx is the optimal solution value.

Theorem 1. *If there are no augmenting paths with respect to a feasible solution x, then cx is maximum.*

Proof. We prove by contradiction. Assume that cx is not optimal. Let x^* be the optimal solution. Let e_i be an edge such that $x_i < x_i^*$ (such an edge must exist). We consider the following three cases:

Case 1: Vertices v_i and v_{i+1} are not saturated with respect to x. This implies the existence of an augmenting path (of Type I), which is a contradiction.

Case II: Either vertex v_i or vertex v_{i+1} is saturated (the other is unsaturated). Without loss of generality, let v_{i+1} be saturated (and v_i be unsaturated). $x_i < x_i^*$ implies that $x_{i+1} > x_{i+1}^*$. We now construct an even length augmenting path of Type I which starts at unsaturated vertex v_i goes through vertex v_{i+1}, and ends at vertex v_{i+2}, leading to a contradiction. In other words, we can use this augmenting path to construct a solution vector which is closer to the optimal solution vector.

Case III: Both vertex v_i and vertex v_{i+1} are saturated. $x_i < x_i^*$ implies that $x_{i+1} > x_{i+1}^*$ and $x_{i-1} > x_{i-1}^*$. In this case, we construct an odd length augmenting path of Type II which starts at vertex v_{i-1}, goes through vertices v_i and v_{i+1}, and ends at vertex v_{i+2}. Again this leads to a contradiction.

To derive an efficient algorithm for the b-matching problem, we next derive some properties of augmenting paths. In the next section, we present a modified algorithm for the b-matching problem. This algorithm chooses at each stage an augmenting path of largest cost. We show that such an augmenting path is an odd length augmenting path of Type I.

5   Properties of Augmenting Paths

Given any augmenting path p of Type I from vertex v_i to vertex v_j, edges at odd position are called *plus* edges, and edges at even positions are called *minus* edges. Similarly, given any augmenting path p of Type II from vertex v_i to vertex v_j, edges at even position are called *plus* edges, and edges at odd positions are called *minus* edges. The cost of an augmenting path p, denoted $c(p)$ is given by $\sum_{e \text{ is a plus edge}} c_e - \sum_{e \text{ is a minus edge}} c_e$. Note that our objective is to maximize the total cost of the matching, and therefore using an augmenting path p with cost $c(p)$ to modify the matching increases the objective function value by $c(p)$. We now describe a modified algorithm which at each stage chooses only an augmenting path which has the largest cost.

Algorithm 2. Algorithm Large-cost Augmentation

1: Initialize $x_i = 0$ for $i = 0, 1, \ldots, n-1$.
2: **while** there is an augmenting path **do**
3:     Search for an augmenting path p with the largest cost. Let path p start at vertex v_i and end at vertex v_j. p is an odd length augmenting path of Type I. Let $\delta = \min\{x_e | e \text{ is odd edge}, d_i - x_i, d_j - x_j\}$. Augment solution by adding δ to each odd edge, and subtracting δ from each even edge.
4: **end while**

We will now show that an augmenting path of largest cost is always an odd length augmenting path of Type I. In Lemma 1 below, we show that at any stage of Algorithm Large-cost Augmentation, an augmenting path of largest cost cannot be an even length augmenting path (either of Type I or of Type II).

Lemma 1. *At any stage of Algorithm Large-cost Augmentation, an even length augmenting path (either of Type I or of Type II) cannot have the largest cost.*

Proof. We provide a proof by induction.

Base Case:
We show by contradiction that a path of length 2 cannot have the largest cost. Without loss of generality, let the path start at vertex v_i and end at vertex v_{i+2}, and go through vertex v_{i+1}. Also, let $c_i > c_{i+1}$. Its existence implies that $x_{i+1} > 0$. Since vertex v_{i+1} is saturated, the edge e_{i+1} must have been augmented in an earlier iteration, which is a contradiction given that edge e_i has higher cost and would have been augmented before edge e_{i+1} gets augmented. Also, the even length augmenting path would not exist because edge e_i would get augmented until either vertex v_i or vertex v_{i+1} get saturated.

Induction Hypothesis:
There are no even length augmenting paths of length at most k.

Induction Step:
Consider an even length augmenting path p of Type I with length $k+1$ from vertex v_i to vertex v_j. Path p arose because an odd length augmenting path p_1 of

Type I from vertex v_j to an intermediate vertex v_l was augmented in an earlier iteration. In addition, the cost of odd length path p_2 from vertex v_i to vertex v_l exceeds the cost of path p_1 from vertex v_l to vertex v_j (otherwise the composite path p cannot be augmenting). This is a contradiction since path p_2 from vertex v_i to vertex v_l would get augmented until either v_i or v_l get saturated, or a minus edge in path p_2 gets empty. In each of the above cases, the even length path p would not exist.

In Lemma 2 below, we show that at any stage of Algorithm Large-cost Augmentation, an augmenting path of largest cost cannot be an odd length augmenting path of Type II.

Lemma 2. *At any stage of Algorithm Large-cost Augmentation, an odd length augmenting path of Type II cannot have the largest cost.*

Proof. We provide a proof by induction.

Base Case:
We show by contradiction that an odd length augmenting path of Type II with length 3 cannot have the largest cost. Without loss of generality, let the path start at vertex v_i and end at vertex v_{i+3}, and go through vertices v_{i+1} and v_{i+2}. Also, let $c_{i+1} > c_i + c_{i+2}$. Its existence implies that $x_i > 0$ and $x_{i+2} > 0$. Since vertices v_i and v_{i+3} are saturated, the edges e_i and e_{i+2} must have been augmented in earlier iterations, which is a contradiction given that edge e_{i+1} has higher cost and would have been augmented before edges e_i and e_{i+2} get augmented. Also, the odd length augmenting path would not exist because edge e_{i+1} would get augmented until either one or both of vertices v_{i+1} and v_{i+2} get saturated. If v_{i+1} gets saturated, then $x_i = 0$ and if v_{i+3} gets saturated, then $x_{i+2} = 0$. In either case, the odd length augmenting path of Type II from vertex v_i to v_{i+2} would not exist.

Induction Hypothesis:
There are no odd length augmenting paths of Type II of length k or less.

Induction Step:
Consider an odd length augmenting path p of Type II with length $k+1$. Without loss of generality, let the path start at vertex v_i and end at vertex v_l. Its existence implies that odd length subpaths p_1 of Type I from v_i to an intermediate vertex v_j and p_3 from vertex v_l to an intermediate vertex v_k were augmented in earlier iterations. In addition, the cost of subpath p_2 from intermediate vertex v_j to intermediate vertex v_k exceeds the costs of subpaths p_1 and p_3. Thus, subpath p_2 would get augmented before subpaths p_1 and p_3. Also, subpath p_2 would get augmented until either one or both of vertices v_j and v_k get saturated, or a minus edge gets empty. If vertex v_j gets saturated, then subpath p_1 would not get augmented and if vertex v_k gets saturated, then subpath p_3 would not get augmented. In either of these two cases, or in the case when a minus edge gets empty, the odd length augmenting path of Type II from vertex v_i to vertex v_l would not exist.

We are now ready to show that the total number of augmentations is $O(n)$. Intuitively, each time an augmentation occurs, either one or both of the end vertices get saturated (and stay saturated after that), or one or more even edges (between saturated vertices) get empty. These even edges participate as minus edges in the augmentation. These edges either stay empty for the duration of the algorithm, or participate at least once as an odd edge (as a plus edge) after getting empty. However, to participate as a plus edge, the augmenting path has to get longer, which implies that at least one of the end vertices of the earlier augmenting path is saturated. In either case, we either run out of vertices, or run out of edges.

Theorem 2. *The total number of augmentations is $O(n)$.*

Proof. We keep a count of the number of saturated vertices in the graph in $saturated_v$. We also count the number of augmentations which empty one or more edges in an augmenting path in $empty_e$.

Note that because we use odd length augmenting paths of Type I, once a vertex gets saturated, it stays saturated for the duration of the algorithm. These vertices are intermediate vertices in an augmenting path, and continue to be saturated after the augmentation. Thus the variable $saturated_v$ is a monotonically increasing integer in the range $[0, n]$. (n is the total number of vertices.)

The variable $empty_e$ keeps a count of the number of augmentations that empty one or more edges in a path. Each time the empty edge (or edges) in an augmenting path p_1 get augmented as a plus edge (plus edges) in a later augmentation, the variable $empty_e$ gets decremented. But this must be preceded by at least one of the two unsaturated vertices in the earlier augmentation p_1 getting saturated (and staying saturated for the duration of the algorithm).

It is easy to see that the variable $empty_e$ is bounded from above by m (the number of edges) because if all edges (between saturated vertices) get empty then there can be no augmenting paths. Also, the net decrease in $count_e$ is no more than n because each decrease results in an increase in $count_v$, whose upper bound is n. From this it follows that the total number of augmentations is $O(m + n) = O(2n) = O(n)$.

We are now ready to present the main result of the paper. We will show in the next section that a max-heap data structure, as well as a balanced binary tree may be used to extract the largest cost augmenting path, as well as construct the solution in $O(\log n)$ time at each stage. We also show that there are at most $O(n)$ stages for the algorithm. This gives us an $O(n \log n)$ algorithm for the b-matching problem.

6   An Efficient Algorithm

We now show that we can extract the augmenting path of largest cost by maintaining all even and odd length augmenting paths of Type I in a max-heap data structure (however the largest cost ones will only be odd length augmenting paths of Type I). After extracting such a path, we use a balanced binary tree to determine the extent to which we can augment the path, as well as maintain the solution. Both these operations can be done in $O(\log n)$ time. We provide an informal description of the algorithm below.

6.1 Informal Description

The algorithm starts with an initial solution $x_i = 0, i = 1, 2, \ldots n - 1$. Corresponding to this solution, there are at most n augmenting paths of Type I. Note that each augmenting path is between successive pairs of unsaturated vertices. We insert these paths into a max-heap. Each path is represented as an interval (l_i, r_i) where l_i is the left unsaturated vertex, and r_i is the right unsaturated vertex (in the clockwise direction). Also associated with each interval is its cost (the key), and two pointers or indices (left and right neighbor pointers): one an index to the location of its left adjacent neighbor interval (left with respect to the actual cyclic graph) and the other an index to its right adjacent neighbor interval. We start by storing each interval in the heap-array in the order they occur in the instance. We then call heapify to maintain the heap structure. Each time an interval is placed in the proper position in the heap-array during heapify, we go to its left (right) neighbor pointer (using the index to reach the left adjacent interval in the heap-array) and modify the index of its right (left) neighbor pointer, reflecting the current position of the interval in the heap-array. This takes at most $O(n \log n)$ time. The above procedure is called Heap-Build.

As long as an augmenting path exists, Algorithm Efficient Large-cost Augmentation picks the augmenting path of largest cost from the max-heap using procedure Heap-Extract. This takes $O(1)$ time. Heap-Extract returns (l_i, r_i), the interval corresponding to the largest augmenting path. This interval is given as input to Algorithm Augment-Interval. Algorithm Augment-Interval keeps track of the saturated and unsaturated vertices, as well as the current solution. Algorithm Augment-Interval returns an integer corresponding to the extent to which the path (l_i, r_i) can be augmented, as well as a vector of three elements, denoting which of the following (not necessarily exclusive) events occur on augmentation: left vertex gets saturated, right vertex gets saturated, one or more edges in the even position gets empty. Algorithm Augment-Interval runs in $O(\log n)$ time each time it is called. This is discussed in Section 7.

Next, we call procedure Heap-Adjust with the output of Algorithm Augment-Interval as its input. If the output of Algorithm Augment-Interval denotes that the left interval is saturated after augmentation, then the interval (l_i, r_i) as well as its left adjacent interval (l_{i-1}, r_{i-1}) (note that $l_i = r_{i-1}$) are removed from the heap-array, and a new interval (l_{i-1}, r_i) is inserted into the heap-array. Note that the above operation corresponds to a merging of the two intervals. Similarly, if the output of Algorithm Augment-Interval denotes that the right interval is saturated after augmentation, then the interval (l_i, r_i) as well as its right adjacent interval (l_{i+1}, r_{i+1}) (note that $r_i = l_{i+1}$) are removed from the heap-array, and a new interval (l_i, r_{i+1}) is inserted into the heap-array. If both the left and the right intervals are saturated, then interval (l_i, r_i) as well its left and right adjacent interval are removed from the heap-array, and a new interval (l_{i-1}, r_{i+1}) corresponding to a merging of all three intervals is inserted. Finally, if one or more edges get empty, then the newly inserted interval is removed from the heap-array and inserted into the bottom of the heap-array. The left and right neighbor pointers of its adjacent intervals, as well as its own left and right neighbor pointers are also suitably updated to reflect its new position. All this takes $O(\log n)$ time.

The pseudocode for Algorithm Efficient Large-cost Augmentation is given in the Appendix.

6.2 Running Time of Algorithm Efficient Large-Cost Augmentation

We now show that the running time of Algorithm Efficient Large-cost Augmentation is $O(n \log n)$.

Theorem 3. *Algorithm Efficient Large-cost Augmentation runs in time $O(n \log n)$.*

Proof. The initialization for Algorithm Efficient Large-cost Augmentation takes time at most $O(n \log n)$, determined by Heap-Build in Statement 8. Initializing capacity-tree for Algorithm Augment-Interval takes $O(n)$ time. Algorithm Efficient Large-cost Augmentation executes Statement 10 through Statement 33 repeatedly until a maximum cost augmenting path has cost 0 (until there is no augmenting path). From Theorem 2, it follows that there are at most $O(n)$ augmentations. We will now determine the time for each statement in this loop which runs $O(n)$ times. Statement 10 takes $O(1)$ time. Statement 10 calls Algorithm Augment-Interval which takes $O(\log n)$ time (Lemma 3). Statements 12 through 23 perform Heap-Extract which takes $O(\log n)$ time. Statements 24 through 28 perform Heap-Insert which takes $(O \log n)$ time. Thus, Algorithm Efficient Large-cost Augmentation runs in $O(n \log n)$ time.

7 Details of Algorithm Augment-Interval

In the next subsection we provide an informal description of Algorithm Augment-Interval. Recall that the input to Algorithm Augment-Interval is an interval, and its output is the vector (left_vertex, right_vertex, edge_empty, cost_zero).

7.1 Informal Description

We construct a balanced binary tree where the leaf nodes represent the edges. Any internal node u represents an interval $[v_i, v_j]$ and the edges it spans are either

$\{(v_i, v_{i+1}), (v_{i+1}, v_{i+2}), ...(v_{j-1}, v_j)\}$ or $\{(v_i, v_{i-1}), (v_{i-1}, v_{i-2}), ... (v_{j+1}, v_j)\}$.
Initially, every leaf edge e is assigned the capacity 0. This corresponds to the
initial solution of Algorithm Efficient Large-cost Augmentation. Every internal
node, such as u, stores a 3-tuple $\{m_{even}(u), m_{odd}(u), \text{offset}\}$. $m_{even}(u)$ is the
bottleneck when the leftmost edge in subtree T_u is an even edge, $m_{odd}(u)$ is the
bottleneck when the leftmost edge in subtree T_u is an odd edge , and the offset
keeps how much we add or subtract to the leftmost edge of T_u in each augmentation. Obviously each leaf has a value only for m_{odd} and offset (corresponding
to a path of length 1). Before we consider the effect of any augmenting path on
the capacities assigned to the edges, note that the initial information stored at
each internal node can be computed in $O(n)$ time.

Moreover, any query range $[v_k, v_{k'}]$ can be represented by the union of $O(\log n)$
sub-ranges where each sub-range is represented by a subtree of the balanced
binary tree built on the edges. The subranges are called canonical subranges of
$[v_k, v_{k'}]$ and the subtrees are call canonical subtrees. For any canonical subtree
T_u spanning $[v_i, v_j]$, the first leaf edge is an odd edge. Consider what happens
when Algorithm Augment-Interval is called with an augmenting path P from v_k
to $v_{k'}$. We need to find the minimum capacity assigned to the even edges of P.
We denote this value by $t_e(P)$. This can be found in $O(\log n)$ time as follows:
$m_{odd}(u)$ ($m_{even}(u)$) is with respect to the edges of T_u where the leftmost edge
is considered to be an odd (even) edge. However, with respect to P, the first
edge of T_u could be an even edge. We need to find for each canonical subtree
T_u whether the first edge is an odd edge or an even edge with respect to P.
After that the minimum of the capacities of all the even edges needs to be
computed.

We also need to find the amount by which the first edge and the last edge
may be augmented. These are denoted by $t_f(P)$ and $t_l(P)$ respectively. Let $t(P)$
be the minimum of $t_e(P), t_f(P), t_l(P)$. Our updating step must decrease the
capacity assigned to each of the even edges by $t(P)$ and increase the capacity
assigned to each of the odd edges by $t(P)$. We update the values stored at the
root node u_i (the root of T_i) for each subtree $T_i, i = 1, 2, ..., m$. We also update the values of the nodes along the path from the root node u_i of T_i for
each i, to the root node of the entire tree. The updating steps can be implemented in $O(\log n)$ time. After the updates, each internal node maintains the
correct minimum capacity information of the edges in its subtrees. For each
subtree T_i, if its leftmost edge is an even edge of the augmenting path P, we
add $-t(p)$ to the offset value stored at its root. But if its leftmost edge is an
odd edge of augmenting path P, we add $t(p)$ to the offset value stored at its
root.

Consider another augmenting path $P' = [v_a, v_b]$ and a canonical subtree $T_{u'}$
of $[v_a, v_b]$. Suppose that $T_{u'}$ is contained in T_{u_i}, a canonical subtree of $[v_k, v_{k'}]$.
In order to determine $t(P')$, we need to compute the minimum capacity of the
edges in $T_{u'}$ that takes into consideration the increase/decrease of the capacity
due to P. This is easily done by traversing the path from u' to the root of the
tree. When visiting u_i, which lies along the path, $t(P)$ is increased or decreased

depending on whether the minimum edge is an odd or an even edge of P'. This way we can compute $t(P')$ in $O(\log n)$ time also. After that we update the $O(\log n)$ canonical subtrees contained by the interval $[v_a, v_b]$ in a similar fashion as described before. The value $t(P)$ is either added to (or subtracted from) the current offset of u'. Therefore, we can handle each augmenting path in $O(\log n)$ time.

Lemma 3. *For each augmenting path, the algorithm Augment-Interval takes* $O(\log n)$ *time.*

Proof. Since any augmenting path $[v_i, v_j]$ can be represented by the union of $O(\log n)$ canonical subtrees, $t_e([v_i, v_j])$ can be obtained in $O(\log n)$ (taking the minimum of $O(\log n)$ values corresponding to each canonical subtree). To compute $t_f([v_i, v_j])$ and $t_l([v_i, v_j])$, we have to start from a leaf and traverse the tree to the root which is also $O(\log n)$. Then, the entire process takes $O(\log n)$.

The pseudocode for Algorithm Augment-Interval, as well as an example to illustrate its details are provided in the Appendix.

8   Conclusion

We provide an $O(n \log n)$ algorithm for the b-matching problem on a cycle. The solution to this problem may be used to solve the $(n - 2, n)$ cyclic scheduling problem in $O(n(\log n)(\log nb_{max})$ time. An interesting open problem would be to provide a strongly polynomial algorithm for the $(n - 3, n)$ cyclic scheduling problem, or more generally, the (k, n) cyclic scheduling problem efficiently. One way to accomplish this is to adapt Megiddos parametric search for the b-matching problem on a hypergraph.

Acknowledgments. The authors would like to thank the anonymous reviewers for their useful comments and suggestions.

References

1. Alfares, H.K.: An efficient two-phase algorithm for cyclic days-off scheduling. Computers and Operations Research 25, 913–923 (1998)
2. Alfares, H.K.: Flexible 4-day workweek scheduling with weekend work frequency constraints. Computers and Industrial Engineering 44, 325–338 (2003)
3. Bartholdi III, J.J., Ratliff, H.D.: Unnetworks with applications to idle time scheduling. Management Science 24(8), 850–858 (1978)
4. Bartholdi, J.J., Orlin, J.B., Ratliff, H.D.: Cyclic scheduling via integer programs with circular ones. Operations Research 28(5), 1074–1085 (1980)
5. Hochbaum, D.S., Levin, A.: Cyclical scheduling and multi-shift scheduling: complexity and approximation algorithms. Discrete Optimization 3, 327–340 (2006)

6. Hochbaum, D.S., Levin, A.: Optimizing over Consecutive 1's and Circular 1's Constraints. SIAM Journal on Optimization 17(2), 311–330 (2006)
7. Tibrewala, R., Phillippe, D., Browne, J.: Optimal scheduling of two consecutive idle periods. Management Science 19(1), 71–75 (1972)

A Appendix

A.1 Pseudocode and Example for Algorithm Efficient Large-Cost and Algorithm Augment-Interval

Example 1. We provide an example for Algorithm Augment-Interval when $n = 7$. We show how the values in the leaves and internal nodes change after each augmentation. The vector of capacities for vertices is $\{v_0, v_1, v_2, v_3, v_4, v_5, v_6\} = \{2, 5, 3, 9, 10, 14, 5\}$. Figure 1 shows the initial values of each node in the tree. The first augmenting path is $\{v_4, v_5\}$ and the the leaf corresponding to the edge e_4 is the only subtree of this augmenting path. Here, we only have to consider the values for $t_f(\{v_4, v_5\})$ and $t_l(\{v_4, v_5\})$ which are 10 and 14 respectively. Therefore, 10 is the bottleneck for this path. We also add 10 to the current offset value in the leaf corresponding to e_4 and increase the m_{odd} value by 10. All the nodes along the path from the leaf to the root node of the entire tree are also updated (Figure 2). $\{v_6, v_0\}$, $\{v_5, v_6\}$, and $\{v_1, v_2\}$ are the next 3 augmenting paths. Finding their bottlenecks and performing the updating is similar to $\{v_4, v_5\}$. Figures 3,4,5 show the tree after each augmentation. The next augmenting path which is depicted in Figure 6 is $\{v_1, v_0, v_6, v_5\}$. The subtrees for this path are T_1, T_2. The leftmost edge in subtree T_1, namely e_6, is an even edge with respect to this path, so $m_{even} = 2$ is selected in this subtree. Then we have $t_e = 2$, $t_f = 2$, and $t_l = 1$. Therefore the bottleneck is 1 for this path. We have to decrease 1 from m_{even} in both T_1 and T_2 (here only T_1 since there is no such value for leaves in the tree) and add 1 to m_{odd}. Since we decrease from the leftmost edge in subtree T_1, its offset value is decreased by one. The offset value in subtree T_2 is increased by·1 as we add to its leftmost edge . The next path is $\{v_1, v_0, v_6, v_5, v_4, v_3\}$. The bottleneck for this path is 1 (Figure 7).

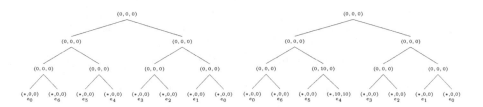

Fig. 1. initial state   **Fig. 2.** path $e_4 = \{v_4, v_5\}$

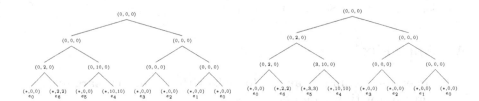

Fig. 3. path $e_6 = \{v_6, v_0\}$

Fig. 4. path $e_5 = \{v_5, v_6\}$

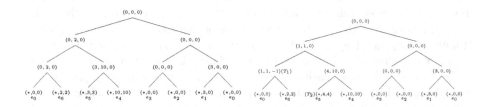

Fig. 5. path $e_1 = \{v_1, v_2\}$

Fig. 6. path $\{e_0, e_6, e_5\} = \{v_1, v_0, v_6, v_5\}$

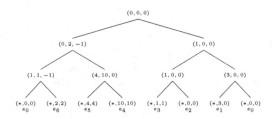

Fig. 7. path $\{e_0, e_6, e_5, e_4, e_3\} = \{v_1, v_0, v_6, v_5, v_4, v_3\}$

Algorithm 3. Algorithm Efficient Large-cost Augmentation

1: Initialize $x_i = 0$ for $i = 0, 1, \ldots, n - 1$ {the solution is initialized}
2: **for** $i = 0$ to n **do**
3:     $heap - array[i].int = (v_{i-1}, v_i)$
4:     $heap - array[i].cost = c_i$
5:     $heap - array[i].left = (v_{i-2}, v_{i-1})$
6:     $heap - array[i].right = (v_i, v_{i+1})$
7: **end for**
    {heap-array is initialized}
8: Heap-Build(heap-array)
    {heap-array is sorted in decreasing order of cost of intervals. when an interval is moved during
    the sorting, the right and left pointers of its left and right intervals are changed to reflect its
    new position.}
9: Initialize capacity-tree {capacity-tree is a balanced binary tree used by Algorithm Augment-
    Interval}
10: max_cost_int = max(heap-array)
    {the maximum cost interval is returned and stored in max_cost_int}
11: (left_vertex,right_vertex,edge_empty,cost_zero) = Augment-Interval(max_cost_int)
    {Algorithm Augment-Interval augments the path by the maximum extent possible Algorithm
    Augment-Interval also modifies the solution and returns whether the left end vertex and or the
    right end vertex get saturated and or one or more minus edges get empty}
12: **if** $l = true$ **then**
13:     left_int = Heap-Extract(max_cost_int(left))
14:     int = merge(left_int, max_cost_int)
    {the left interval of max_cost_int is removed from heap-array and merged with interval
    max_cost_int}
15:     cost = left_int.cost + max_cost_int.cost
    {cost of new interval is computed}
16: **else**
17:     int = max_cost_int;
    {if l = false, int is max_cost_int}
18: **end if**
19: **if** $r = true$ **then**
20:     right_int = Heap-Extract(max_cost_int(right))
21:     int = merge(right_int, int)
    {the right interval of max_cost_int is removed from heap-array and merged with interval
    max_cost_int}
22:     cost = right_int.cost + int.cost
    {cost of new interval is computed}
23: **end if**
24: **if** $e = true$ **then**
25:     interval int is placed at the bottom of heap-array
    {interval int is inserted into the bottom of heap-array}
26: **else**
27:     Heap-Insert(int)
    {or in sorted order in both cases, the left and right pointers of its left and right intervals are
    adjusted to reflect its current position}
28: **end if**
29: **if** $cost\_zero = true$ **then**
30:     stop
31: **else**
32:     go to step 9
33: **end if**
    {If Algorithm Augment-Interval returns cost of interval is zero then stop else continue }

Algorithm 4. Algorithm Augment-Interval

1: **Input:** A path $[v_i, v_j]$
2: $left\_vertex, right\_vertex, edge\_empty = false$
3: Find all the canonical subtrees $T_1, T_2, ..., T_m$ of path $[v_i, v_j]$
4: **for** $i \leftarrow 1$ to m **do**
5:     check if the leftmost edge in T_i is located in even or odd position with respect to path $[v_i, v_j]$
6:     **if** the leftmost edge in T_i is located in even position **then**
7:         $minimum\_even\_edge[i] \leftarrow m_{even}(i)$
8:     **else**
9:         $minimum\_even\_edge[i] \leftarrow m_{odd}(i)$
10:    **end if**
11: **end for**
12: $t_e([v_i, v_j]) \leftarrow min(minimum\_even\_edge)$
13: $t_f([v_i, v_j]) \leftarrow$ maximum amount that can be added to the leftmost edge
      { We start from the leaf nodes corresponding to (v_{i-1}, v_i) and (v_i, v_{i+1}), traverse the tree to the root and update the amounts according to the offsets stored on the nodes along the paths}
14: $t_l([v_i, v_j]) \leftarrow$ maximum amount that can be added to the rightmost edge
      {similar to t_f procedure}
15: $bottleneck \leftarrow min(t_e([v_i, v_j]), t_f([v_i, v_j]), t_l([v_i, v_j]))$
16: Update the offset value for each $T_1, T_2, ..., T_m$
      {if the leftmost edge in T_u is in even position according to the path (v_i, v_j), subtract the bottleneck from the offset in T_u, otherwise add the bottleneck to the offset}
17: Update the values $\{m_{even}(u), m_{odd}(u)\}$ for each $T_1, T_2, ..., T_m$ and also update these values in each node along the path from root node u_i of T_i to the root of the entire tree for each subtree
      {if the leftmost edge in subtree T_u is in even position according to path (v_i, v_j), subtract the bottleneck from $m_{even}(u)$ and add the bottleneck to $m_{odd}(u)$, but if the leftmost edge in subtree T_u is in odd position , subtract the bottleneck from $m_{odd}(u)$ and add the bottleneck to $m_{even}(u)$ }
18: **if** $bottleneck = t_e([v_i, v_j])$ **then**
19:    $edge\_empty = true$
20: **end if**
21: **if** $bottleneck = t_f([v_i, v_j])$ **then**
22:    $left\_vertex = true$
23: **end if**
24: **if** $bottleneck = t_l([v_i, v_j])$ **then**
25:    $right\_vertex = true$
26: **end if**
27: **Output:** $\{left\_vertex, right\_vertex, edge\_empty, bottleneck\}$

Generalized Rainbow Connectivity of Graphs[*]

Kei Uchizawa[1], Takanori Aoki[2], Takehiro Ito[2], and Xiao Zhou[2]

[1] Graduate School of Science and Engineering, Yamagata University,
Jonan 4-3-16, Yonezawa-shi, Yamagata 992-8510, Japan
uchizawa@yz.yamagata-u.ac.jp
[2] Graduate School of Information Sciences, Tohoku University,
Aoba-yama 6-6-05, Sendai, Miyagi 980-8579, Japan
{takanori,takehiro,zhou}@ecei.tohoku.ac.jp

Abstract. Let $C = \{c_1, c_2, \ldots, c_k\}$ be a set of k colors, and let $\boldsymbol{\ell} = (\ell_1, \ell_2, \ldots, \ell_k)$ be a k-tuple of nonnegative integers $\ell_1, \ell_2, \ldots, \ell_k$. For a graph $G = (V, E)$, let $f : E \to C$ be an edge-coloring of G in which two adjacent edges may have the same color. Then, the graph G edge-colored by f is $\boldsymbol{\ell}$-rainbow connected if every two vertices of G have a path P such that the number of edges in P that are colored with c_j is at most ℓ_j for each index $j \in \{1, 2, \ldots, k\}$. Given a k-tuple $\boldsymbol{\ell}$ and an edge-colored graph, we study the problem of determining whether the edge-colored graph is $\boldsymbol{\ell}$-rainbow connected. In this paper, we characterize the computational complexity of the problem with regards to certain graph classes: the problem is NP-complete even for cacti, while is solvable in polynomial time for trees. We then give an FPT algorithm for general graphs when parameterized by both k and $\ell_{\max} = \max\{\ell_j \mid 1 \leq j \leq k\}$.

1 Introduction

Graph connectivity is one of the most fundamental graph-theoretic properties. In the literature, several measures for graph connectivity have been studied, such as requiring hamiltonicity, edge-disjoint spanning trees, or edge- or vertex-cuts of sufficiently large size. Recently, there has been some interest in studying problems on colored graphs, due to their applications in areas such as computational biology, transportation and telecommunications [9]. In this paper, we generalize an interesting concept of graph connectivity on colored graphs, called the *rainbow connectivity*, which was introduced by Chartrand *et al.* [6] and has been extensively studied in the literature [2, 4–8, 11, 12].

Let $G = (V, E)$ be a graph with vertex set V and edge set E; we often denote by $V(G)$ the vertex set of G and by $E(G)$ the edge set of G. Let $C = \{c_1, c_2, \ldots, c_k\}$ be a set of k colors, and let $\boldsymbol{\ell} = (\ell_1, \ell_2, \ldots, \ell_k)$ be a k-tuple of nonnegative integers $\ell_1, \ell_2, \ldots, \ell_k$. Consider a mapping $f : E \to C$, called an *edge-coloring* of G. Note that f is not necessarily a proper edge-coloring, that is, f may assign a same color to two adjacent edges. We denote by $G(f)$ the graph G

---

[*] This work is partially supported by JSPS KAKENHI Grant Numbers 22700001, 23500001 and 23700003.

S.K. Ghosh and T. Tokuyama (Eds.): WALCOM 2013, LNCS 7748, pp. 233–244, 2013.
© Springer-Verlag Berlin Heidelberg 2013

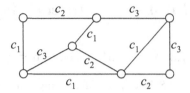

Fig. 1. An ℓ-rainbow connected graph, where $\ell = (1, 3, 2)$

edge-colored by f. Then, a path P in $G(f)$ connecting two vertices u and v in V is called an ℓ-*rainbow path* between u and v if the number of edges in P that are colored with c_j is at most ℓ_j for every index $j \in \{1, 2, \ldots, k\}$. The edge-colored graph $G(f)$ is ℓ-*rainbow connected* if $G(f)$ has an ℓ-rainbow path between every two vertices in V. Note that these ℓ-rainbow paths are not necessarily edge-disjoint for pairs of vertices. For example, the edge-colored graph $G(f)$ in Fig. 1 is ℓ-rainbow connected for $\ell = (1, 3, 2)$.

The concept of ℓ-rainbow connectivity was originally introduced by Chartrand *et al.* [6] for the special case where $\ell = (1, 1, \ldots, 1)$. Chakraborty *et al.* [4] defined the RAINBOW CONNECTIVITY problem which asks whether a given edge-colored graph is $(1, 1, \ldots, 1)$-rainbow connected, and showed that the problem is NP-complete in general. Then, Uchizawa *et al.* [12] characterized the computational complexity of the problem with regards to certain graph classes, and also settled it with regards to graph diameters. (Remember that the *diameter* of a graph G is the maximum number of edges in a shortest path between any two vertices in G.)

In this paper, we introduce and study the generalized problem, defined as follows: Given a k-tuple ℓ and an edge-coloring f of a graph G, the GENERALIZED RAINBOW CONNECTIVITY problem is to determine whether $G(f)$ is ℓ-rainbow connected. Thus, (ordinary) RAINBOW CONNECTIVITY is a specialization of GENERALIZED RAINBOW CONNECTIVITY. We first give precise complexity analyses for GENERALIZED RAINBOW CONNECTIVITY with regards to certain graph classes. We then give an FPT algorithm for the problem on general graphs when parameterized by both $k = |C|$ and $\ell_{\max} = \max\{\ell_j \mid 1 \leq j \leq k\}$. Below we explain our results more precisely, together with comparisons with known results [12].

[Graph classes]
From the viewpoint of graph classes, we clarify a boundary on graph classes between tractability and NP-completeness: GENERALIZED RAINBOW CONNECTIVITY is NP-complete even for cacti, while there is a polynomial-time algorithm for trees. Note that trees and cacti are very close to each other in the following sense: trees form a graph class which is a subclass of cacti, and the treewidth of cacti is two [3]. It is remarkable that the boundary is different from the known one for RAINBOW CONNECTIVITY [12]: it is NP-complete for outerplanar graphs, and is solvable in polynomial time for cacti. Therefore, the NP-complete proof given by [12] does not imply our result. We also remark that our polynomial-time algorithm for trees is always faster than a naive one, as discussed in Section 3.1.

[FPT algorithm]

In Section 3.2, we give an algorithm which solves GENERALIZED RAINBOW CON-NECTIVITY for general graphs in time $O(k2^{k\ell_{\max}}mn)$ using $O\big(kn2^{k\ell_{\max}}\log(\ell_{\max}+1)\big)$ space, where n and m are the numbers of vertices and edges in a graph, respectively. Therefore, the problem can be solved in polynomial time for the following two cases: (a) $k = O(\log n)$ and $\ell_{\max}$ is a fixed constant; and (b) k is a fixed constant and $\ell_{\max} = O(\log n)$. We remark that our FPT algorithm generalizes the known one [12]: the same running time and space complexity of the known FPT algorithm for RAINBOW CONNECTIVITY [12] can be obtained from our result as the special case where $\ell_{\max} = 1$.

2   NP-Completeness for Cacti

A graph G is a *cactus* if every edge is part of at most one cycle in G [3]. (See Fig. 2 as an example of cacti.) The main result of this section is the following theorem.

Theorem 1. GENERALIZED RAINBOW CONNECTIVITY *is* NP-*complete even for cacti and* $\boldsymbol{\ell} = (2, 2, \ldots, 2)$.

Let $G(f)$ be a given edge-colored graph. We can clearly cheek in polynomial time whether a given path in $G(f)$ is an $\boldsymbol{\ell}$-rainbow path, and hence GENERALIZED RAINBOW CONNECTIVITY belongs to NP. We below show that the problem is NP-hard even for cacti and $\boldsymbol{\ell} = (2, 2, \ldots, 2)$ by a reduction from the 3-OCCURRENCE 3SAT problem defined as follows: Given a 3CNF formula ϕ such that each variable appears at most three times in ϕ, determine whether ϕ is satisfiable. 3-OCCURRENCE 3SAT is known to be NP-complete [10].

Suppose that the formula ϕ consists of n variables $x_1, x_2, \ldots, x_n$ and m clauses $C_1, C_2, \ldots, C_m$. Without loss of generality, we can assume that any literal of a variable x_i, $1 \le i \le n$, appears at most twice in ϕ; otherwise, ϕ contains only positive (or negative) literals of x_i, and hence we can safely fix its assignment. In what follows, we construct a cactus G_ϕ, an edge-coloring f_ϕ of G_ϕ and a k-tuple $\boldsymbol{\ell} = (2, 2, \ldots, 2)$, as a corresponding instance. We then prove that ϕ is satisfiable if and only if the edge-colored graph $G_\phi(f_\phi)$ is $\boldsymbol{\ell}$-rainbow connected.

[Graph G_ϕ]

We first construct a gadget G_j for each clause C_j, $1 \le j \le m$, as follows: G_j is a cycle consisting of four vertices p_j, u_j, p'_j, u'_j embedded in clockwise order

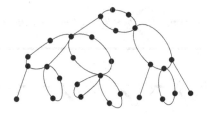

Fig. 2. Cactus G

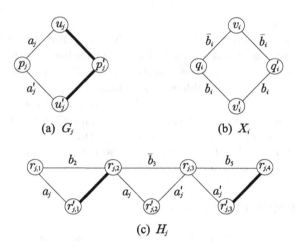

Fig. 3. (a) Gadget G_j for a clause C_j, (b) gadget X_i for a variable x_i, and (c) gadget H_j for the clause $C_j = (x_2 \vee \overline{x_3} \vee x_5)$

on the plane. (See Fig. 3(a).) We then construct a gadget X_i for each variable x_i, $1 \le i \le n$, as follows: X_i is a cycle consisting of four vertices q_i, v_i, q'_i, v'_i embedded in clockwise order on the plane. (See Fig. 3(b).) We lastly construct a gadget H_j for each clause C_j, $1 \le j \le m$, as follows: (i) make a path of four vertices $r_{j,1}$, $r_{j,2}$, $r_{j,3}$, $r_{j,4}$; (ii) add three vertices $r'_{j,1}$, $r'_{j,2}$, $r'_{j,3}$; and (iii) connect $r'_{j,1}$ to both $r_{j,1}$ and $r_{j,2}$, connect $r'_{j,2}$ to both $r_{j,2}$ and $r_{j,3}$, and connect $r'_{j,3}$ to both $r_{j,3}$ and $r_{j,4}$. (See Fig. 3(c).)

Using $G_1, G_2, \ldots, G_m$, $X_1, X_2, \ldots, X_n$ and $H_1, H_2, \ldots, H_m$ given above, we construct the corresponding graph G_ϕ as follows. (See Fig. 4.) We connect p'_j to p_{j+1} for every $j \in \{1, 2, \ldots, m-1\}$, and connect p'_m to q_1. We then connect q'_i to q_{i+1} for every $i \in \{1, 2, \ldots, n-1\}$, and connect q'_n to $r_{1,1}$. We complete the construction of G_ϕ by connecting $r_{j,4}$ to $r_{j+1,1}$ for every $j \in \{1, 2, \ldots, m-1\}$. Since each gadget consists of either a single cycle or consecutive three cycles, G_ϕ is clearly a cactus.

Before constructing the edge-coloring f_ϕ of G_ϕ, we define some terms. For each gadget G_j, $1 \le j \le m$, we call the path $p_j u_j p'_j$ the j-th upper path, and

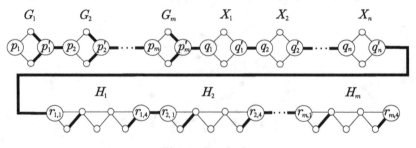

Fig. 4. Graph G_ϕ

call the path $p_j u'_j p'_j$ the j-th *lower path*. For each gadget X_i, $1 \le i \le n$, we call the path $q_i v_i q'_i$ the i-th *positive path*, and call the path $q_i v'_i q'_i$ the i-th *negative path*; the i-th positive path corresponds to $x_i = 1$, and the i-th negative path corresponds to $x_i = 0$. Let $\phi = \wedge^m_{j=1}(l_{j,1} \vee l_{j,2} \vee l_{j,3})$ be the given formula, where $l_{j,1}$, $l_{j,2}$ and $l_{j,3}$ are literals of $x_1, x_2, \ldots, x_n$ contained in the clause C_j.

[Edge-coloring f_ϕ of G_ϕ]

We construct f_ϕ as in the following four steps (i)–(iv).

   (i) Let $a_1, a_2, \ldots, a_m$ and $a'_1, a'_2, \ldots, a'_m$ be $2m$ distinct colors. For each $j \in \{1, 2, \ldots, m\}$, we assign a_j to (p_j, u_j), and a'_j to (p_j, u'_j). (See Fig. 3(a).)

  (ii) Let $b_1, b_2, \ldots, b_n$ and $\bar{b}_1, \bar{b}_2, \ldots, \bar{b}_n$ be $2n$ distinct (new) colors. For each $i \in \{1, 2, \ldots, n\}$, we assign $\bar{b}_i$ to both (q_i, v_i) and (v_i, q'_i), and b_i to both (q_i, v'_i) and (v'_i, q'_i). (See Fig. 3(b).)

 (iii) For each clause $C_j = l_{j,1} \vee l_{j,2} \vee l_{j,3}$, $1 \le j \le m$, we assign some of $b_1, b_2, \ldots, b_n$ and $\bar{b}_1, \bar{b}_2, \ldots, \bar{b}_n$ to the edges $(r_{j,1}, r_{j,2}), (r_{j,2}, r_{j,3})$ and $(r_{j,3}, r_{j,4})$ in the gadget H_j, as follows: For each $k \in \{1, 2, 3\}$, we assign b_i to $(r_{j,k}, r_{j,k+1})$ if the k-th literal $l_{j,k}$ is a positive literal of x_i; and assign $\bar{b}_i$ to $(r_{j,k}, r_{j,k+1})$ if the k-th literal $l_{j,k}$ is a negative literal of x_i. Moreover, we assign a_j to both $(r_{j,1}, r'_{j,1})$ and $(r_{j,2}, r'_{j,2})$, and assign a'_j to both $(r'_{j,2}, r_{j,3})$ and $(r_{j,3}, r'_{j,3})$. (See Fig. 3(c).)

  (iv) Let U be the set of the edges that are not assigned colors in (i)–(iii) above. We assign a new distinct color for each edge in U. (See Figs. 3 and 4, where the edges in U are depicted by thick lines.)

Remember that any literal of a variable x_i, $1 \le i \le n$, appears at most twice in ϕ. Therefore, in the step (iii), each of the colors $b_1, b_2, \ldots, b_n, \bar{b}_1, \bar{b}_2, \ldots, \bar{b}_n$ is assigned to at most two edges in $H_1, H_2, \ldots, H_m$.

We finally set $\ell = (2, 2, \ldots, 2)$, and complete the construction of the corresponding instance.

The following two lemmas for $G_\phi(f_\phi)$ clearly imply Theorem 1.

Lemma 1. $G_\phi(f_\phi)$ *is ℓ-rainbow connected if and only if $G_\phi(f_\phi)$ has an ℓ-rainbow path between p_1 and $r_{m,4}$.*

Lemma 2. $G_\phi(f_\phi)$ *has an ℓ-rainbow path between p_1 and $r_{m,4}$ if and only if ϕ is satisfiable.*

In the rest of the section, we prove Lemmas 1 and 2.

[Proof of Lemma 1]

It is trivially true that, if $G_\phi(f_\phi)$ is ℓ-rainbow connected, then $G_\phi(f_\phi)$ has an ℓ-rainbow path between p_1 and $r_{m,4}$. Below we prove that $G_\phi(f_\phi)$ is ℓ-rainbow connected if $G_\phi(f_\phi)$ has an ℓ-rainbow path between p_1 and $r_{m,4}$.

Let s and t be an arbitrary pair of vertices in G_ϕ. We consider a partition of the vertex set $V(G_\phi)$ into the following three groups: $V^1 = \bigcup^m_{j=1} V(G_j)$, $V^2 = \bigcup^n_{i=1} V(X_i)$ and $V^3 = \bigcup^m_{j=1} V(H_j)$. In any subgraph induced by only one of V^1, V^2 and V^3, every color is assigned to at most two edges in the subgraph. Similarly, in the subgraph induced by V^1 and V^2, every color is assigned to at most two edges; in the subgraph induced by V^2 and V^3, when we remove the edges $(r_{j,1}, r_{j,2}), (r_{j,2}, r_{j,3}), (r_{j,3}, r_{j,4})$ for every $j \in \{1, 2, \ldots, m\}$, every color is assigned to at most two edges. Thus, it suffices to verify the case where $s \in V^1$ and $t \in V^3$. Let P be the ℓ-rainbow path between p_1 and $r_{m,4}$ in $G_\phi(f_\phi)$, and let j_1 and j_2 be the indices satisfying $s \in V(G_{j_1})$ and $t \in V(H_{j_2})$. Then, we can easily construct an ℓ-rainbow path P' between s and t, as follows: P' consists of a subpath of P between p_{j_1} and $r_{j_2,4}$ together with some of the five edges $(u_{j_1}, p'_{j_1}), (u'_{j_1}, p'_{j_1}), (r'_{j_2,1}, r_{j_2,2}), (r'_{j_2,2}, r_{j_2,3}), (r'_{j_2,3}, r_{j_2,4})$. $\square$

[Proof of Lemma 2]

Necessity. We prove that, if $G_\phi(f_\phi)$ has an ℓ-rainbow path between p_1 and $r_{m,4}$, then ϕ is satisfiable. Let P be an ℓ-rainbow path in $G_\phi(f_\phi)$ between p_1 and $r_{m,4}$. For each gadget G_j, $1 \le j \le m$, we denote by $P \cap G_j$ the graph (path) induced by $E(P) \cap E(G_j)$. Then, each subpath $P \cap G_j$, $1 \le j \le m$, is either j-th upper path or j-th lower path. Similarly, for each gadget X_i, $1 \le i \le n$, we denote by $P \cap X_i$ the graph (path) induced by $E(P) \cap E(X_i)$. Then, each subpath $P \cap X_i$, $1 \le i \le n$, is either i-th positive path or i-th negative path. Consider the following truth assignment $z = (z_1, z_2, \ldots, z_n) \in \{0, 1\}^n$: for each index $i \in \{1, 2, \ldots, n\}$,

$$z_i = \begin{cases} 1 & \text{if } P \cap X_i \text{ is the } i\text{-th positive path;} \\ 0 & \text{if } P \cap X_i \text{ is the } i\text{-th negative path.} \end{cases} \tag{1}$$

We now show that z is a satisfying truth assignment for ϕ. Clearly, any ℓ-rainbow path must go through either (p_j, u_j) with the color a_j or (p_j, u'_j) with the color a'_j for each $j \in \{1, 2, \ldots, m\}$. Then, since $\ell = (2, 2, \ldots, 2)$, P must pass through at least one of the edges $(r_{j,1}, r_{j,2})$, $(r_{j,2}, r_{j,3})$ and $(r_{j,3}, r_{j,4})$ in each clause gadget H_j, $1 \le j \le m$. Let $(r_{j,k}, r_{j,k+1})$ be such an edge. We show that the literal $l_{j,k}$ corresponding to the edge $(r_{j,k}, r_{j,k+1})$ is true by z, and hence the clause C_j is satisfied; consequently, z is satisfying, as required. Consider the case where the edge $(r_{j,k}, r_{j,k+1})$ receives the color $\bar{b}_i$ for some i, $1 \le i \le n$. (The proof is similar for the other case where $(r_{j,k}, r_{j,k+1})$ receives the color b_i.) Then, by the construction of f_ϕ, the literal $l_{j,k}$ corresponding to $(r_{j,k}, r_{j,k+1})$ is a negative literal of the variable x_i. Since the color $\bar{b}_i$ is assigned to each of the two edges in the i-th positive path in X_i, P must go through the i-th negative path in X_i. Consequently, by Eq. (1), we have $z_i = 0$ in z, and hence the literal $l_{j,k}$ is true by z.

Sufficiency. We prove that $G_\phi(f_\phi)$ has an ℓ-rainbow path between p_1 and $r_{m,4}$ if ϕ is satisfiable. Let $z = (z_1, z_2, \ldots, z_n)$ be a satisfying truth assignment for ϕ. For each $j \in \{1, 2, \ldots, m\}$, we denote by l_{j,k_j}, $1 \le k_j \le 3$, a literal satisfied by z in C_j.

Consider the following path P^X from q_1 to q'_n: For each gadget X_i, $1 \le i \le n$, take the i-th positive path if $z_i = 1$, and otherwise take the i-th negative path. Clearly, for each $i \in \{1, 2, \ldots, n\}$, both of the edges in $P^X \cap E(X_i)$ receive $\bar{b}_i$ if $z_i = 1$, and receive b_i if $z_i = 0$. Consider then the following path P^H_j from $r_{j,1}$ to $r_{j,4}$ for each $j \in \{1, 2, \ldots, m\}$: make a path consisting of (r_{j,k_j}, r_{j,k_j+1}) together with the four edges $(r_{j,\alpha}, r'_{j,\alpha}), (r'_{j,\alpha}, r_{j,\alpha+1}), (r_{j,\beta}, r'_{j,\beta})$ and $(r'_{j,\beta}, r_{j,\beta+1})$, where $\alpha, \beta \in \{1, 2, 3\} \backslash \{k_j\}$ and $\alpha < \beta$. We obtain the path P from q_1 to $r_{m,4}$ by connecting P^X and $P_1^H, P_2^H, \ldots, P_m^H$ in this order. Clearly, every color is assigned to at most two edges in P. Moreover, for each $j \in \{1, 2, \ldots, m\}$, one of a_j and a'_j is assigned to only one edge in P; such a color is said to be *available*. Then, we can obtain a path P^G from p_1 to p'_m such that, for each $j \in \{1, 2, \ldots, m\}$, it takes the j-th upper path if a_j is available, and otherwise takes the j-th lower path. Finally, we can obtain an ℓ-rainbow path between p_1 and $r_{m,4}$ by connecting the paths P^G and P.   □

3   Algorithms

As we have shown in Theorem 1, GENERALIZED RAINBOW CONNECTIVITY is NP-complete even for cacti and hence it cannot be solved in polynomial time unless P = NP. However, we give two algorithms in this section: in Section 3.1, we give an efficient polynomial-time algorithm for trees; in Section 3.2, we give an FPT algorithm for general graphs when parameterized by both k and ℓ_{max}.

3.1   Polynomial-Time Algorithm for Trees

The main result of this subsection is the following theorem.

Theorem 2. GENERALIZED RAINBOW CONNECTIVITY *can be solved for a tree* T *in time* $O(kn)$, *where* $k = |C|$ *and* n *is the number of vertices in* T.

In the remainder of this subsection, we give an $O(kn)$-time algorithm as a proof of Theorem 2. It is obvious that the problem is in P for trees, because a naive $O(n^3)$-time algorithm exists: for each pair of vertices in a tree, it determines whether the unique path between the pair is an ℓ-rainbow path. We remark that our $O(kn)$-time algorithm is always faster than the naive one; our algorithm runs in linear time if k is a constant, and in time $O(n^2)$ even if $k = O(n)$; notice that $k \le n - 1$.

[**Terms and ideas**]

Let $T = (V, E)$ be a given tree. One may assume without loss of generality that T is a rooted tree with an arbitrarily chosen root r. Let u be a vertex of T, and we denote by $d(u)$ the number of children of u. For each $i \in \{1, 2, \ldots, d(u)\}$, let u_i be a child of u ordered arbitrarily, and let e_i be the edge joining u and u_i, as illustrated in Fig. 5. Let T_{u_i} be the subtree of T which is rooted at u_i and is induced by all descendants of u_i in T. We denote by T_u^i the subtree of T which consists of the vertex u, the edges $e_1, e_2, \ldots, e_i$ and the subtrees $T_{u_1}, T_{u_2}, \ldots, T_{u_i}$. In Fig. 5, T_u^i is indicated by a dotted closed curve. Clearly $T_u = T_u^{d(u)}$. For the sake of notational convenience, we denote by T_u^0 the tree consisting of a single vertex u.

Let $C = \{c_1, c_2, \ldots, c_k\}$ be the color set, and let $f : E \to C$ be a given edge-coloring of T. Note that any path P in T must be an ℓ-rainbow path; otherwise

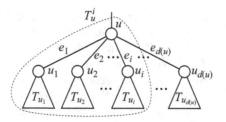

Fig. 5. Tree T_u

there is no ℓ-rainbow path between the end-vertices of P. For a pair of vertices $v, w \in V(T)$ and an index $j \in \{1, 2, \ldots, k\}$, we denote by $t_j(v, w)$ the number of edges in the (unique) path between v and w that are colored with c_j by f.

Consider the subtree T_u^i for a vertex u of T and an integer $i \in \{0, 1, \ldots, d(u)\}$. We classify the paths in T_u^i into two subclasses, and check whether every path in T_u^i is an ℓ-rainbow path. For an index $j \in \{1, 2, \ldots, k\}$, we define $a_j(T_u^i)$ as follows:

$$a_j(T_u^i) = \max\{t_j(u, w) \mid w \in V(T_u^i)\}.$$

Therefore, by the values $a_j(T_u^i)$, $1 \le j \le k$, we can check whether all paths between the root u of T_u^i and vertices of T_u^i are ℓ-rainbow paths; more specifically, such paths are all ℓ-rainbow paths if $a_j(T_u^i) \le \ell_j$ for all indices $j \in \{1, 2, \ldots, k\}$. Similarly, for an index $j \in \{1, 2, \ldots, k\}$, we define $b_j(T_u^i)$ as follows:

$$b_j(T_u^i) = \max\{t_j(v, w) \mid v, w \in V(T_u^i)\}.$$

Then, by the values $b_j(T_u^i)$, $1 \le j \le k$, we can check whether all paths that do not necessarily contain u are ℓ-rainbow paths; indeed, T_u^i is ℓ-rainbow connected if and only if $b_j(T_u^i) \le \ell_j$ for all indices $j \in \{1, 2, \ldots, k\}$.

Our algorithm computes $a_j(T_u^i)$ and $b_j(T_u^i)$ for each vertex u of T and all indices $i \in \{0, 1, \ldots, d(u)\}$ and $j \in \{1, 2, \ldots, k\}$ from the leaves to the root r of T by means of dynamic programming. Then, the edge-colored tree $T(f) = T_r(f)$ is ℓ-rainbow connected if and only if $b_j(T_r) \le \ell_j$ for all indices $j \in \{1, 2, \ldots, k\}$.

[Algorithm]

We first compute $a_j(T_u^0)$ and $b_j(T_u^0)$ for each vertex u of T and all indices $j \in \{1, 2, \ldots, k\}$. Since T_u^0 consists of a single vertex u, there is no edge in T_u^0. Therefore, we clearly have $a_j(T_u^0) = 0$ and $b_j(T_u^0) = 0$.

We then compute $a_j(T_u^i)$ and $b_j(T_u^i)$, $1 \le i \le d(u)$, for each internal vertex u of T from the counterparts of T_u^{i-1} and T_{u_i}. (See Fig. 5.) Remember that $T_u = T_u^{d(u)}$, and that T_u^i is obtained from T_u^{i-1} and T_{u_i} by joining u and u_i. For an edge e in T, let

$$h_j(e) = \begin{cases} 1 & \text{if } f(e) = c_j; \\ 0 & \text{otherwise.} \end{cases}$$

We first compute $a_j(T_u^i)$, that is, check whether all paths between the root u of T_u^i and vertices of T_u^i are ℓ-rainbow paths. Consider an arbitrary path P between u and some vertex v of T_u^i. Then, there are the following two cases:

(i) v is in T_u^{i-1}, and hence P is a path in T_u^{i-1}; and

(ii) v is in T_{u_i}, and hence P consists of e_i and the path between u_i and v.

Therefore, the value $a_j(T_u^i)$ can be computed as follows:

$$a_j(T_u^i) = \max\{a_j(T_u^{i-1}),\ a_j(T_{u_i}) + h_j(e_i)\}. \tag{2}$$

We then compute $b_j(T_u^i)$, that is, check whether all paths that do not necessarily contain u are ℓ-rainbow paths. Consider an arbitrary path P between two vertices v and w in T_u^i. Then, there are the following three cases:

(i) both v and w are in T_u^{i-1}, and hence P is a path in T_u^{i-1};

(ii) both v and w are in T_{u_i}, and hence P is a path in T_{u_i}; and

(iii) one of v and w is in T_u^{i-1} and the other is in T_{u_i}, and hence P is a path starting from v, passing through u and e_i, and ending with w.

Therefore, the value $b_j(T_u^i)$ can be computed as follows:

$$b_j(T_u^i) = \max\{b_j(T_u^{i-1}),\ b_j(T_{u_i}),\ a_j(T_u^{i-1}) + h_j(e_i) + a_j(T_{u_i})\}. \tag{3}$$

[Proof of Theorem 2]

We now show that our algorithm above runs in time $O(kn)$.

For each vertex u of T and all indices $j \in \{1, 2, \ldots, k\}$, we can compute $a_j(T_u^0)$ and $b_j(T_u^0)$ in time $O(k)$. Therefore, the initialization can be done in time $O(kn)$ for all vertices in T and all indices $j \in \{1, 2, \ldots, k\}$.

For an internal vertex u of T and all indices $j \in \{1, 2, \ldots, k\}$, as described in Eqs. (2) and (3), we can compute $a_j(T_u^i)$ and $b_j(T_u^i)$, $i \geq 1$, from the counterparts of T_u^{i-1} and T_{u_i} in time $O(k)$. Since there are $n-1$ edges in T, the computation occurs $n-1$ times for each of $a_j(T_u^i)$ and $b_j(T_u^i)$. Therefore, for the root r of T, we can compute the values $a_j(T_r)$ and $b_j(T_r)$ for all indices $j \in \{1, 2, \ldots, k\}$ in time $O(kn)$.

Then, from the values $b_j(T_r)$, $1 \leq j \leq k$, we can determine in time $O(k)$ whether the edge-colored tree $T(f)$ is ℓ-rainbow connected.

In this way, our algorithm solves GENERALIZED RAINBOW CONNECTIVITY for a tree in time $O(kn)$ in total. This completes the proof of Theorem 2.

3.2 FPT Algorithm for General Graphs

In this subsection, we give an FPT algorithm for GENERALIZED RAINBOW CONNECTIVITY when parameterized by both k and $\ell_{\max}$.

Theorem 3. *For a k-tuple ℓ and an edge-coloring f of a graph G, one can determine whether the edge-colored graph $G(f)$ is ℓ-rainbow connected in time $O(k 2^{k\ell_{\max}} mn)$ using $O(kn 2^{k\ell_{\max}} \log(\ell_{\max} + 1))$ space, where n and m are the numbers of vertices and edges in G, respectively.*

As a proof of Theorem 3, we give an algorithm to determine whether $G(f)$ has ℓ-rainbow paths from a vertex s to all the other vertices. It suffices to design such an algorithm which runs in time $O(k 2^{k\ell_{\max}} m)$ using $O(kn 2^{k\ell_{\max}} \log(\ell_{\max} + 1))$ space. Then, Theorem 3 clearly holds.

[Terms and ideas]

We first introduce some terms. For a vertex v of a graph $G = (V, E)$, we denote by $N(v)$ the set of all neighbors of v (which does not include v itself), that is, $N(v) = \{w \in V \mid (v, w) \in E\}$. We remind the reader that a *walk* in a graph is a sequence of adjacent vertices and edges, each of which may appear more than once; while a *path* is a walk in which each vertex appears exactly once. The *length* of a walk is defined as the number of edges in the walk. A walk W in $G(f)$ is called an ℓ-*rainbow walk* if the number of edges in W that are colored with c_j is at most ℓ_j for every index $j \in \{1, 2, \ldots, k\}$.

We extend ideas in [1, 12]. For a graph $G = (V, E)$ and a color set C with $|C| = k$, let $f : E \to C$ be a given edge-coloring of G. We choose an arbitrary vertex $s \in V$. We indeed give an algorithm which determines whether the edge-colored graph $G(f)$ has an ℓ-rainbow *walk* W from s to each vertex $v \in V \setminus \{s\}$; one can obtain an ℓ-rainbow path between s and v, as the sub-walk of W. Since $|C| = k$ and $\ell_{\max} = \max\{\ell_j \mid 1 \leq j \leq k\}$, every ℓ-rainbow walk is of length at most $k\ell_{\max}$. Therefore, our algorithm is based on a dynamic programming approach with respect to the lengths of walks from s: $G(f)$ has an ℓ-rainbow walk from s to a vertex v with length exactly i if and only if there exists at least one vertex w in $N(v)$ such that $G(f)$ has an ℓ-rainbow walk from s to w with length exactly $i - 1$ in which the color $c_{j'} = f((w, v))$ is assigned to at most $(\ell_{j'} - 1)$ edges.

Based on the idea above, for an integer $i \in \{1, 2, \ldots, k\ell_{\max}\}$ and a vertex $v \in V$, we define a set $\Gamma_s(i, v)$ of k-tuples $\boldsymbol{\ell}' = (\ell_1', \ell_2', \ldots, \ell_k')$ of nonnegative integers $\ell_1', \ell_2', \ldots, \ell_k'$, as follows:

$$\Gamma_s(i, v) = \{(\ell_1', \ell_2', \ldots, \ell_k') \mid G(f) \text{ has an } \ell\text{-rainbow walk } W \text{ between } s \text{ and } v$$
$$\text{such that } \ell_1' + \ell_2' + \cdots + \ell_k' = i \text{ and } c_j \text{ is assigned}$$
$$\text{to exactly } \ell_j' \text{ edges in } W \text{ for all } j \in \{1, 2, \ldots, k\}\}.$$

Note that $\Gamma_s(i, v) = \emptyset$ if $G(f)$ has no ℓ-rainbow walk between s and v of length exactly i. Clearly, $G(f)$ has an ℓ-rainbow path from s to a vertex $v \in V \setminus \{s\}$ if and only if $\Gamma_s(i, v) \neq \emptyset$ for some integer $i \in \{1, 2, \ldots, k\ell_{\max}\}$. By a dynamic programming approach, we compute the sets $\Gamma_s(i, v)$ from $i = 1$ to $k\ell_{\max}$ for all vertices $v \in V$. Then, using the sets $\Gamma_s(i, v)$, it can be determined in time $O(k\ell_{\max}n)$ whether $G(f)$ has ℓ-rainbow paths from s to all vertices $v \in V \setminus \{s\}$.

[Algorithm]

We first compute the set $\Gamma_s(1, v)$ for each vertex $v \in V$. Clearly, the walks with length exactly 1 from s are only the edges (s, v) for the vertices v in $N(s)$. Therefore, if $v \in N(s)$ and $\ell_{j'} \geq 1$ where $c_{j'} = f((s, v))$, then we have

$$\Gamma_s(1, v) = \{(\ell_1', \ell_2', \ldots, \ell_k')\}, \tag{4}$$

where $\ell_{j'}' = 1$ and $\ell_j' = 0$ for the other indices $j \in \{1, 2, \ldots, k\} \setminus \{j'\}$; otherwise

$$\Gamma_s(1, v) = \emptyset. \tag{5}$$

We then compute the set $\Gamma_s(i, v)$ for an integer $i \geq 2$ and each vertex $v \in V$. Suppose that we have already computed $\Gamma_s(i - 1, w)$ for all vertices $w \in V$. Obviously, $G(f)$ has an ℓ-rainbow walk from s to a vertex v with length exactly i if and only if, for some vertex $w \in N(v)$, there exists a k-tuple $(\ell'_1, \ell'_2, \ldots, \ell'_k) \in \Gamma_s(i - 1, w)$ such that $\ell'_{j'} \leq \ell_{j'} - 1$ for the color $c_{j'} = f((w, v))$. Therefore, we can compute $\Gamma_s(i, v)$ for a vertex $v \in V$, as follows:

$$\Gamma_s(i, v) = \Big\{ (\ell'_1, \ldots, \ell'_{j'-1}, \ell'_{j'} + 1, \ell'_{j'+1}, \ldots, \ell'_k) \mid$$
$$w \in N(v), \, (\ell'_1, \ell'_2, \ldots, \ell'_k) \in \Gamma_s(i - 1, w),$$
$$\text{and } \ell'_{j'} \leq \ell_{j'} - 1 \text{ for the color } c_{j'} = f((w, v)) \Big\}. \quad (6)$$

[Proof of Theorem 3]

Using Eqs. (4)–(6) one can correctly compute $\Gamma_s(i, v)$, $1 \leq i \leq k\ell_{\max}$, for all vertices $v \in V$. Thus, we now show that our algorithm runs in time $O(k2^{k\ell_{\max}}m)$ and uses $O(kn2^{k\ell_{\max}} \log(\ell_{\max} + 1))$ space.

We first claim that $|\Gamma_s(i, v)| \leq \binom{k\ell_{\max}}{i}$ for a vertex v in T and each integer $i \in \{1, 2, \ldots, k\ell_{\max}\}$. Consider an arbitrary k-tuple $(\ell'_1, \ell'_2, \ldots, \ell'_k) \in \Gamma_s(i, v)$. Then, $0 \leq \ell'_j \leq \ell_j \leq \ell_{\max}$ holds for each index $j \in \{1, 2, \ldots, k\}$, and $i = \ell'_1 + \ell'_2 + \cdots + \ell'_k \leq k\ell_{\max}$. Thus, the number of k-tuples in $\Gamma_s(i, v)$ can be bounded by the number of combinations which choose i elements from $k\ell_{\max}$ elements, and hence $|\Gamma_s(i, v)| \leq \binom{k\ell_{\max}}{i}$.

We now show that our algorithm uses $O(kn2^{k\ell_{\max}} \log(\ell_{\max} + 1))$ space. For a vertex v and an integer i, each k-tuple $\ell' \in \Gamma_s(i, v)$ can be represented by $O(k \cdot \log(\ell_{\max} + 1))$ bits, and hence the set $\Gamma_s(i, v)$ can be represented by using $O\big(\binom{k\ell_{\max}}{i} \cdot k \log(\ell_{\max} + 1)\big)$ space. Therefore, we can represent the sets $\Gamma_s(i, v)$ for all vertices $v \in V$ and all integers $i \in \{1, 2, \ldots, k\ell_{\max}\}$ using the space of

$$\sum_{i=1}^{k\ell_{\max}} \sum_{v \in V} O\Big(k \cdot \binom{k\ell_{\max}}{i} \cdot \log(\ell_{\max} + 1) \Big) = O\big(kn2^{k\ell_{\max}} \log(\ell_{\max} + 1)\big).$$

We finally estimate the running time of our algorithm. By Eqs. (4) and (5) the sets $\Gamma_s(1, v)$ can be computed in time $O(kn)$ for all vertices $v \in V$. By Eq. (6) the set $\Gamma_s(i, v)$ for a vertex v and an integer i can be computed in time $O\Big(|N(v)| \cdot \binom{k\ell_{\max}}{i-1} \cdot k \Big)$, because $|\Gamma_s(i - 1, w)| \leq \binom{k\ell_{\max}}{i-1}$, the condition $\ell'_{j'} \leq \ell_{j'} - 1$ for the color $c_{j'} = f((w, v))$ can be checked in time $O(1)$, and $O(k)$ time is required to store the obtained k-tuple $(\ell'_1, \ldots, \ell'_{j'-1}, \ell'_{j'} + 1, \ell'_{j'+1}, \ldots, \ell'_k)$ into $\Gamma_s(i, v)$. Therefore, the sets $\Gamma_s(i, v)$ can be computed for all vertices $v \in V$ and all integers $i \in \{2, 3, \ldots, k\ell_{\max}\}$, in time

$$\sum_{i=2}^{k\ell_{\max}} \sum_{v \in V} O\Big(k \cdot \binom{k\ell_{\max}}{i-1} \cdot |N(v)| \Big) = O(k2^{k\ell_{\max}} m).$$

Using the sets $\Gamma_s(i, v)$, $1 \leq i \leq k\ell_{\max}$, it can be determined in time $O(k\ell_{\max} n)$ whether $G(f)$ has ℓ-rainbow paths from s to all vertices $v \in V \setminus \{s\}$. Since G

can be assumed to be a connected graph, $n - 1 \leq m$ and hence our algorithm takes time $O(k2^{k\ell_{\max}}m)$ in total.

This completes the proof of Theorem 3.   □

4   Conclusion

In this paper, we introduced GENERALIZED RAINBOW CONNECTIVITY. We proved that the problem is NP-complete even for cacti, while is solvable in polynomial time for trees. We also gave an FPT algorithm for general graphs when parameterized by both k and $\ell_{\max} = \max\{\ell_j \mid 1 \leq j \leq k\}$.

References

1. Alon, N., Yuster, R., Zwick, U.: Color-coding. J. ACM 42, 844–856 (1996)
2. Ananth, P., Mande, M., Sarpatwar, K.: Rainbow connectivity: hardness and tractability. In: Proc. of IARCS Annual Conference on Foundations of Software Technology and Theoretical Computer Science, FSTTCS 2011, pp. 241–251 (2011)
3. Brandstädt, A., Le, V.B., Spinrad, J.P.: Graph Classes: A Survey. Society for Industrial and Applied Mathematics, Philadelphia (1999)
4. Chakraborty, S., Fischer, E., Matsliah, A., Yuster, R.: Hardness and algorithms for rainbow connection. J. Combinatorial Optimization 21, 330–347 (2011)
5. Chandran, L.S., Das, A., Rajendraprasad, D., Varma, N.M.: Rainbow connection number and connected dominating sets. J. Graph Theory 71, 206–218 (2012)
6. Chartrand, G., Johns, G.L., McKeon, K.A., Zhang, P.: Rainbow connection in graphs. Mathematica Bohemica 133, 85–98 (2008)
7. Chartrand, C., Johns, G.L., McKeon, K.A., Zhang, P.: The rainbow connectivity of a graph. Networks 54, 75–81 (2009)
8. Caro, Y., Lev, A., Roditty, Y., Tuza, Z., Yuster, R.: On rainbow connectivity. The Electronic J. Combinatorics 15, R57 (2008)
9. Fellows, M.R., Guo, J., Kanj, I.: The parameterized complexity of some minimum label problems. J. Computer and System Sciences 76, 727–740 (2010)
10. Garey, M.R., Johnson, D.S.: Computers and Intractability: A Guide to the Theory of NP-Completeness. Freeman, San Francisco (1979)
11. Krivelevich, M., Yuster, R.: The rainbow connection of a graph is (at most) reciprocal to its minimum degree. J. Graph Theory 63, 185–191 (2010)
12. Uchizawa, K., Aoki, T., Ito, T., Suzuki, A., Zhou, X.: On the rainbow connectivity of graphs: complexity and FPT algorithms. To appear in Algorithmica, doi:10.1007/s00453-012-9689-4

Fixed-Parameter Tractability of Error Correction in Graphical Linear Systems

Peter Damaschke[1], Ömer Eğecioğlu[2], and Leonid Molokov[1]

[1] Department of Computer Science and Engineering,
Chalmers University, 41296 Göteborg, Sweden
{ptr,molokov}@chalmers.se
[2] Department of Computer Science, University of California,
Santa Barbara, CA 93106-5110
omer@cs.ucsb.edu

Abstract. In an overdetermined and feasible system of linear equations $Ax = b$, let vector b be corrupted, in the way that at most k entries are off their true values. Assume that we can check in the restricted system given by any minimal dependent set of rows, the correctness of all corresponding values in b. Furthermore, A has only coefficients 0 and 1, with at most two 1s in each row. We wish to recover the correct values in b and x as much as possible. The problem arises in a certain chemical mixture inference application in molecular biology, where every observable reaction product stems from at most two candidate substances. After formalization we prove that the problem is NP-hard but fixed-parameter tractable in k. The FPT result relies on the small girth of certain graphs.

Keywords: sparse system of linear equations, error correction, girth, even cycle matroid, parameterized algorithm.

1 Introduction

Let $Ax = b$ be a system of m linear equations in n variables, over the real numbers. Suppose that $Ax = b$ was obtained from some feasible linear system $Ax = b'$ by changing at most k of the coefficients in the vector b'. That is, b differs from the unknown true vector b' in at most k positions, but we are not told which. The maximum number k of errors may or may not be known. Our goal is to recover the correct values in b (and x) as far as possible, using a certain correctness criterion for entries of b that will be introduced below.

We were led to the problem by an application where we wish to infer the amounts of chemical compounds in an unknown mixture. We can only measure amounts of products of chemical split reactions each of which can stem from one or more candidate substances. This is modelled by a system of linear equations $Ax = b$. Each row (equation) corresponds to a measured substance, and each column of A (resp. variable of x) corresponds to a candidate compound. A with entries a_{ij} is the incidence matrix, that is, $a_{ij} = 1$ if split product i appears in compound j, and $a_{ij} = 0$ else. The a_{ij} are known, and b_i is the measured amount of product i. Typically every split product is contained in very few candidate

S.K. Ghosh and T. Tokuyama (Eds.): WALCOM 2013, LNCS 7748, pp. 245–256, 2013.
© Springer-Verlag Berlin Heidelberg 2013

compounds, hence the rows of A contain very few 1s. We are particularly interested in the reconstruction of protein mixtures after enzymatic digestion into peptides which can be identified and measured. Most peptides come from only one or two candidate proteins, and simulated protein digestion data suggest that already equations with at most two variables suffice to infer most of the protein amounts, provided that all measurements are correct [2]. A practical issue is that, as a result of experimental errors, some of the measured values in b may be corrupted. Without errors we would merely have to solve the linear system, which is even overdetermined. But in the presence of errors it is clear that any inference algorithm needs some assumptions about the number or the nature of errors, as well as the manner by which they can be detected. Here we will adopt what we call the *independent errors assumption*; see below.

For any set R of rows, let $A[R]$ be the matrix A restricted to R, and let $b[R]$ be the vector b restricted to the corresponding entries, in the rows of R. Our systems are overdetermined. Consider any subset C of rows that is linearly dependent and minimal with this property, that is, every proper subset of C is linearly independent. Following the terminology of matroid theory we call such C a *circuit*. (Note that circuits can be of any size up to $rank(A) + 1$.) Every row of a circuit C is a unique linear combination of the other rows of C. It follows that, if $b[C]$ has exactly one false entry, then the system $A[C]x = b[C]$ is not feasible. However, if $b[C]$ has several false entries, these errors are unlikely to cancel out each other such that $A[C]x = b[C]$ remains feasible by chance. Since the false entries in b result from independent measurement errors, deviations follow some continuous probability distribution and are uncorrelated. Hence errors in a circuit lead almost surely to infeasibility. This motivates the following

Independent Errors Assumption: *Whenever C is a circuit and $A[C]x = b[C]$ is feasible (we also call the circuit C balanced), then all values in $b[C]$ are true.*

Note that minimality of C is essential here. The assumption trivially extends to unions of circuits, but not to arbitrary dependent sets. If some row in a dependent set D is not contained in any circuit with other rows of D, the row is independent of the rest of D, and then arbitrary changes of the corresponding entry of vector b will keep the system feasible. Our assumption is of similar spirit as the very common "general position" assumptions in computational geometry (e.g., no three points are on the same line, no three lines intersect in a point). The setting also resembles the combinatorial group testing problem where elements of a set can be faulty or clean, and one can test pools of elements, with the result that either the entire pool is clean, or some faults are present. But when faults are present, the test does not tell us what the faulty elements are. In our variant however, pools are restricted to circuits of some matroid. For ease of presentation we assume an idealized computational model with precise real numbers, in practice we must allow small tolerances when we check two numbers for equality. Our problem is now preliminarily stated as follows:

BALANCED CIRCUITS RECOVERY: Given a linear system $Ax = b$ where some unknown subset of the entries in b are faulty, identify all entries of vector b that can be confirmed true under the independent errors assumption.

After that, it only remains to solve the linear system restricted to the rows with confirmed entries of b. Two questions arise: Which entries of b can we recover, if we can recover any at all, and how difficult is this algorithmically? We focus on the case where all coefficients in A are 0 or 1, and every row of A contains at most two entries that are 1. This problem is not as limited and specialized as it might seem. The restriction to $0, 1$-matrices with sparse rows is immediately motivated by applications as above, and the independent errors assumption is not restrictive at all; loosely speaking it just says that random errors will not collectively appear correct by pure chance.

Organization of the Paper: First we have to put some work into the formal problem statement and terminology. In Section 2 we characterize the entries in b that are recoverable under the independent errors assumption. In the case of two variables per equation, our problem can be formalized as a graph labeling problem where the entries of x and b are turned into vertices and edges, respectively. We introduce some notions and facts about matroids from graphs, which are needed in the following (see also [11]). The graph formulation is essential for our algorithm. In Section 3 we prove NP-hardness of BALANCED CIRCUITS RECOVERY in graphs. Section 4 gives a preparation for an FPT algorithm, with the number k of faulty edges as the parameter: we show that, after some pre-processing, we always find a circuit of logarithmic size. This extends the known fact that graphs of constant minimum degree have logarithmic girth, but since circuits and cycles are different objects, the matter requires some care. What we have here is the *even cycle matroid* of the associated graph, for which the circuits are different than the cycles of the ordinary cycle matroid. Based on the girth we give our FPT result in Section 5, by constructing a kernel of $O(k \log k)$ vertices. In Section 6 we obtain an $O^*((\log k)^k)$ time bound. Section 7 lists some open questions. Due to space limitations some proofs and straightworward algorithm details are omitted.

Related Literature: In [2] we considered an error model where all b_i are changed by at most some small ϵ, and we gave graph-theoretic and LP methods for controlling the error in the solution vector x. The motivation for the present study is that, besides general measurement inaccuracies, a number k of measured amounts may be totally wrong and should be detected first. Only for ease of presentation we assume here that all non-faulty b_i are accurate. If they are slightly disturbed, then unbalanced circuits remain unbalanced, and "nearly" balanced circuits within some tolerance have to be considered as balanced.

Approximability and parameterized complexity of finding maximal feasible subsystems of linear systems is studied, e.g., in [4,5]. Like ours, this problem is of special interest in the case of graphs (e.g., for some models in statistical physics), and it can be generalized to so-called gain graphs where vertices and edges are labeled with elements of a group. The minimum number of unsatisfied edges is known as the *frustration index*, and its computation is NP-hard already in special cases. We refer to the extensive annotated bibliography in [15]. However, BALANCED CIRCUITS RECOVERY differs from this suite of problems in that we made an additional mild assumption on the confirmation of correct edge labels. Next, the matroid circuits we have to deal with are even cycles and connected pairs of odd cycles (see details in Section 2), which loosely relates our problem to both feedback set problems [12,7,3] and odd cycle transversals (OCT) [13,7,9] whose parameterized complexity is well investigated. Remarkably, [8] uses matroids for kernelization, too. LP techniques as applied to OCT and other problems in [10] do not seem to be immediately applicable to our problem. In [1] we enumerated solutions with minimal support in linear systems with a constant number of nonzeros per row, however in an error-free setting.

2   Characterizations and Formalization

Remember that our input is a linear system $Ax = b$, where some entries of the observed b are faulty but obey the independent errors assumption.

Definition 1. *A row i is correct if b_i has its true value, otherwise it is faulty. A set of rows is correct if every row in that set is correct. A row i is recoverable if a unique value b_i is consistent with the independent errors assumption.*

Note that we cannot directly "see" which rows are correct, rather, we learn them only by checking circuits for being balanced. Row i being recoverable means informally that we could infer the true b_i, given enough computation time. Clearly, rows in balanced circuits are recoverable, and an obvious question is whether there exist more recoverable rows.

Definition 2. *We inductively define reachable rows as follows. A row in a balanced circuit is reachable; a single row in a circuit where all other rows are reachable is reachable, too; and no other rows are reachable.*

As mentioned, all rows in balanced circuits are recoverable. Next, any row i that appears in some circuit C where all other rows are recoverable, is recoverable, too. This follows from the fact that row i of matrix A is a unique linear combination of the other rows of C: since the true values in $b[C]$, perhaps except b_i, can be determined, we can finally determine the true b_i as well, thereby even ignoring the given value. In summary, all reachable rows are recoverable. The converse is also true, but due to space limitations we skip the proof.

Theorem 1. *The recoverable rows are exactly the reachable rows.*   □

From now on we deal with the announced "graphical" case.

Definition 3. *Let $Ax = b$ be a system of m linear equations in n variables, with at most two variables per equation, that appear with coefficient 1. We represent it as a graph with n vertices and m edges as follows. Its vertices are the variables in x. For every equation (row) $x_u + x_v = b_i$, the graph has an edge uv with edge label $b_{uv} := b_i$. For every trivial equation $2x_u = b_i$ with only one variable, the graph has a loop with edge label $b_{uu} := b_i$. Variable x_v is also called the vertex label of v. The graph may comprise parallel edges and also several loops at the same vertex, since the given matrix A may have identical rows.*

We have multiplied the trivial equations by 2 (and doubled b_i) to give all equations the same form. Despite possible parallel edges the notation b_{uv} will not cause confusion, as it will be clear from context which edge we refer to.

Due to the correspondence established in the Definitions and Theorem 1 we can use the terms *row* and *edge* interchangeably and speak of *recoverable (reachable) edges*. Moreover we can state the following graph problem, whose complexity with respect to parameter k will be studied here.

BALANCED CIRCUITS RECOVERY in graphs: Given a graph $G = (V, E)$, possibly with parallel edges and loops, and an edge labeling b, identify all reachable edges. As for the parameter k, the following is assumed: There exists a labeling b' that differs from b on at most k edges called the faulty edges; G with labeling b' has only balanced circuits; and in G with labeling b, no circuit containing faulty edges is balanced.

Of course, it is essential to know which edge sets in the graph correspond to the circuits, i.e., minimal dependent sets of rows in the linear system.

Definition 4. *A path or cycle in a graph is simple if it does not cross itself, that is, every vertex appears at most once on it. The length of a path or cycle is the number of edges. We consider a loop as a simple odd cycle of length 1. A bow tie is either a pair of vertex-disjoint simple odd cycles connected by a simple path whose inner vertices do not appear in any of the two cycles, or a pair of simple odd cycles with exactly one common vertex.*

The following is implicit in earlier literature [14,6]. (In [6] one can also find an interesting treatment of the algebra of the even cycle matroid.)

Theorem 2. *The circuits are exactly the simple even cycles and bow ties.* □

The two types of circuits behave differently when it comes to the vertex labels. In a bow tie, the vertex labels are uniquely determined. This is because the incidence matrix of a simple odd cycle has a nonzero determinant. As opposed to this, the incidence matrix of a simple even cycle has determinant zero, and the possible vectors of vertex labels form a 1-dimensional space: We can alternatingly add/subtract some free value to/from the vertex labels.

For our algorithm we will need some "technical" generalization of graphs (which is well established in matroid theory, cf. signed graphs and gain graphs).

Definition 5. *A signed graph is a graph where every edge also has a sign, besides the edge label. A sign is even or odd. Signs can be added modulo 2 where even=0 and odd=1. Vertex and edge labels are related as follows. The label b_{uv} of an odd edge uv satisfies $b_{uv} = x_u + x_v$; note that $b_{uv} = b_{vu}$. The label d_{uv} of an even edge uv satisfies $d_{uv} = x_u - x_v$. Note that $d_{uv} = -d_{vu}$, that is, the orientation of an even edge matters.*

The operation of merging edges in a signed graph works as follows. Let w be some vertex of degree 2 with neighbors u and v, where possibly $u = v$. We replace w and edges wu and wv with a new edge uv whose sign is the sum of signs of wu and wv. The label of uv is built according to these rules:

If both wu and wv are odd, then uv is even, and
$$d_{uv} = x_u - x_v = b_{uw} - x_w + x_w - b_{vw} = b_{uw} - b_{wv}.$$
If both wu and wv are even, then uv is even, and
$$d_{uv} = x_u - x_v = d_{uw} + x_w - x_w - d_{vw} = d_{uw} + d_{wv}.$$
If wu is odd and wv is even, then uv is odd, and
$$b_{uv} = x_u + x_v = b_{uw} - x_w + x_w + d_{vw} = b_{uw} - d_{wv}.$$
If wu is even and wv is odd, then uv is odd, and
$$b_{uv} = x_u + x_v = b_{vw} - x_w + x_w + d_{uw} = b_{vw} - d_{wu}.$$

Note that:

(1) In terms of the linear system, merging just means to eliminate the variable x_w by combining the equations for the labels of wu and wv.

(2) We can consider the original graph as a signed graph where all signs are odd. After a sequence of mergings, an odd (even) edge can represent a path of odd (even) length with inner vertices of degree 2 in the original graph. It is straightforward to prove associativity: The label of an edge does not depend on the order the merge steps are applied to consecutive edges in a path. Also the notions of circuit and balanced circuit can now be lifted to signed graphs in a straightforward way. In particular we have:

Corollary 1. *The circuits in signed graphs are exactly the simple even cycles and bow ties, with the modification that the sign of a cycle (odd or even) is now the sum of signs of its edges.*   □

3   NP-Hardness of Determining the Recoverable Edges

Theorem 3. BALANCED CIRCUITS RECOVERY *in graphs is NP-hard.*

Proof. We will demonstrate that the following decision problem is NP-complete: Given a graph with edge signs and edge labels and a specific edge, is this edge recoverable? Then the Theorem follows, because even edges are only used as a shorthand for a path of two odd edges. (An instance of BALANCED CIRCUITS RECOVERY has odd edges only.)

The well-known NP-complete SUBSET SUM takes as input $n+1$ real numbers $a_1, \ldots, a_n; s$ and asks whether $\sum_{i \in A} a_i = s$ holds for some subset $A \subseteq \{1, \ldots, n\}$. Given an instance of SUBSET SUM, we construct in linear time an instance of BALANCED CIRCUIT RECOVERY as follows (consult Fig.1 for an example of the reduction graph):

Create two vertices u and v, each with a loop with odd sign. The loop at u gets label 0, and the loop at v gets label $2s$. For $i = 1$ create vertices u_1, v_1 and two edges uu_1 and uv_1. For each item a_i, $i > 1$, from the sequence, create two vertices u_i, v_i and four edges $u_{i-1}u_i$, $u_{i-1}v_i$, $v_{i-1}u_i$, $v_{i-1}v_i$. Also create two edges $u_n v$, $v_n v$. All these non-loop edges are even. Edges get the following labels. Every edge $u_i u_{i+1}$ and $v_i u_{i+1}$ gets label 0, and every edge $u_i v_{i+1}$ and $v_i v_{i+1}$ gets label $-a_{i+1}$. Similarly, edge uu_1 gets label 0, and edge uv_1 gets label $-a_1$. Both $u_n v$ and $v_n v$ get label 0.

The idea is that every even edge, so to speak, shifts the vertex label by either a_{i+1} or 0 when we proceed from u to v. Based on this, we will show the following equivalence (remember what *recoverable* means, from Definition 1).

Claim. The two loops are recoverable if and only if the SUBSET SUM instance is a YES instance.

Proof of Claim. Trivially, the graph without the loops at u and v contains only even cycles. Due to Theorem 2, the circuits containing a loop are exactly the bow ties connecting the loops at u and v by a simple path. That is, the two loops can only appear together in circuits. By the inductive definition of reachable edges (rows) and Theorem 1, we can only infer edge labels in balanced circuits, and one further true edge label at a time, in a circuit where all other edges are already recovered. But we cannot infer two new edge labels in a circuit simultaneously. Hence the loops are recoverable if and only if they appear in a balanced circuit.

The 2^n shortest simple paths from u to v go through the vertices u_i or v_i strictly in the order of indices $i = 1, \ldots, n$. On a simple path from u to v we may also go from index $i+1$ back to index i, but then we have to return immediately to $i+1$, to avoid repeated visits of a vertex. Any such zig zag path of three edges can be replaced with the one edge between its end vertices with indices i and $i+1$. The shift of labels on the zig zag path and the single edge is the same; this is easy to verify by our choice of even edge labels and their effect on the vertex labels. (Note that the direct edge is the result of merging the three edges in the zig zag path.) Hence it suffices to consider only bow ties with shortest paths from u to v. Now, going through v_i means to add a_i to the solution A, and going through u_i means not to add a_i to A. Since the loop at v has label $2s = s + s$, the total shift must be exactly s, and the claimed equivalence is established.   □

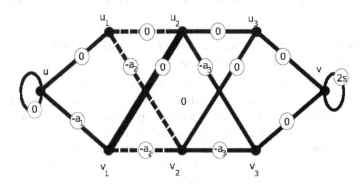

Fig. 1. The reduction graph (Theorem 3) for 3 items. The dashed lines show a zig zag path that may be replaced with the bold edge in a bow tie.

4   Girth of Reduced Signed Graphs

In view of the NP-hardness result in Theorem 3 it is natural to study the parameterized version of BALANCED CIRCUITS RECOVERY in graphs, with the maximum number k of faulty edges as the parameter. The following is a building block of our parameterized algorithm.

Given any signed graph with real-valued edge labels we construct the *reduced* signed graph by repeatedly applying the following steps as long as possible:

0-rule: If vertex w has a single loop with odd sign, and w is not adjacent with further edges, delete w and the loop.

1-rule: If w is a vertex of degree 1 with neighbor u, delete w and the edge wu.

2-rule: If w is a vertex of degree 2 with neighbors u and v (where possibly $u = v$), merge the edges wu and wv.

It is easy to establish that these rules can be applied in any order and yield a uniquely determined reduced signed graph. We also need:

Lemma 1. *In a signed graph G with n vertices and minimum degree $d \geq 3$ we find a circuit with at most $4 \log_{d-1} n + 2$ edges in polynomial time. Consequently, in a reduced signed graph with n vertices we find a circuit with at most $4 \log_2 n + 2$ edges in polynomial time.*

Proof. We only sketch the construction, details are omitted due to space limitations. By breadth-first search we easily find a cycle C of logarithmic length. If C is even, we are done. If C is odd, we shrink C to a super vertex and run BFS again, to find another cycle and a path connecting it to C, both of logarithmic length. Finally note Corollary 1.   □

5   Parameterized Algorithm for the Recoverable Edges

Suppose that we have identified a balanced circuit C. Then the edges in C are recovered, due to the independent errors assumption. Furthermore we can *contract* C by successively contracting its edges.

Edge Contraction: Once an edge uv is recognized as correct, we can *contract* it and obtain an equivalent smaller problem instance. The principle is simply to eliminate one of the variables, say x_v, which is possible since we know that the edge label is correct. For clarity we discuss all details as needed later.

If the considered edge is a loop ($u = v$), we have the following cases. If the loop is even, then $d_{uu} = 0$ (otherwise the loop would not be correct), and we can simply delete the loop. If the loop is odd, then $x_u = b_{uu}/2$ is enforced, thus we keep the loop as an indicator that the x_u has been determined. In the following we suppose $u \neq v$ and discuss the real edge contraction. We keep vertex u and eliminate v; this choice is arbitrary. Let w denote any further vertex. When we contract uv, all edges uw ($w \neq v$) are unchanged. We merge uv with every edge vw ($w \neq v$) exactly as described earlier in Section 2. Note that at least one such edge exists, since otherwise uv is in no circuit (by Theorem 2), hence it would never be confirmed as correct. It remains to consider odd loops at v (whereas even loops are meaningless, as seen before). Any odd loop at v, with label b_{vv}, is transformed into an odd loop at u with the following label b_{uu}.

If uv is odd, then $b_{uu} = 2x_u = 2b_{uv} - 2x_v = 2b_{uv} - b_{vv}$.
If uv is even, then $b_{uu} = 2x_u = 2d_{uv} + 2x_v = 2d_{uv} + b_{vv}$.

It may be helpful to notice what the final result of *contracting a balanced circuit* C is. If C is a simple even cycle, there remains one vertex without loop, that is, the label of that vertex is not "internally" determined by C, corresponding to the 1-dimensional solution space of a simple even cycle. If C is a bow tie, we eventually get one vertex with two odd loops attached, however they have equal edge labels as C is balanced, hence one of them is redundant. – We are ready to give a high-level description of our algorithm for BALANCED CIRCUITS RECOVERY in graphs.

Preprocessing: In the following we work with signed graphs. Initially, every edge gets an odd sign. First we apply the 1-rule as long as possible. From Theorem 1 and Corollary 1 we get: The removed edges do not appear in any circuit, hence they are not recoverable nor can they contribute to recovery of other edge labels. Clearly, we can ignore them henceforth. Next we also apply the 2-rule and 0-rule as long as possible, hence we obtain the reduced signed graph.

Main Loop of the Algorithm: We apply Lemma 1 to find a circuit C with at most $4\log_2 n + O(1)$ edges. We check whether C is balanced, by testing whether the linear system induced by the edges in C is feasible.

If C is balanced, we contract C. Contraction does not alter the degrees of vertices outside C, only some edges incident with C may get new end vertices and adjusted labels. (Remember that parallel edges are allowed, they are *not* removed from our graphs. While contraction of edges may render other edges parallel, these edges are still kept, hence the degrees of vertices outside C are *not* diminished.) In particular, the graph is still a reduced signed graph after contraction, hence no merging takes places among the remaining edges.

If C is not balanced, clearly C contains *some* faulty edge, but it is important to notice that we do not know which edges are faulty. Therefore, in this case we delete *all* edges of C (but not their vertices) from the graph, and then we reduce the remaining graph again, by exhaustively applying the three rules.

We iterate the process of circuit detection, contraction, deletion and reduction, until the remaining graph has some constant size $O(1)$.

Re-inserting Edges: As said in the beginning of this section, contraction only yields equivalent problem instances, and vertices eliminated during the contraction of balanced circuits have labels uniquely determined by the labels of the vertices we keep, hence we need not consider them any more. Roughly speaking, no information gets lost by contraction.

The situation is different for faulty circuits. Since we have removed edges that are only "potentially" faulty, we must eventually re-insert them one by one to guarantee an equivalent problem instance. The details of this step need some more discussion. In the following, G denotes the input graph after initial exhaustive application of the 1-rule (hence G has minimum degree 2), and H denotes the graph obtained by our processing so far. Let $e = uv$ be some edge that we want to put back next. Recall that any vertex of H may stand for a vertex of G or represent a subset of original vertices of G identified due to edge contractions. Similarly, any edge of H may stand for an original edge of G or a path of merged edges from G. Thus, any end of e, say u, may lie in a vertex of H or on an edge of H, or u may even lie outside H, because the edges adjacent to u have been removed as well (e.g., by the 1-rule). If e is already on a path represented by an edge of H, clearly we need not re-insert e.

Now we treat the other cases. If u is in a vertex of H, we simply attach e to this vertex in the obvious sense. If u is on an edge f of H, we insert vertex u in H and subdivide f. If u is not in H, then due to minimum degree 2 another path in G different from edge uv starts in u and ends somewhere in H, or in v. We also re-insert such a path, but then exhaustively apply the 2-rule again. A few simple case inspections show that, in either case regarding the ends u and v of e, re-insertion of an edge e adds at most 2 vertices and 2 further edges to H. With at most k faulty edges, the size of H thus increases by $O(k \log n)$ in total, due to Lemma 1. (After each re-insertion we can also update the edge labels in H straightforwardly, since the original edge labels from G are still known.)

Analysis: By the last observation we can reduce a graph with n vertices to a kernel of $k \log n$ vertices (with some fixed logarithm base) in polynomial time. Thus an upper bound n on the kernel size is implicitly given by $n = k \log n$. In order to bound n in terms of k only, observe that $n = k \log n$ implies $k > \log n$ (for large enough n). Therefore $n = k \log n = k(\log k + \log \log n) = O(k \log k)$.

Work on the Kernel: The above procedure has computed, in polynomial time, a kernel of $O(k \log k)$ vertices, with the property that every edge outside the kernel is already recovered or not recoverable at all. Thus it only remains to solve the problem on the kernel. Even if we naively enumerate all circuits in the kernel, test them, and get the other recoverable edges inductively (Theorem 1), the time depends only on k. Thus we have finally shown:

Theorem 4. BALANCED CIRCUITS RECOVERY *in graphs parameterized by the number of faulty edges is in FPT.*   □

6   Branching Strategies

Theorem 4 establishes our FPT result, however exhaustive enumeration of all circuits in the kernel would be wasteful. We consider more efficient strategies for this last phase separately. Before we solve the remaining problem on the kernel, we reduce this graph once more: Isolated odd loops removed by the 0-rule and edges removed by the 1-rule are not recoverable anyhow, and in every path P of edges merged by the 2-rule, either all or none of the edges in P are recoverable, such that we need not distinguish them. Also remember that contracting balanced circuits only removes edges that are already recovered (since the independent errors assumption holds, and edges that became parallel are kept in the graph). In parameterized time bounds we use the O^* notation that omits polynomial factors, and log means the logarithm with a suitable base.

Theorem 5. BALANCED CIRCUITS RECOVERY *in graphs with n vertices and m edges, at most k of them faulty, is solvable within the minimum of the following time bounds: $O^*(2^m)$; $O^*(m^k/k!)$; and $O^*((\log n)^k)$.*

Proof. We may guess the subset of faulty edges and delete them. Clearly, these are at most 2^m and at most $m^k/k!$ subsets. In every branch we run the algorithm from Section 5, with the difference that we abort it when a faulty circuit is detected. Every branch needs polynomial time.

Alternatively, we may run the algorithm from Section 5, now with the difference that, when a faulty circuit C is detected, we branch on C, thereby guessing only one faulty edge and deleting it. Due to Lemma 1, C has $\log n$ edges. Since we can apply this step at most k times, we obtain a bounded search tree of size $(\log n)^k$. Clearly, if some branch is successful, we have found all recoverable edges, otherwise we report that more than k faulty edges were present.

A side remark is that the reachable edges that are not in balanced circuits (see the definition of reachable rows/edges and Theorem 1) are eventually detected, as they are linearly dependent from already recovered edges. □

Since we have $n = O(k \log k)$ as said above, we obtain $O^*((\log n)^k) = O^*((\log k)^k) = O^*(c^{k \log \log k})$ for some constant $c > 1$. Hence the latter method is faster unless $m = O(k \log \log k)$.

Corollary 2. BALANCED CIRCUITS RECOVERY *in graphs with at most k faulty edges is solvable in $O^*((\log k)^k)$ time.* □

7   Conclusions

We proved NP-hardness and gave an FPT algorithm for the problem of recovering the entries of vector b in a linear system $Ax = b$, where A is a 0,1-matrix with at most two 1s per row, assuming that b has at most k errors which are uncorrelated in a sense. The problem is motivated by the inference of chemical mixtures under measurement errors, and can be rephrased as a graph problem. Already membership in FPT is not trivial. Some obvious open questions are: Can we improve the FPT time bound, possibly by using stronger relations between edge number and girth? What about matrices with general nonzero coefficients, and with some more than two nonzeros per row? (Do the resulting hypergraph problems inherit some of the useful graph structure?) Can we generalize the FPT approach to gain graphs [15], provided that there exist natural applications? The complexity of approximation might be interesting, too, but this was not the scope of this paper. Finally, it would be worthwhile to apply the algorithms to real protein quantitation data and validate the error assumptions made in the parameterization.

Acknowledgment. This work has been supported by the Swedish Research Council (Vetenskapsrådet), grant no. 2010-4661, "Generalized and fast search strategies for parameterized problems". Early stages of the third author's work have also been supported by Devdatt Dubhashi through a Chalmers Bioscience Initiative grant. The work was done while the second author was visiting Chalmers during his sabbatical 2011–2012.

References

1. Damaschke, P.: Sparse Solutions of Sparse Linear Systems: Fixed-Parameter Tractability and an Application of Complex Group Testing. In: Marx, D., Rossmanith, P. (eds.) IPEC 2011. LNCS, vol. 7112, pp. 94–105. Springer, Heidelberg (2012); Extended version to appear in Theor. Comp. Sci.
2. Damaschke, P., Molokov, L.: Error Propagation in Sparse Linear Systems with Peptide-Protein Incidence Matrices. In: Bleris, L., Măndoiu, I., Schwartz, R., Wang, J. (eds.) ISBRA 2012. LNCS, vol. 7292, pp. 72–83. Springer, Heidelberg (2012)
3. Dehne, F.K., Fellows, M.R., Langston, M.A., Rosamond, F.A., Stevens, K.: An $O(2^{O(k)}n^3)$ FPT Algorithm for the Undirected Feedback Vertex Set Problem. Theory Comput. Syst. 41, 479–492 (2007)
4. Feige, U., Reichman, D.: On the Hardness of Approximating Max-Satisfy. Info. Proc. Lett. 97, 31–35 (2006)
5. Giannopoulos, P., Knauer, C., Rote, G.: The Parameterized Complexity of Some Geometric Problems in Unbounded Dimension. In: Chen, J., Fomin, F.V. (eds.) IWPEC 2009. LNCS, vol. 5917, pp. 198–209. Springer, Heidelberg (2009)
6. Grossman, J.W., Kulkarni, D.M., Schochetman, I.E.: Algebraic Graph Theory Without Orientation. Lin. Algebra and its Appl. 212/213, 289–307 (1994)
7. Guo, J., Gramm, J., Hüffner, F., Niedermeier, R., Wernicke, S.: Compression-Based Fixed-Parameter Algorithms for Feedback Vertex Set and Edge Bipartization. J. Comput. Syst. Sci. 72, 1386–1396 (2006)
8. Kratsch, S., Wahlström, M.: Compression via Matroids: A Randomized Polynomial Kernel for Odd Cycle Transversal. In: Rabani, Y. (ed.) SODA 2012, pp. 94–103. SIAM (2012)
9. Lokshtanov, D., Saurabh, S., Sikdar, S.: Simpler Parameterized Algorithm for OCT. In: Fiala, J., Kratochvíl, J., Miller, M. (eds.) IWOCA 2009. LNCS, vol. 5874, pp. 380–384. Springer, Heidelberg (2009)
10. Narayanaswamy, N.S., Raman, V., Ramanujan, M.S., Saurabh, S.: LP can be a Cure for Parameterized Problems. In: Dürr, C., Wilke, T. (eds.) STACS 2012. LIPIcs, vol. 14, pp. 338–349 (2012)
11. Oxley, J.: Matroid Theory, 2nd edn. Oxford Univ. Press (2011)
12. Raman, V., Saurabh, S., Subramanian, C.R.: Faster Fixed Parameter Tractable Algorithms for Finding Feedback Vertex Sets. ACM Trans. Algor. 2, 403–415 (2006)
13. Reed, B.A., Smith, K., Vetta, A.: Finding Odd Cycle Transversals. Oper. Res. Lett. 32, 299–301 (2004)
14. Tutte, W.T.: On Chain-Groups and the Factors of Graphs. Coll. Math. Societatis János Bolyai 25 (Algebraic Methods in Graph Theory, Szeged), 793–818 (1978)
15. Zaslavsky, T.: A Mathematical Bibliography of Signed and Gain Graphs and Allied Areas. Electron. J. Comb., Dynamic Surveys in Combinatorics, no. DS8 (1999)

Lower Bounds for Ramsey Numbers for Complete Bipartite and 3-Uniform Tripartite Subgraphs

Tapas Kumar Mishra and Sudebkumar Prasant Pal*

Department of Computer Science and Engineering
Indian Institute of Technology Kharagpur, 721302, India
tap1cse@gmail.com, spp@cse.iitkgp.ernet.in

Abstract. In this paper we establish lower bounds of the number $R'(a, b)$ so that any bicoloring of the edges of the complete undirected graph K_n with $n \geq R'(a, b)$ vertices, always admits a monochromatic complete bipartite subgraph $K_{a,b}$, where a and b are natural numbers. We show that $R'(2, b) > 2b + 1$ for $b \geq 4$. We establish a lower bound for $R'(a, b)$ using the probabilistic method that improves the lower bound given by Chung and Graham [4]. Further, we also use Lovász' local lemma to derive a better lower bound for $R'(a, b)$. We define $R'(a, b, c)$ be the minimum number n such that any n-vertex 3-uniform hypergraph $G(V, E)$, or its complement $G'(V, E^c)$ contains a $K_{a,b,c}$. Here, $K_{a,b,c}$ is defined as the complete tripartite 3-uniform hypergraph with vertex set $A \cup B \cup C$, where the A, B and C have a, b and c vertices respectively, and $K_{a,b,c}$ has abc 3-uniform hyperedges $\{u, v, w\}$, $u \in A$, $v \in B$ and $w \in C$. We derive lower bounds for $R'(a, b, c)$ using probabilistic methods.

Keywords: Ramsey number, biparite graph, local lemma, probabilistic method, r-uniform hypergraph.

1 Introduction

We define $R'(a, b)$ as the minimum number n of vertices so that any n-vertex simple undirected graph G or its complement G' must contain the complete bipartite graph $K_{a,b}$. Equivalently, $R'(a, b)$ is the minimum number n of vertices such that any bicoloring of the edges of the n-vertex complete undirected graph K_n would contain a monochromatic $K_{a,b}$. The significance of such a number is that it gives us the minimum number of vertices needed in a graph so that two mutually disjoint subsets of vertices with cardinalities a and b can be guaranteed to have the complete bipartite connectivity property as mentioned. In the analysis of social networks it may be worthwhile knowing whether all persons in some subset of a persons share b friends, or none of the a persons of some other subset share friendship with some set of b persons. This can also be helpful in the analysis of dependencies, where there are many entities in one partite set, which are all dependent on entities in the other partite sets; we need to achieve consistencies that either all dependencies exist between a pair of partite sets, or none of the dependencies exist between possibly another pair of partite sets. These Ramsey numbers are different from the usual Ramsey numbers $R(a, b)$, where $R(a, b)$ is the

---

* Corresponding author.

S.K. Ghosh and T. Tokuyama (Eds.): WALCOM 2013, LNCS 7748, pp. 257–264, 2013.

smallest number such that any undirected graph G with n or more vertices contains either a K_a or an independent set of size b. We know that $R(3,3) = 6$, $R(3,4) = 9$, $R(3,8) = 28$ and $R(3,9) = 36$, $R(4,4) = 18$, $R(4,5) = 25$ (see [11,12]).

1.1 Existing Results

From the definition of $R'(a,b)$, it is clear that $R'(1,1)=2$ and $R'(1,2)=3$. To see that $R'(1,3) \geq 6$, observe that we need at least 4 vertices and neither a 4-cycle nor its complement has a $K_{1,3}$. Further, observe that neither a 5-cycle in K_5, nor its complement (also a 5-cycle) has a $K_{1,3}$. The numbers $R'(1,b)$ are however known exactly, and are given by Burr and Roberts [2] as $R'(1,b) = 2b - 1$ for even b, and $2b$, otherwise. Chvátal and Harary [3] were the first to show that $R(C_4, C_4) = 6$, where C_4 is a cordless cycle of four vertices. We use $R(G_1, G_2)$ to denote the minimum number such that for every undirected graph G with $R(G_1, G_2)$ or more vertices, either (i) G contains G_1 as subgraph, or (ii) the complement graph G' of G contains G_2 as subgraph. As $K_{2,2}$ is identical to C_4, $R'(2,2) = 6$. Note that $R'(2,3) = 10$ [1], $R'(2,4) = 14$ [6], and $R'(2,5) = 18$ [6]. There are many such results for $R'(a,b)$ for various values of a and b in [11]. Harary [7] proved that $R(K_{1,n}, K_{1,m}) = n + m - x$, where $x = 1$ if both n and m are even and $x = 0$ otherwise.

1.2 Our Contribution

In the next section we establish a lower bound of $2b + 1$ for $R'(2,b)$ for all $b \geq 2$. We provide an explicit construction and use combinatorial arguments. Note that Lortz and Mengersen [8] conjectured that $R'(2,b) \geq 4b - 3$, for all $b \geq 2$. Exoo et al. [6] proved that $R'(2,b) \leq 4b - 2$ for all $b \geq 2$, where the equality holds if and only if a strongly regular $(4b - 3, 2b - 2, b - 2, b - 1)$-graph exists. A k-regular graph G with n vertices is called strongly regular (n, k, p, q)-graph if every pair of adjacent vertices share exactly p neighbours and every non-adjacent pair of vertices share exactly q neighbours. In Section 3, we consider Ramsey numbers $R'(a,b)$ for integers a and b and establish lower bounds using probabilitic methods. In Section 4 we extend similar methods for 3-uniform tripartite hypergraphs, deriving lower bounds for the Ramsey numbers $R'(a,b,c)$; we are unaware of any literature concerning such lower bounds for such hypergraphs. Here, $R'(a,b,c)$ is the minimum number n such that any n-vertex 3-uniform hypergraph $G(V, E)$, or its complement $G'(V, E^c)$ contains a $K_{a,b,c}$. Here, $K_{a,b,c}$ is defined as the complete tripartite 3-uniform hypergraph with vertex set $A \cup B \cup C$, where the A, B and C have a, b and c vertices respectively, and $K_{a,b,c}$ has abc 3-uniform hyperedges $\{u, v, w\}$, $u \in A$, $v \in B$ and $w \in C$. In Section 5 we conclude with a few remarks and future research directions.

2 A Constructive Lower Bound for $R'(2, b)$

The following lower bound for $R'(2, b)$ involves an explicit construction as follows. We are not aware of better lower bounds for $R'(2, b)$ in the literature, to the best of our knowledge.

Theorem 1. $R'(2, b) > 2b + 1$, for all integers $b \geq 2$.

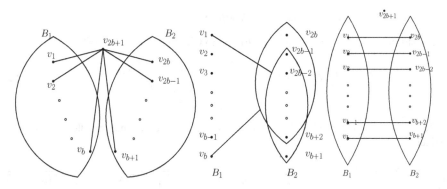

Fig. 1. Construction of G(left two): generation of B_1, B_2 and addition of edges. Resulting G'(rightmost): In G', B_1 and B_2 become K_b, and B_1 and B_2 have a perfect matching.

Proof. For $b \geq 4$, we show that with $2b + 1$ vertices, there always exist a graph G such that both G and its complement G' do not contain $K_{2,b}$. The entire construction is illustrated in Figure 1. Let the vertices be labelled v_1, v_2, ..., v_{2b+1}. Connect v_{2b+1} to every other vertex. In order to avoid $K_{2,b}$, no other vertex out of v_1, v_2, ..., v_{2b} should be connected to b or more vertices in the set v_1, v_2, ..., v_{2b}. So, any of these $2b$ vertices can have a maximum of $b - 1$ neighbours other than v_{2b+1}. We distribute these $2b$ vertices into two groups, keeping v_1, v_2, ..., v_b in one group B_1, and $v_b + 1$, $v_b + 2$, ..., v_{2b} in the other group B_2. Now every vertex from B_1 can be connected to at most $b - 1$ vertices from B_2 such that we can still avoid $K_{2,b}$. There are $\binom{b}{b-1} = b$ such distinct groups of size $b - 1$ in B_2. Now each vertex of B_1 is connected to one such distinct group of size $b - 1$ from B_2. We claim the degree of every vertex except v_{2b+1} is b. Firstly, every vertex of B_1 is connected to $b - 1$ vertices of B_2, and the single vertex v_{2b+1}. Secondly, every vertex of B_2 (i) is connected to v_{2b+1}, and (ii) also present in exactly $b - 1$ separate groups, where each group is connected to exactly one vertex of B_1. So, every vertex of $B_1 \cup B_2$ has degree b. Therefore G is $K_{2,b}$-free.

Now consider G'. Since v_{2b+1} is connected to every other vertex in G, it is isolated in G'. So, the number of possible neighbours for vertices v_1, v_2, ..., v_{2b+1} becomes $2b - 1$. Since each vertex in G is connected to $b - 1$ vertices other than v_{2b+1}, the number of possible neighbours for each vertex is restricted to $(2b - 1) - (b - 1) = b$, as illustrated in Figure 1. Now we argue that such neighbouring sets of b vertices of any two vertices differ in at least one vertex. Observe that in G', B_1 and B_2 include complete graphs K_b, and the edges between B_1 and B_2 form a perfect matching. As a result the neighbouring sets of any two vertices differ by at least one vertex in G. Since the number of common neighbours between any two vertices is no more than $b - 1$, G' is also $K_{2,b}$-free.   □

3   Probabilistic Lower Bounds for $R'(a, b)$

In the first Section 3.1 we use the probabilistic method that Erdós applied to prove lower bounds on the original Ramsey numbers [5]. In the Section 3.2, we demonstrate improved lower bounds using the Lovász' local lemma.

3.1   Application of the Probabilistic Method

The best known lower bound on $R'(a, b)$ due to Chung and Graham [4] is

$$R'(a, b) > \left(2\pi\sqrt{ab}\right)^{\left(\frac{1}{a+b}\right)} \left(\frac{a+b}{e^2}\right) 2^{\frac{ab-1}{a+b}} \tag{1}$$

Table 1. Lower bounds for $R'(a, b)$ from Inequality 3.1(left), Theorem 2 (middle) and Theorem 3 (right)

b	4	5	6	7	8	14	15	16
a								
1	2,3,4	3,4,5	3,5,6	3,5,7	3,6,8	5,10,17	5,11,18	6,12,19
2	3,5,6	4,6,7	5,7,9	5,8,10	6,9,12	9,17,23	10,18,24	10, 19, 26
3	5,7,8	6,8,9	7,10,12	8,12,14	9,14,16	16,26,32	17,29,35	18,31,37
4	6,9,10	8,11,12	10,14,15	12,16,18	14,19,22	26,41,46	28,45,50	30,49,55
5		11,14,16	13,18,20	16,22,24	19,27,29	40,60,65	43,67,72	47,74,80
6			17,23,25	21,29,31	26,35,38	59,87,93	66,98,104	72,109,116
7				27,37,39	34,46,48	86,123,129	96,139,147	106,156,165
8					43,58,61	119,168,178	136,193,204	152,219,232
14						556,755,820	678,922,1005	817,1113,1219
15							836,1136,1246	1019,1385,1525
16								1254,1704,1886

We derive a tighter lower bound using the probabilistic method as follows.

Theorem 2. *For natural numbers a and b,* $R'(a, b) > \dfrac{\left(a^a b^b 2\pi\sqrt{ab}\right)^{\left(\frac{1}{a+b}\right)} 2^{\left(\frac{ab-1}{a+b}\right)}}{e}$.

Proof. First we find the probability p of existence of a particular monochromatic $K_{a,b}$ and then sum that probability over all possible distinct $K_{a,b}$ to get the probability of existence of some monochromatic $K_{a,b}$. To get a lower bound on $R'(a, b)$, we choose the largest value of n, keeping the probability p stictly less than unity. This would ensure the existence of some graph G with n vertices such that both G and G' are free from any monochromatic $K_{a,b}$. Let n be the number of vertices of graph G. Then the total number of distinct $K_{a,b}$'s possible is $\binom{n}{a}\binom{n-a}{b}$. Each $K_{a,b}$ has exactly ab edges. Each edge can be either of two colors with equal probability. The probability that a particular $K_{a,b}$ will have all ab edges of a specific color is $\left(\frac{1}{2}\right)^{ab}$. So, the probability that a particular $K_{a,b}$ is monochromatic is $2\left(\frac{1}{2}\right)^{ab} = 2^{1-ab}$. The probability p that some $K_{a,b}$ is monochromatic is $\binom{n}{a}\binom{n-a}{b}2^{1-ab}$. Our objective is to choose as large n as possible with $p < 1$. So, choosing n as in Theorem 2 and using Stirling's approximation (replacing $a!$ by $\sqrt{2\pi}\frac{a^{a+\frac{1}{2}}}{e^a}$ and $b!$ by $\sqrt{2\pi}\frac{b^{b+\frac{1}{2}}}{e^b}$), we get $p < 1$. The details of the derivation shown in [9] are omitted in this version. This guarantees the existence of an n-vertex graph for which some edge bicoloring would not result in any monochromatic $K_{a,b}$.   □

See Table 1 for the first two lower bounds for $R'(a,b)$ for each pair (a,b), due to Inequality 3.1 and Theorem 2, respectively. The ratio of our lower bound in Inequality 2 and that of Chung and Graham's as in Inequality 3.1 is close to 1.359 for values of a close to b, and close to Euler's number e for values of a very small compared to b. So our lower bound gives an improvement that varies between 1.35 to e depending upon the values of a and b.

3.2 A Lower Bound for $R'(a,b)$ Using Lovász' Local Lemma

We are interested in the question of existence of a monochromatic $K_{a,b}$ in any bi-colouring of the edges of K_n. Since the same edge may be present in many distinct $K_{a,b}$'s, the colouring of any particular edge may effect the monochromaticity in many $K_{a,b}$'s. This gives the motivation of use of Lovász' local lemma (see [10]) in this context. We use the following Corollary 1 of Lovász's local lemma to account for such dependencies.

Corollary 1. *If every event E_i, $1 \leq i \leq m$ is dependent on at most d other events and $Pr[E_i] \leq p$, and if $ep(d+1) \leq 1$, then $Pr\left[\bigcap_{i=1}^{n} \overline{E_i}\right] > 0$.*

Theorem 3. *If $e\left(2^{1-ab}\right)\left(ab\binom{n-2}{a+b-2}\binom{a+b-2}{b-1}+1\right) \leq 1$ then $R'(a,b) > n$.*

Proof. We consider a random bicolouring of the complete graph K_n in which each edge is independently coloured red or blue with equal probability. Let S be the set of edges of an arbitrary $K_{a,b}$, and let E_S be the event that all edges in this $K_{a,b}$ are coloured monochromatically. For each such S, the probability of E_S is $P(E_S) = 2^{1-ab}$. We enumerate the sets of edges of all possible $K_{a,b}$'s as $S_1, S_2, ..., S_m$, where $m = \binom{n}{a}\binom{n-a}{b}$ and each S_i is the set of all the edges of the ith $K_{a,b}$. Clearly, each event E_{S_i} is mutually independent of all the events E_{S_j} from the set $\{E_{S_j} : |S_i \cap S_j| = 0\}$. For each E_{S_i}, the number of events outside this set satisfies the inequality $|\{E_{S_j} : |S_i \cap S_j| \geq 1\}| \leq ab\binom{n-2}{a+b-2}\binom{a+b-2}{b-1}$; every S_j in this set shares at least one edge with S_i, and therefore such an S_j shares at least two vertices with S_i. We can choose the rest of the $a + b - 2$ vertices of S_j from the remaining $n-2$ vertices of K_n, out of which we can choose $b-1$ for one partite of S_j, and the remaining $a - 1$ to form the second partite of S_j, yielding a $K_{a,b}$ that shares at least one edge with S_i. We apply Corollary 1 to the set of events $E_{S_1}, E_{S_2}, ..., E_{S_m}$, with $p = 2^{1-ab}$ and $d = ab\binom{n-2}{a+b-2}\binom{a+b-2}{b-1}$, enforcing the premise $ep(d+1) \leq 1$, resulting in the lower bound for n, so that $Pr\left[\bigcap_{i=1}^{m} \overline{E_{S_i}}\right] > 0$. This non-zero probability (of none of the events E_{S_i} occuring, for $1 \leq i \leq m$) implies the existence of some bicoloring of the edges of K_n with no monochromatic $K_{a,b}$, thereby establishing the theorem.   □

Solving the inequality in the statement of Theorem 3, we can compute lower bounds for $R'(a,b)$, for natural numbers a and b. Such lower bounds for some larger values of a and b show significant improvements over the bounds computed using Theorem 2 (see Table 1).

4   Lower Bounds for Ramsey Numbers for Complete Tripartite 3-Uniform Subgraphs

Let $R'(a, b, c)$ be the minimum number n such that any n-vertex 3-uniform hypergraph $G(V, E)$, or its complement $G'(V, E)$ contains a $K_{a,b,c}$. An r-uniform hypergraph is a hypergraph where every hyperedge has exactly r vertices. (Hyperedges of a hypergraph are subsets of the vertex set. So, usual graphs are 2-uniform hypergraphs.) Here, $K_{a,b,c}$ is defined as the complete tripartite 3-uniform hypergraph with vertex set $A \cup B \cup C$, where the A, B and C have a, b and c vertices respectively, and $K_{a,b,c}$ has abc 3-uniform hyperedges $\{u, v, w\}$, $u \in A$, $v \in B$ and $w \in C$. It is easy to see that $R'(1, 1, 1) = 3$; with 3 vertices, there is one possible 3-uniform hyperedge which either is present or absent in G.

Theorem 4. $R'(1, 1, 2) = 4$.

Proof. Consider the complete 3-uniform hypergraph with vertex set $V = \{1, 2, 3, 4\}$ and set of exactly four hyperedges $H = \{\{1, 2, 3\}, \{1, 2, 4\}, \{1, 3, 4\}, \{2, 3, 4\}\}$. Since vertex 1 is present in 3 hyperedges, any (empty or non-empty) subset S of H, or its complement $H \setminus S$ must contain at least two hyperedges containing the vertex 1. Observe that any such set of two hyperedges must have a $K_{1,1,2}$ as subset.   □

We have checked the facts that $R'(1, 1, 3) = 5$ and $R'(1, 1, 4) = 6$ by exhaustive search; the codes for these programs are available on request from the authors. Determining such Ramsey numbers for higher parameters by exhaustive searching using computer programs is computationally very expensive in terms or running time. We state our conjecture for $R'(1, 1, b)$ as follows.

Conjecture 1. $R'(1, 1, b) = b + 2$.

Table 2. Lower bounds for $R'(a, a, a)$ by Theorem 5 (left) and Theorem 6 (right)

a	3	4	5	6	7	8
$R'(a, a, a)$	14,19	84,138	800,1765	11773,35167	269569,1073543	9650620,50616072

Table 3. Lower bounds for $R'(a, b, c)$ by Theorem 5 (left) and Theorem 6 (right)

	a=2	a=3	a=3	a=3	a=4	a=4	a=5	a=6	a=6	a=6	a=6
c	5	3	4	5	4	5	5	2	3	4	5
b											
2	9,13	8,11	11,16	16,22	18,25	26,36	40,58	11,16	21,29	36,52	59,87
3	16,22	14,19	23,32	35,50	41,61	68,107	124,208		50,74	107,175	209,371
4	26,36		41,61	68,107	84,138	159,281	334,653			277,521	643,1354
5	40,58			124,208		334,653	800,1765				1740,4194

4.1 Probabilistic Lower Bound for $R'(a, b, c)$

Theorem 5. $R'(a, b, c) > \dfrac{\left(a^a b^b c^c \sqrt{(2\pi)^3 abc}\right)^{\left(\frac{1}{a+b+c}\right)} 2^{\left(\frac{abc-1}{a+b+c}\right)}}{e}.$

Proof. Consider the probability of existence of a particular $K_{a,b,c}$ in G or G', where G is a 3-uniform hypergraph and G' is its complement. The sum p of such probabilities over all possible distinct $K_{a,b,c}$'s is an upper bound on the probability that some $K_{a,b,c}$ exists in G or G'. Let n be the number of vertices of hypergraph G. As in the proof of Theorem 2, we observe that the number of $K_{a,b,c}$'s is no more than $\binom{n}{a}\binom{n-a}{b}\binom{n-a-b}{c}$. Each $K_{a,b,c}$ has exactly abc hyperedges. Each hyperedge can be present in G or G' with equal probability. So, the probability that all hyperedges of a particular $K_{a,b,c}$ are in G is $\left(\frac{1}{2}\right)^{abc}$. Therefore, the probability that a particular $K_{a,b,c}$ is present in either G or G' is $2\left(\frac{1}{2}\right)^{abc} = 2^{1-abc}$. So, the probability p that some $K_{a,b,c}$ is either in G or in G', is $\binom{n}{a}\binom{n-a}{b}\binom{n-a-b}{c}2^{1-abc}$. Using the lower bound from Theorem 5 for $R'(a, b, c)$ as the value for the number n of vertices, we can verify that $p < 1$. This establishes the theorem since $p < 1$ implies the existence of a hypergraph G of n vertices such that neither G nor G' has a $K_{a,b,c}$. For details, see [9]. □

See Tables 2 and 3 for some computed lower bounds based on Theorem 5.

4.2 A Lower Bound for $R'(a, b, c)$ Using Lovász' Local Lemma

Theorem 6. *If* $e\left(2^{1-abc}\right)\left(abc\binom{n-3}{a+b+c-3}\binom{a+b+c-3}{b-1}\binom{a+c-2}{c-1} + 1\right) \leq 1$ *then* $R'(a, b, c) > n$.

Proof. Here we perform analysis as done earlier in Section 3.2. Consider a random bicoloring of the hyperedges of the complete 3-uniform hypergraph of n vertices, in which each hyperedge is independently colored red or blue with equal probability. Let S be the set of hyperedges of an arbitrary $K_{a,b,c}$, and let E_S be the event that the $K_{a,b,c}$ is coloured monochromatically. For each such S, $P(E_S) = 2^{1-abc}$. If we enumerate all possible $K_{a,b,c}$'s as $S_1, S_2, ..., S_m$, where $m = \binom{n}{a}\binom{n-a}{b}\binom{n-a-b}{c}$ and each S_i is the set of all the hyperedges of the ith $K_{a,b,c}$, each event E_{S_i} is mutually independent of all the events from the set $\{E_{S_j} : |S_i \cap S_j| = 0\}$. For each E_{S_i}, the number of events outside this set satisfies the inequality $\{E_{S_j} : |S_i \cap S_j| \geq 1\} \leq abc\binom{n-3}{a+b+c-3}\binom{a+b+c-3}{b-1}\binom{a+c-2}{c-1}$, as every S_j in this set shares at least one of the abc hyperedges of S_i, and therefore S_j shares at least three vertices with S_i. We can choose the rest of the $a + b + c - 3$ vertices of S_j from the remaining $n - 3$ vertices, out of which we can choose $b - 1$ for the second partite of S_j, and the remaining $c - 1$ for the third partite of S_j, thereby yielding a $K_{a,b,c}$ which shares at least one hyperedge edge with S_i. We can apply Corollary 1 to the set of events $E_{S_1}, E_{S_2}, ..., E_{S_m}$, with $p = 2^{1-abc}$, $d = abc\binom{n-3}{a+b+c-3}\binom{a+b+c-3}{b-1}\binom{a+c-2}{c-1}$, yielding $ep(d + 1) \leq 1 \Rightarrow Pr\left[\bigcap_{i=1}^{m} \overline{E_{S_i}}\right] > 0$. Since no event E_{S_i} occurs for some random bicoloring of the hyperedges, no monochromatic $K_{a,b,c}$ exists in that bicoloring. This establishes the theorem. □

See Tables 2 and 3 for some computed lower bounds based on Theorem 6; the values based on Theorem 6 to the right in each cell of these tables are much better than those based on Theorem 5, to the left in the respective cells.

5   Concluding Remarks

The probabilistic method is useful in establishing lower bounds for Ramsey numbers. It is worthwhile studying the application of Lovász' local lemma, possibly more effectively and accurately, so that higher lower bounds may be determined. In our work we have considered the bicoloring of K_n and the existence of a monochromatic $K_{a,b}$ in arbitrary bicolorings of the edges of K_n; some authors consider complete bipartite graphs $K_{n,n}$ instead of complete graphs like K_n and derive bounds for corresponding Ramsey numbers. If one uses four parameters, a, b, c and d, then one can consider the existence of a monochromatic $K_{a,b}$ or a monochromatic $K_{c,d}$, in the bicolorings of the edges of K_n (or $K_{n,n}$). For values and bounds on such Ramsey numbers see [11]. For computing the lower bounds in Tables 1, 2 and 3, we have used computer programs. The code for these programs are available from the authors on request. As the sizes of the complete bipartite graphs (tripartite 3-uniform hypergraphs) grow, the computation time required for computing the lower bounds becomes prohibitive.

References

1. Burr, S.A.: Diagonal Ramsey Numbers for Small Graphs. Journal of Graph Theory 7, 57–69 (1983)
2. Burr, S.A., Roberts, J.A.: On Ramsey numbers for stars. Utilitas Mathematica 4, 217–220 (1973)
3. Chvátal, V., Harary, F.: Generalized Ramsey Theory for Graphs, II. Small Diagonal Numbers. Proceedings of the American Mathematical Society 32, 389–394 (1972)
4. Chung, F.R.K., Graham, R.L.: On Multicolor Ramsey Numbers for Complete Bipartite Graphs. Journal of Combinatorial Theory (B) 18, 164–169 (1975)
5. Erdős, P., Spencer, J.: Paul Erdős: The Art of Counting. The MIT Press (1973)
6. Exoo, G., Harborth, H., Mengersen, I.: On Ramsey Number of $K_{2,n}$. In: Alavi, Y., Chung, F.R.K., Graham, R.L., Hsueds, D.F. (eds.) Graph Theory, Combinatorics, Algorithms, and Applications, pp. 207–211. SIAM, Philadelphia (1989)
7. Harary, F.: Recent Results on Generalized Ramsey Theory for Graphs. In: Alavi, Y., et al. (eds.) Graph Theory and Applications, pp. 125–138. Springer, Berlin (1972)
8. Lortz, R., Mengersen, I.: Bounds on Ramsey Numbers of Certain Complete Bipartite Graphs. Results in Mathematics 41, 140–149 (2002)
9. Mishra, T.K., Pal, S.P.: Lower bounds for Ramsey numbers for complete bipartite and 3-uniform tripartite subgraphs (manuscript, 2012)
10. Motwani, R., Raghavan, P.: Randomized Algorithms, pp. 115–120. Cambridge University Press, New York (1995)
11. Radziszowski, S.P.: Small Ramsey Numbers. The Electronic Journal on Combinatorics (2011)
12. West, D.B.: Introduction to Graph Theory, 2nd edn. Pearson Prentice Hall (2006)

Improved Fixed-Parameter Algorithm
for the Minimum Weight 3-SAT Problem

Venkatesh Raman and Bal Sri Shankar

The Institute of Mathematical Sciences, Chennai 600113, India
{vraman,balsri}@imsc.res.in

Abstract. The problem of finding a satisfying assignment for a CNF formula that minimizes the weight (the number of variables that are set to 1) is NP-complete even if the formula is a 2-CNF formula. It generalizes the well-studied problem of finding the smallest hitting set for a family of sets, which can be modeled using a CNF formula with no negative literals. The natural parameterized version of the problem asks for a satisfying assignment of weight at most k.

It is known that when the input instance is a 2-CNF formula, the problem actually is equivalent (in terms of parameterized and exact complexity) to the vertex cover (or 2-hitting set) problem. In this paper, we present the first non-trivial fixed-parameter algorithm for the problem when the given input is a 3-CNF formula.

We give an $2.85^k n^{O(1)}$ algorithm for determining whether a given 3-CNF formula on n variables has a satisfying assignment with weight at most k. This improves the trivial $3^k n^{O(1)}$ time algorithm for the problem and answers a question asked in an earlier paper. This implies that within the same time, we can test whether a given 3-CNF formula has a *weak backdoor* on k variables, to a 0-valid formula – i.e. whether there are k variables such that there exists an assignment to these variables that results in a 0-valid formula (formulas that are satisfiable by an all 0 assignment). This improves the naive $6^k n^{O(1)}$ algorithm for the problem.

1 Introduction and Motivation

Satisfiability is a fundamental problem that encodes several computational problems. Variations of the problem appear as canonical complete problems for several complexity classes. While it is well known that the satisfiability of a formula in CNF form is a canonical NP-complete problem, testing whether a CNF formula has a satisfying assignment with weight[1] at least k is a canonical complete problem for the parameterized complexity class $W[2]$ [2]. If the number of variables in each clause is bounded, it is a canonical $W[1]$-complete problem [2]. These results imply that it is unlikely that these problems are *fixed parameter tractable* (FPT). In other words, it is unlikely that they have an algorithm with running time $O(f(k)n^{O(1)})$ on input formulas of size n.

---

[1] The *weight* of an assignment is the number of variables assigned 1 by the assignment.

S.K. Ghosh and T. Tokuyama (Eds.): WALCOM 2013, LNCS 7748, pp. 265–273, 2013.

On the other hand, if the question is whether a d-CNF formula (for a fixed d) has a satisfying assignment with weight *at most* k, then this generalizes the well-studied d-hitting set problem and independently, turns out to be fixed parameter tractable with the weight as a parameter ([7,5], cf. Section 2). When we restrict our attention to 2-CNF formulas (MINONES 2-SAT) this problem generalizes the well-studied VERTEX COVER problem. For, given a graph $G = (V, E)$, a satisfying assignment of weight at most k on the formula $\wedge(u \vee v)$, where the $\wedge$ runs over all edges (u, v) in E, where u and v are variables corresponding to vertices u and v of G, corresponds to a vertex cover of size at most k in G. However, notice that we do not require negated literals to encode VERTEX COVER using 2-CNF formulas, and thus it appears that MINONES 2-SAT is a more general version of the vertex cover problem. In [6], the authors give a polynomial time parameter preserving reduction from MINONES 2-SAT to VERTEX COVER thereby showing that the parameterized and exact optimization versions of MINONES 2-SAT problem can be solved in the same time (with a polynomial time overhead) as the corresponding versions of VERTEX COVER. They ask, in general, whether finding a weight at most k satisfying assignment can be done in better than $O^*(c^k)^2$ time for a c-SAT formula. We answer the question affirmatively in this paper when $c = 3$.

A natural question, due to the equivalence between minimum weight 2 hitting set problem and minimum weight 2 CNF SAT is whether this equivalence continues for higher valus of c. One evidence in the negative, has been shown in [4] where the authors show that when the input instance is a 3 CNF formula, the problem of finding a weight k satisfying assignment is unlikely to have a polynomial (in k) kernel (see Section 2 for the definitions). This is in contrast to the fact that 3-hitting set has a kernel on $O(k^2)$ elements [1]. But this does not rule out equivalence (between 3-hitting set and weight k 3 CNF SAT) in terms of fixed-parameter tractable algorithms. While $O^*(2.27^k)$ algorithms exist [8] for 3-hitting set (and hence for weight k satisfying assignment in monotone 3-CNF formulae), ours is the first step in beating the $3^k n^{O(1)}$ algorithm for finding weight k satisfying assignment in general 3-CNF formula.

Another motivation for studying this problem stems from the need to understand the role of 'backdoors' to a formula. Given a boolean formula F, a *weak backdoor* [3] to a family $\mathcal{F}$ of boolean CNF formulae is a subset S of variables of the formula such that there exists an assignment to the variables of S, that after simplification in F, results in a formula that is in $\mathcal{F}$ and is satisfiable. If F has a k sized weak backdoor, and if satisfiability of a formula in $\mathcal{F}$ is testable in polynomial time, then the satisfiability of the entire formula can be checked in $O^*(2^k)$ time. Our result implies that one can test whethere a given 3 CNF formula has a k-sized weak backdoor to the family of what are called 0-valid formulae (formulae which are satisfiable by setting all variables to 0), in $O^*(2.85^k)$ time improving on the earlier known bound of $O^*(6^k)$ [3].

---

[2] O^* notation ignores polynomial factors.

2   Preliminaries

A parameterized problem is denoted by a pair $(Q, k) \subseteq \Sigma^* \times \mathbb{N}$. The first component Q is a classical language, and the number k is called the parameter. Such a problem is *fixed–parameter tractable* (FPT) if there exists an algorithm that decides it in time $O(f(k)n^{O(1)})$ on instances of size n. A *kernelization algorithm* takes an instance (x, k) of the parameterized problem as input, and in time polynomial in $|x|$ and k, produces an equivalent instance (x', k') such that $|x'|$ is a function purely of k. The output x' is called the *kernel* of the problem and its size is $|x'|$. We refer the reader to [2,7] for more details on the notion of fixed-parameter tractability.

Let P be an arbitrary set, whose elements we shall refer to as *variables*. A *literal* is either a variable or its negation. An assignment for P is a function $t : P \to \{0, 1\}$. Sometimes, we also refer to an assignment setting (mapping) a variable to 'true' or 'false' when we mean to say 1 or 0 respectively.

A formula is in *conjunctive normal form* (CNF) if it is a conjunction of clauses, where a clause is a disjunction of literals. A c-SAT formula has at most c literals in any clause. We assume that no clause is repeated and all variables in a clause are distinct. The *weight* of an assignment is the number of variables that are set to one by that assignment. We refer to the problem of finding a smallest weight satisfying assignment for c-SAT formulae as MINONES c-SAT.

The main algorithm we give is a 'branching algorithm'. By this, the algorithm at any point performs one of several choices of partial assignments, and then at that branch recursively checks for an assignment with fewer weight (fewer by the number of variables already set to 1 in that branch). A branching vector $(b_1, b_2, ...b_l)$ for an l-way branch means that the step has set respectively $b_1, b_2, ...b_l$ variables to 1. And the runtime of the branching step is the one obtained by solving the recursive equation $T(k, n) \leq \sum_{i=1}^{l} T(k - b_i, n)$ at that step. The solution for the recursive equations we obtain can be solved using an appropriate generating function (see [7]) and can be verified using induction or any symbolic algebra package (for example, Mathematica).

Simple FPT algorithm for weight at most k assignments. The natural parameterized version of MINONES c-SAT is FPT for any fixed c, when parameterized by the weight: pick a clause that contains only positive literals (as long as one exists) and branch by setting each of the variables to 1. This results in a c-way branch of depth at most k. If at any leaf every clause has at least one negated literal, then the assignment that sets all the remaining variables to 0 satisfies all such clauses. Otherwise (if all leaves have at least one clause containing all positive literals), then we return the answer that there is no assignment with weight at most k. This results in an $O(c^k m)$ algorithm where m is the number of clauses in the formula.

3   Improved Algorithm

3.1   The Algorithm

The following branching algorithm solves the problem in time $O^*(2.85^k)$.

Step 1.1. Consider the formula and find a pair of clauses of the form $(a \vee b \vee c)$ and $(a \vee d \vee e)$, where a, b, c, d and e are variables, all distinct from each other.

Branch as follows :

Branch1 Set $a = 1$
Branch2 Set $a = 0, b = 1, d = 1$
Branch3 Set $a = 0, c = 1, d = 1$
Branch4 Set $a = 0, c = 1, e = 1$
Branch5 Set $a = 0, b = 1, e = 1$

The branching vector of this step is $(1, 2, 2, 2, 2)$ resulting in a runtime of $O^*(2.6^k)$.

Step 1.2. Consider the formula and find a pair of clauses of the form $(a \vee b \vee c)$ and $(a \vee b \vee e)$, where a, b, c, and e are variables, all distinct from each other.

Branch as follows :

Branch1 Set $a = 1$
Branch2 Set $a = 0, b = 1$
Branch3 Set $a = 0, b = 0, c = 1, e = 1$

The branching vector of this step is $(1, 1, 2)$ resulting in a runtime of $O^*(2.414^k)$.

We branch as above as long as any pair of clauses as mentioned can be found.

Step 2. Find in the formula a pair of clauses of the form $(a \vee b \vee c)$ and $(\neg a \vee d \vee e)$, where b, c, d and e are variables, not necessarily distinct, but they are all distinct from a. Branch as follows :

Branch1 Set $a = 0, b = 1$
Branch2 Set $a = 0$ and $c = 1$
Branch3 Set $a = 1$ and $d = 1$
Branch4 Set $a = 1$ and $e = 1$

The branching vector is $(1, 1, 2, 2)$ with time $O^*(2.731^k)$. Branch as above as long as any pair of clauses as mentioned can be found.

Step 3. Find in the formula, a triplet of clauses of the form $(a \vee b \vee c)$, $(\neg a \vee \neg d \vee e)$ and $(d \vee f \vee g)$, where a, b, c, d, e, f and g are variables, not necessarily distinct. Note, however that as we have already branched based on Step 1, a, b and c are distinct from d, f and g. Also, within a clause, all variables are distinct. Now, we branch as follows:

Branch1 Set $a = 0$ and $b = 1$
Branch2 Set $a = 0$ and $c = 1$
Branch3 Set $a = 1, d = 0$ and $f = 1$
Branch4 Set $a = 1, d = 0$ and $g = 1$
Branch5 Set $a = 1, d = 1$ and $e = 1$

The branching vector is $(1, 1, 2, 2, 3)$ with a time complexity of $O^*(2.84^k)$. Branch as above as long as any pair of clauses as mentioned can be found.

Step 4. Find in the formula, a set of clauses of the form $\{(a_1 \vee a_2 \vee a_3), (\neg a_1 \vee \neg a_4 \vee \neg a_5), (a_4 \vee a_6 \vee a_7), (a_5 \vee a_8 \vee a_9)\}, \{(\neg a_2 \vee \neg a_{10} \vee \neg a_{11}), (a_{10} \vee a_{12} \vee a_{13}), (a_{11} \vee a_{14} \vee a_{15})\}, \{(\neg a_3 \vee \neg a_{16} \vee \neg a_{17}), (a_{16} \vee a_{18} \vee a_{19})$ and $(a_{17} \vee a_{20} \vee a_{21})\}$, where a_i, $i = 1$ to 21 are variables, not necessarily distinct. Branch as follows:

Branch1 Set $a_1 = 1, a_4 = 0$, and $a_6 = 1$

Branch2 Set $a_1 = 1, a_4 = 0$ and $a_7 = 1$

Branch3 Set $a_1 = 1, a_4 = 1, a_5 = 0$ and $a_8 = 1$

Branch4 Set $a_1 = 1, a_4 = 1, a_5 = 0$ and $a_9 = 1$

Branches 5 to 8   Same as branches 1 to 4 except that the variable a_1 is replaced by a_2 and variables a_i is replaced respectively by the variables a_{i+6} for $i = 4$ to 9.

Branches 9 to 12   Same as branches 1 to 4 except that the variable a_1 is replaced by a_3 and variables a_i is replaced respectively by the variables a_{i+12} for $i = 4$ to 9.

The branching vector of the step is $(2, 2, 3, 3, 3, 2, 2, 3, 3, 2, 2, 3, 3)$ having time $O^*(2.85^k)$. We branch as above as long as any set of clauses as mentioned can be found.

Step 5. If in the formula no set of clauses as mentioned in the steps 1 to 4 can be found, then :

if the number of clauses of the form $(a \vee b \vee c)$ where a, b and c are variables is $\leq k$, then return YES otherwise return NO.

3.2   Proof of Correctness

It is clear that the branches in Steps 1 to 4 are exhaustive and exclusive and cover all satisfying assignments of the set of clauses found in that branch.

The running time is dominated by the runtime in Step 4 and is $O^*(2.85^k)$ as claimed.

To complete the proof of correctness, we essentially have to prove the correctness of the last step. In order to do this, it is sufficient to prove the following two claims:

Claim 1. Given a formula F' in 3-CNF that results after application of the branching steps 1 to 4 repeatedly until none of those steps can be applied, the original formula F has a satisfying assignment with weight $\leq k$ if and only if the formula F' has a satisfying assignment with weight $\leq k'$. Here k' is the value of the parameter, passed as an argument along with F' after applications of the repeated application of the branching steps.

Claim 2. The (above mentioned) formula F' obtained after repeated applications of Steps 1 to 4 (until these steps are no longer applicable) has a satisfying assignment of weight $\leq k'$ if and only if this formula contains $\leq k'$ clauses of the form $(a \vee b \vee c)$, where a, b, and c are variables.

These claims are proved below.

Proof of Claim 1. Claim1 is clearly true if the branching has been made using the first 3 branching steps. For the 4th branching step, if a set of clauses as mentioned is found in the formula, then any satisfying assignment must set all of these clauses to the value 1. Now in order for the first clause $(a_1 \lor a_2 \lor a_3)$ to be satisfied, at least one among a_1, a_2 and a_3 must be set to 1.

Now, $a_1 = 1 \implies a_4 = 0$ or $(a_4 = 1$ and $a_5 = 0)$ (in order to satisfy the clause $(\neg a_1 \lor \neg a_4 \lor \neg a_5)$).

Also, $a_4 = 0 \implies (a_6 = 1$ or $a_7 = 1)$ (in order to satisfy the clause $(a_4 \lor a_6 \lor a_7)$),

and $a_4 = 1$ and $a_5 = 0 \implies (a_8 = 1$ or $a_9 = 1)$ (in order to satisfy the clause $(a_5 \lor a_8 \lor a_9)$).

These implications give the branches 1 to 4 of this step. Similar implications obtained by replacement of the variables a_1 by a_2, and a_i by a_{i+6} for $i = 4$ to 9 give the branches 5 to 8, and, replacing a_1 by a_3 and a_i by a_{i+12} for $i = 4$ to 9 give the branches 9 to 12 of this step.

Proof of Claim 2. First, let F' have a satisfying assignment A of weight $\leq k'$. Now suppose that F' contains $> k'$ clauses of the form $(a \lor b \lor c)$. Since the assignment A satisfies F', hence it must satisfy each of these $> k'$ clauses. This implies that these clauses contain at least one pair of the form $(a \lor b \lor c)$ and $(a \lor d \lor e)$, where the variables b, c, d and e are not necessarily distinct. However, such a pair can not exist, as it must have been eliminated during branching in step 1, thus contradicting our supposition.

Now, let F' contain $\leq k'$ clauses of the form $(a \lor b \lor c)$, where a, b and c are variables. We first construct an assignment A' as follows :

1. For any clause of the form $(a \lor b \lor c)$ in F', where a, b and c are variables such that none amongst a, b and c (denoted by v) occurs in F' in any set of clauses of the form $(\neg v \lor \neg d \lor \neg e), (d \lor f \lor g)$ and $(e \lor h \lor i)$, the assignment A' arbitrarily sets one variable amongst a, b and c to 1.
2. For any clause of the form $(a \lor b \lor c)$ in F', where a, b and c are variables such that there exists a set of clauses in F' of the form $(\neg a \lor \neg d \lor \neg e), (d \lor f \lor g)$, and $(e \lor h \lor i)$, one variable amongst b and c is chosen which does not occur in any such set of clauses. This chosen variable is assigned the value 1 by the assignment A'. One such variable must exist, as if both b and c also occur in similar sets, then the entire set of 9 clauses would have been eliminated in the branching step 3.
3. After all clauses having all positive variables have been thus eliminated, the assignment sets all variables occuring in the remaining clauses to 0.

By construction, the weight of the assignment A' is $\leq$ the number of clauses in F' of the form $(a \lor b \lor c)$, and is thus $\leq k'$.

We now prove that the assignment A' satisfies F'. This is proved as follows: Consider any clause C present in the formula F'. This clause must be of one of the following forms:

- $C = (a \vee b \vee c)$, where a, b and c are variables. Such a clause must be satisfied by A', as by definition, it sets atleast one amongst the variables a, b and c to 1.
- $C = (\neg a \vee b \vee c)$, where a, b and c are variables. Now the variable a cannot occur in any clause of the form $(a \vee d \vee e)$ in F', as any such pair of clauses would have been eliminated during branching Step 2. Thus a must be assigned the value 0 by the assignment A', and hence the clause C is satisfied by it.
- $C = (\neg a \vee \neg b \vee c)$, where a, b and c are variables. Here, the variables a and b can not both occur in clauses of the form $(\neg a \vee d \vee e)$ and $(\neg b \vee f \vee g)$ in the formula F', as any such triplet of clauses would have been eliminated during branching Step 3. Hence, at least one of the 2 variables a and b must be assigned the value 0 by A'. This shows that A' must satisfy the clause C.
- $C = (\neg a \vee \neg b \vee \neg c)$, where a, b and c are variables. Now if the variables a, b and c all occur in clauses that contain only positive literals, i.e. in clauses of the form $(a \vee d \vee e), (b \vee f \vee g)$ and $(c \vee h \vee i)$, then during the construction of the assignment A', atleast one of the variables amongst a, b and c must be set to 0 (during Step 3 of the construction). This is because depending upon which clause amongst the 3 mentioned clauses is encountered first during Step 2 of the construction, this clause will be eliminated by setting a variable other than a, b or c occuring in this clause to 1. Now since this variable amongst a, b and c cannot occur in any other clause with all positive literals, it will be eventually set to 0 in Step 3.

    Hence this clause C is satisfied by the assignment A'.

Thus we have

Theorem 1. *Given a 3-CNF formula, and an integer k, we can determine in $O^*(2.85^k)$ time, whethere it has a satisfying assignment with weight at most k.*

3.3   Applications to Weak Backdoor

A family $\mathcal{F}$ of boolean CNF formulae is called 0-valid if for each formula in the family, the assignment that sets all variables to 0 satisfies it. Given a boolean formula F, a weak backdoor to the family of 0-valid formulae is a subset S of variables of F such that there is an assignment A to the variables of S such that F with the partial assignment A simplifies to a 0 valid formula.

The following branching idea gives an easy $O^*(6^k)$ algorithm (see [3]) to determine whether there is a weak backdoor on k variables in the given formula to a 0-valid formula. For each clause, branch by choosing each of the three variables to be in the backdoor and try either of the two setting for the chosen variable. At the leaf level, we simply check whether the given formula is 0-valid. We osberve below that Theorem 1 gives an improved algorithm.

Notice that our algorithm in trying to determine whether there is a weight at most k satisfying assignment to the formula finds a weight at most k partial assignment that results in a formula with every clause containing at least one negative literal, i.e. a formula which is 0-valid. I.e. it finds a weak backdoor set of at most k variables.

Conversely suppose that there is a weak backdoor set S to 0-valid formulae, with at most k variables, and the partial assignment A to the variables (that reduces the formula to a 0-valid formula) sets some variable v to 0. We claim that $S - v$ is still a weak backdoor set to 0-valid formula with at most $k - 1$ variables. Suppose the formula resulting after administering the partial assignment (dictated by A) to the variables of $S - v$ is not 0-valid. As F with the partial assignment A to the variables of S is 0-valid, this means that some clause containing variable v is not satisfied by setting any of the variables in the clause to 0. But this is a contradiction as v was part of the weak backdoor in which the satisfying assignment set v to 0.

Essentially, we have argued that to obtain a resulting formula that is 0-valid, a weak backdoor need not set any variable to 0. I.e. there exists a weak backdoor to 0-valid with at most k variables if and only if there is a weight at most k satisfying assignment. Thus, from Theorem 1, we have

Corollary 1. *Given a 3-CNF formula F, we can determine in $O^*(2.85^k)$ time whether there is a weak backdoor on at most k variables, to 0-valid formulae.*

4   Conclusions and Open Problems

We have given the first algorithm that beats the trivial $O^*(3^k)$ algorithm to determine whether a given 3-CNF formula has a satisfying assignment with weight at most k. This also implies that one can test in the same time whether a given 3-CNF formula has a weak backdoor to a 0-valid formula.

It would be interesting to determine whether such an algorithm exists for determining whether a given 3-CNF formula has a weak backdoor to a null formula. The problem here is to determine whether a given 3-CNF formula has at most k variables such that some assignment to them results in a satisfying assignment to the formula. This also has a trivial $O^*(6^k)$ algorithm, and improving this is an interesting open problem.

Another open problem is to improve our bound further. As our main aim was to beat the $O^*(3^k)$ bound, we developed a simple branching algorithm. We believe that with a few involved branching steps, one should be able to improve the running time bound. Obtaining better than $O^*(c^k)$ algorithm for c-CNF formula for $c > 3$ is another interesting open problem.

Acknowledgement. The first author acknowledges fruitful discussions with Neeldhara Misra. We also thank the anonymous referees for helpful comments that improved the presentation.

References

1. Abu-Khzam, F.N.: A kernelization algorithm for d-hitting set. J. Comput. Syst. Sci. 76, 524–531 (2010)
2. Downey, R.G., Fellows, M.R.: Parameterized Complexity. Springer (November 1999)

3. Gaspers, S., Szeider, S.: Backdoors to Satisfaction. In: Bodlaender, H.L., Downey, R., Fomin, F.V., Marx, D. (eds.) Fellows Festschrift 2012. LNCS, vol. 7370, pp. 287–317. Springer, Heidelberg (2012)
4. Kratsch, S., Wahlström, M.: Two Edge Modification Problems without Polynomial Kernels. In: Chen, J., Fomin, F.V. (eds.) IWPEC 2009. LNCS, vol. 5917, pp. 264–275. Springer, Heidelberg (2009)
5. Mahajan, M., Raman, V.: Parameterizing above guaranteed values: Maxsat and maxcut. J. Algorithms 31, 335–354 (1999)
6. Misra, N., Narayanaswamy, N.S., Raman, V., Shankar, B.S.: Solving MINONES-2-SAT as Fast as VERTEX COVER. In: Hliněný, P., Kučera, A. (eds.) MFCS 2010. LNCS, vol. 6281, pp. 549–555. Springer, Heidelberg (2010)
7. Niedermeier, R.: Invitation to Fixed Parameter Algorithms. Oxford Lecture Series in Mathematics and Its Applications. Oxford University Press, USA (2006)
8. Niedermeier, R., Rossmanith, P.: An efficient fixed-parameter algorithm for 3-hitting set. J. Discrete Algorithms 1, 89–102 (2003)

On Directed Tree Realizations of Degree Sets

Prasun Kumar[1], M.N. Jayalal Sarma[1], and Saurabh Sawlani[2]

[1] Department of Computer Science and Engineering, IIT Madras, Chennai, India
[2] Department of Electrical Engineering, IIT Madras, Chennai, India

Abstract. Given a degree set $D = \{a_1 < a_2 < \ldots < a_n\}$ of non-negative integers, the minimum number of vertices in any tree realizing the set D is known[11]. In this paper, we study the number of vertices and multiplicity of distinct degrees as parameters of tree realizations of degree sets. We explore this in the context of both directed and undirected trees and asymmetric directed graphs. We show a tight lower bound on the maximum multiplicity needed for any tree realization of a degree set. For the directed trees, we study two natural notions of realizability by directed graphs and show tight lower bounds on the number of vertices needed to realize any degree set. For asymmetric graphs, if $\mu_A(D)$ denotes the minimum number of vertices needed to realize any degree set, we show that $a_1 + a_n + 1 \leq \mu_A(D) \leq a_{n-1} + a_n + 1$. We also derive sufficiency conditions on a_i's under which the lower bound is achieved.

We study the following related algorithmic questions. (1) Given a degree set D and a non-negative integer r (as 1^r), test whether the set D can be realized by a tree of exactly $\mu_T(D) + r$ number of vertices. We show that the problem is fixed parameter tractable under two natural parameterizations of $|D|$ and r. We also study the variant of the problem : (2) Given a tree T, and a non-negative integer r (in unary), test whether there exists another tree T' such that T' has exactly r more vertices than T and has the same degree set as T. We study the complexity of the problem in the case of directed trees as well.

1 Introduction

Representation of graphs is an important theme in various algorithmic tasks related to graphs. The standard methods used are adjacency matrix and adjacency list representations. Since many applications require more succinct representation, degree sets and degree sequences have been considered where the uniqueness of the graph being represented can be traded off for succinctness. There is a host of computational [1] and combinatorial problems [11,9,10,3] associated with these representations themselves.

In this context, we study tree realizations[1] of degree sets $D = \{a_1 < a_2 < \ldots < a_n\}$. It is known[11] that the minimum number of vertices necessary and sufficient for a graph to realize any degree set D is exactly $a_n + 1$. If the graph

---

[1] See Section 2 for formal definitions.

S.K. Ghosh and T. Tokuyama (Eds.): WALCOM 2013, LNCS 7748, pp. 274–285, 2013.
© Springer-Verlag Berlin Heidelberg 2013

is restricted to be a tree, this is known [11] to be exactly $(\sum_{i=1}^{n}(a_i - 1)) + 2$ if and only if $a_1 = 1$.

We study the tree-realizability of degree sets under multiplicity constraints on each degree. That is, realizations where the multiplicity of each vertex is upper bounded by a number m. The realization that achieves the minimum number of vertices has exactly one vertex of each degree except for the degree 1. Hence the degree distribution is skewed. Can the degree set be realized by a tree with smaller maximum multiplicity if we are allowed to use more vertices? We answer this in the negative by arguing that the standard construction is indeed optimal in terms of maximum multiplicity.

Theorem 1. *The minimum multiplicity of* pendant vertices *in any tree realization for the degree set* $D = \{1 = a_1 < a_2 < \ldots < a_n\}$ *is* $\sum_{i=1}^{n} a_i - 2n + 3$.

We define the notion of degree multiset where each element repeats atleast m times. We also generalize the above theorem to the case of degree multiset.

We turn to degree set realizations using directed graphs. We study some natural variants. In the first variant, the degree set is said to be realized if there is a directed graph such that every vertex has either the in-degree *or* the out-degree from the set, and every number in the degree set appears as the in-degree of some vertex *or* as the out-degree of some vertex in the graph. We call this the ∨-REALIZATION of D. We observe a connection between this variant and the undirected graph realizations of degree sets in the case of bipartite graphs (in particular trees) and hence derive the minimum multiplicities in this case.

In the second variant, which we call the ∧-REALIZATION of D, the degree set is to be realized by a directed graph such that every vertex has the in-degree *and* the out-degree from the set and every element in the set appears as the in-degree of some vertex *and* as the out-degree of some vertex. We prove the following theorem for directed tree ∧-realizations.

Theorem 2. *The minimum order of any directed tree* ∧-*realizing a degree set* D *is* $2(\sum_{i=2}^{n}(a_i - 1) + 1)$.

Relaxing the tree constraints, we study the ∧ realizability of D in the context of asymmetric directed graphs[2]. These are classes of directed graphs where there are no cycles of length 2. We prove the following:

Theorem 3. *For any degree set* $D = \{a_1 < a_2 < \ldots < a_n\}$, *let* $\mu_A(D)$ *be the minimum number of vertices of any asymmetric directed graph realizing* D, *then:*

$$a_1 + a_n + 1 \leq \mu_A(D) \leq a_{n-1} + a_n + 1$$

We also give sufficient conditions among the a_i's under which the lower bound is achieved.

---

[2] In [2], Chartrand *et al* studied directed asymmetric graph realizations of degree sets D. However, their definition of realization is only with respect to out-degree which differs from our definition.

Turning to algorithmic questions : we consider the following TREE EXTENSION PROBLEM : Given a degree set D and a number r, test whether there is an undirected tree T that realizes the degree set D such that T has exactly $\mu_T(D)+r$ vertices. Here $\mu_T(D)$ denotes the minimum order of any tree realizing D.

From a known characterization of the realizability of D with a given number of vertices using the well-studied Frobenius problem[8], we show that the problem is polynomial-time many-one equivalent to INTEGER KNAPSACK PROBLEM.

We study parametrized versions of the problem, with respect to two parameters - $|D|$, r. We show the following results.

Theorem 4. TREE EXTENSION PROBLEM *is fixed parameter tractable with respect to the parameters $|D|$ and r, when r is presented in unary.*

We study the following variant of the computational question. Given a tree T and a non-negative integer r, test whether there is another tree T' with the same degree set but now having exactly $|T|+r$ number of vertices. We show that this problem can be solved in log-space and hence in polynomial time.

The analogous problems for directed trees turn out to be surprisingly easier. We prove the following characterization. For a degree set D, let $\mu_\wedge(D)$ ($\mu_\vee(D)$) denote the minimum number of vertices required to $\wedge$-realize ($\vee$-realize) the degree set D using a directed tree.

Theorem 5. *Given a set D and a value r, the degree set can always be $\wedge$-realized (resp. $\vee$-realized) using a directed tree of $\mu_\wedge(D) + r$ (resp. $\mu_\vee(D) + r$ vertices.)*

2   Preliminaries

Let $G = (V, E)$ be a graph[3]. For $v \in V$, by $d(v)$ we denote the degree of the vertex v in G. A degree-set of a graph G (first studied by [11]) is a subset of $\mathbb{N}$[4] defined as follows: $D(G) = \{d(v) : v \in G\}$. A set $D \subset \mathbb{N}$ is said to be *realizable* if and only if there is a graph G such that $D(G) = D$.

The degree-sequence of the graph G is the sequence of numbers : $d(G) = (d(v) : v \in G)$. A sequence D with elements from $\mathbb{N}$ is said to be realizable if there is a graph whose degree sequence (up to the ordering) is $d(G)$. Several results are known about characterizing realizability of degree sequences using graphs and various subclasses of graphs[3,10,9].

Let $\mu(D)$ denote the minimum number of vertices that must be present in any realization of D. Let $\mu_T(D)$ denote the minimum number of vertices that must be present in any realization of D when the graph is restricted to be a tree.

In directed graphs, for a vertex v we denote by $d^+(v)$ and $d^-(v)$, the outdegree and indegree respectively. We write the outdegree and indegree for a vertex v_i as an ordered pair (a_i, b_i) which means $d^+(v_i) = a_i, d^-(v_i) = b_i$. A directed graph is said to be *asymmetric* if it does not have cycles of length two. Let $\mu_A(D)$ denote

---

[3] All graphs being considered in this paper are simple.
[4] $\mathbb{N}$ includes 0.

the minimum number of vertices that any asymmetric directed graph realizing D must have. If it is clear from the context, we drop the notation for type of realizability.

An intermediate case between degree sets and degree sequences is that of multiplicity-constrained degree sets, where we restrict the number of times that a term in the degree set appears in the realization. A natural restriction to study is when the multiplicity is bounded from above, given that the degree distribution in the realization of D with trees is highly skew. We also consider the complementary variant, where the multiplicity is bounded from below. In these cases, we can denote the degrees with a multi-set $D_m = \{a_1^{m_1}, a_2^{m_2}, \ldots, a_n^{m_n}\}$ where $a_i^{m_i}$ denotes that a_i is appearing at least m_i times in the multiset and m_is are positive integers. We now focus on a very special case of the degree multiset when $a_1 = 1, m_1 = 1$, which we need later in our construction. Under this assumption, $D_m = \{1, a_2^{m_2}, \ldots, a_n^{m_n}\}$. Since $1 \in D_m$, there exists a tree realization for D_m, we obtain the lower bound for any tree realizing D_m. We state the following proposition. The proof of this is an easy generalization of the argument in [11] which assumes each $m_i = 1$.

Proposition 1. *The minimum order of a tree realizing $D_m = \{1, a_2^{m_2}, \ldots, a_n^{m_n}\}$ is $\sum_{i=2}^{n} m_i(a_i - 1) + 2$.*

We briefly introduce the basics of parametrized complexity that we need in the paper. We refer the reader to a standard textbook[6] for details. A parametrized computational problem instance is denoted by (I, k) where k is the parameter. A problem is fixed parameter tractable (FPT) with respect to the parameter k if there is an algorithm solving the problem in time $f(k).n^{O(1)}$ where n is the size of the instance. In general, the choice of the parameter k is not unique. That is, it is possible to parametrize a problem in more than one way and using more than one parameter.

3   Multiplicity Lower Bounds in Tree-Realizations

In this section, we prove lower bounds for the multiplicities of the pendant vertices (vertices of degree 1) in any realization of a degree set D using trees. We prove Theorem 1.

Theorem 6. *The minimum multiplicity of pendant vertices in any tree realization for the degree set $D = \{1 = a_1 < a_2 < \ldots < a_n\}$ is $\sum_{i=1}^{n} a_i - 2n + 3$.*

Proof. The set $D = \{1 = a_1 < a_2 < \ldots < a_n\}$ can be realized by a tree[11]. Minimum order of such a tree is $\sum_{i=1}^{n}(a_i - 1) + 2$. In minimum order tree construction, each a_i is connected with exactly $a_i - 2$ pendant vertices for $i = 3, 4, \ldots, n - 1$ and for $i = 2$ *and* n, a_i's are connected with $a_i - 1$ pendant vertices and then a_i is connected with a_{i+1} for $i = 2, \ldots, n - 1$. Let m_i be the multiplicity of a_i in a tree realization T. Then, $(1^{m_1}, a_2^{m_2}, \ldots, a_n^{m_n})$ will be the degree sequence of T.

Case 1 : when $a_2 \geq 3$. We recall that, if degree sequence $d = (d_1 \geq d_2 \geq \ldots \geq d_n)$ is being realized by a tree then number of pendant vertices in any tree realization [1] of d is $\sum_{i=1}^{k}(d_i - 2) + 2$ where k is the largest index such that $d_k \geq 3$. Hence, $m_1 = 2 + (a_2 - 2)m_2 + (a_3 - 2)m_3 + \ldots + (a_n - 2)m_n$, $\forall i$ $m_i \geq 1$. m_1 will be minimum if $m_i = 1$ for each $i = 2, 3, \ldots, n$ and the tree construction described above meets exactly this requirement. So minimum value $m_1 = 2 + (a_2 - 2) + (a_3 - 2) + \ldots + (a_n - 2) = \sum_{i=1}^{n} a_i - 2(n-1) + 1$ $= \sum_{i=1}^{n} a_i - 2n + 3$

Case 2 : when $a_2 = 2$. We first construct the tree for the degree set $D_1 = \{1 = a_1 < a_3 < \ldots < a_n\}$ in the way mentioned above and then introduce a vertex v. Now make v adjacent to any one pendant vertex, say u, so that v becomes the new pendant vertex and $d(u) = 2$. Degree set of this modified tree is D and number of pendant vertices is same as that in the tree realization of D_1 which is same as $m_1 = 2 + (a_3 - 2) + (a_4 - 2) + \ldots + (a_n - 2 = 2 + (a_2 - 2) + (a_3 - 2) + \ldots + (a_n - 2) = \sum_{i=1}^{n} a_i - 2(n-1) + 1 = \sum_{i=1}^{n} a_i - 2n + 3$

□

The above lemma can be generalized to the case of multisets and the proofs are quite similar.

Theorem 7. *The minimum multiplicity of pendant vertices in any tree realization for the degree multiset* $D_m = \{1, a_2^{m_2}, \ldots, a_n^{m_n}\}$ *is* $\sum_{i=2}^{n} m_i(a_i - 2) + 2$.

4   Minimum-Order Realizability of Directed Trees

In this section we explicitly compute the minimum number of vertices needed to $\wedge$-realize (resp. $\vee$-realize) the given degree set D using directed trees.

We describe $\vee$-realizability first. We prove the following general upper bound for $\mu_\vee(D)$. Let $\mu_B(D)$ denote the minimum number of vertices for any undirected bipartite graph realizing the degree sequence D. Given any undirected bipartite realization of a degree set by a graph $G = (U, V, E)$ we assign directions from U to V. This gives a $\vee$-realization of the same graph using a directed bipartite graph. Thus, we have the following proposition.

Proposition 2. $\mu_\vee(D) \leq \mu_B(D)$

Indeed, this proposition holds for directed trees as well since the underlying undirected graph is bipartite. We now argue that this upper bound is tight for trees and show the following theorem.

Theorem 8. *For the degree set* $D = \{1 = a_1 < a_2 \ldots < a_n\}$, *minimum order of a directed tree* $T(V, E)$ *so that* $\forall v \in V, d^+(v) \in D$ *or* $d^-(v) \in D$, *and for each* $a_i \in D$ *there is a vertex* $u \in V$ *such that* $d^+(u) = a_i$ *or* $d^-(u) = a_i$, *is same as the minimum order undirected tree realizing* D, *i.e.* $\sum_{i=1}^{n}(a_i - 1) + 2$.

Proof. The upper bound follows from the above proposition through the undirected tree-realization of D with optimal number of vertices.

Now we need to prove a lower bound on the order of a directed tree satisfying the given constraints and then give a realization which meets this bound.

For each i, $a_i \in D$ will appear as both (a_i, a_j) and (a_k, a_i) at least once, where $a_j, a_k \in D$. Thus, $1 \leq a_i + a_j \leq 2a_n$. Let $T(V, E)$ be a directed tree for D satisfying the constraints. We have,

$$\sum_{v \in V} (d^-(v) + d^+(v)) = 2|E| = 2(V - 1) \geq \sum_{i=1}^{n} a_i + (V - n)$$

This implies the lower bound $|V| \geq 2 + \sum_{i=1}^{n}(a_i - 1)$.   □

Now we turn to $\wedge$-realizability of D using directed trees. It can be noted that a necessary condition is $0 \in D$ since the tree has to contain leaf nodes whose in-degree or out-degree has to be 0.

Theorem 9. *For the degree set $D = \{0 < 1 < a_2 < \ldots < a_n\}$, the minimum order of a directed tree which $\wedge$-realizes the degree set D, is $2\left(\sum_{i=1}^{n}(a_i - 1)\right) + 2$.*

Proof. We prove the upper bound by constructing the directed tree. Construct a path with $2(n-1)$ number of vertices, say $u_1, u_2, \ldots, u_{2n-2}$. Now add $(a_2 - 1)$ pendant vertices to u_1. For each $2 \leq i \leq 2n - 1$, add $a_{\lceil \frac{i}{2} \rceil + 1} - 2$ pendant vertices to u_i. Add $a_n - 1$ pendant vertices to the u_{2n-2}.

In this tree, first 2 vertices are having degree a_2, next 2 vertices are having degree a_3 and so on. Now we assign directions. Start with the first vertex u_1 in the path. Direct all edges connected with u_1 towards u_1. For the next vertex in the path u_2 assign directions to all adjacent edges away from u_2. Repeat this process to assign direction to all edges. Since each a_i, for $i = 2, 3, \ldots, n$, appears exactly twice and because of the way we are assigning directions to edges, a_i once appears as $(a_i, 0)$ and once as $(0, a_i)$ in final directed tree. For pendant vertices in undirected graph, indegree and outdegree pair occurs as either $(1, 0)$ or $(0, 1)$.

To prove the minimality, we first observe that the number of vertices in the above construction is $|V| = \sum_{i=2}^{n} 2(a_i - 1) + 2$. Now, consider the corresponding degree multiset $\{1, a_2^2, a_3^2, \ldots, a_n^2\}$. Applying proposition 1 with $m_i = 2 \forall i$ gives a matching lower bound on $|V|$.   □

5   Minimum Order $\wedge$-Realizability of Asymmetric Graphs

In this section we study $\wedge$ realizations of degree sets with asymmetric directed graphs. We introduce a notation for convenience in this section. For a directed graph G, let $\mathcal{A}_G$ denote the set that is $\wedge$-realized by G. Since the realizability is fixed, we drop it from the notation. Recall that $\mu_A(D)$ denotes the minimum order of any asymmetric directed graph realizing D. We start with a simple case which is similar to the starting point in [2].

Lemma 1. *If $D = \{a\}$ where a is a non-negative integer, then $\mu_A(D) = 2a+1$.*

Proof. This case is similar to [2]. When $a = 0$ the graph is an isolated vertex. For $a \geq 1$, all vertices in a directed graph with $\mathcal{A}_G = \{a\}$ must have both indegree and outdegree equal to a. Consider a vertex v, since the graph is asymmetric, v is connected to $2a$ distinct vertices. Accounting for these vertices and v, we have $2a + 1$ vertices. Hence, $\mu_A(D) \geq 2a + 1$. To complete the proof, we need to prove that $\mu_A(D) \leq 2a + 1$. To do this, we will come up with a construction of a directed graph with $\mathcal{A}_G = \{a\}$ and order $2a + 1$.

We define G to be the directed graph with the vertex set $\{v_1, v_2, \ldots, v_{2a+1}\}$. The edges are as follows: $\{(v_i, v_j)|1 \leq i \leq 2a + 1 \text{ and } i+1 \leq j \leq i+a\}$ (where subscripts are modulo $2a + 1$). Clearly, G is asymmetric and has $2a + 1$ vertices with $\mathcal{A}_G = \{a\}$. Hence the proof. $\square$

Theorem 10. *If $D = \{a_1 < a_2 < \ldots < a_n\}, n \geq 2$ is a set of positive integers then*

$$a_1 + a_n + 1 \leq \mu_A(D) \leq a_{n-1} + a_n + 1.$$

Proof. We know that there is at least one vertex v of G with either indegree or outdegree equal to a_n. Without loss of generality, let us assume that $d^+(v) = a_n$. Now, we know that $d^-(v) \geq a_1$. Therefore, $d^+(v) + d^-(v) \geq a_n + a_1$. Since G is also asymmetric, it implies that the order of G is at least $a_1 + a_n + 1$.

To prove that $\mu_A(S) \leq a_{n-1} + a_n + 1$, we proceed by induction. By Lemma 1, we know that $\mu_A(\{a_1\}) = 2a_1 + 1$. Let the graph representing this be G_1. Divide G_1 into three components, C_x, C_y - each containing a_1 vertices, and C_z - containing the remaining vertex. From G_1, we obtain G_2, by adding a new component C_1 containing $a_2 - a_1$ vertices and adding the following edge set $E = \{(v_x, v_1)|v_x \in C_x \wedge v_1 \in C_1\} \cup \{(v_1, v_y)|v_1 \in C_1 \wedge v_y \in C_y\}$. Thus, we have an asymmetric directed graph for the degree set $\{a_1 < a_2\}$ with order $a_1 + a_2 + 1$.

Now consider that there exists an asymmetric directed graph G_{n_0} with $\mathcal{A}_G = \{a_1 < a_2 < \ldots < a_{n_0}\}$, with order $a_{n_0-1} + a_{n_0} + 1$. G_{n_0} contains a total of $2n_0$ components :

- C_{n_0-1}, containing $a_{n_0} - a_{n_0-1}$ vertices with outdegree and indegree equal to a_1.
- C_i, for i from 1 to $n_0 - 2$, each containing $a_{i+1} - a_i$ vertices with outdegree a_1 and indegree a_{n_0-1-i}.
- C'_j, for j from 1 to $n_0 - 2$, each containing $a_{j+1} - a_j$ vertices with outdegree a_{n_0-1-j} and indegree a_1.
- C_x, containing a_1 vertices with outdegree a_{n_0} and indegree a_{n_0-1}.
- C_y, containing a_1 vertices with outdegree a_{n_0-1} and indegree a_{n_0}.
- C_z, containing 1 vertex with outdegree and indegree a_1.

From G_{n_0}, we obtain G_{n_0+1}, by adding two new components - C_{n_0} containing $a_{n_0+1} - a_{n_0}$ vertices, and C'_{n_0-1} containing $a_{n_0} - a_{n_0-1}$ vertices, and adding the edge set $E = E_1 \cup E_2 \cup E_3$, where

- $E_1 = \{(v_x, v_{n_0})|v_x \in C_x \wedge v_{n_0} \in C_{n_0}\} \cup \{(v_{n_0}, v_y)|v_{n_0} \in C_{n_0} \wedge v_y \in C_y\}$

$-\ E_2 = \{(v_y, v_{n_0-1})|v_y \in C_y \wedge v_{n_0-1} \in C'_{n_0-1}\} \cup \{(v_{n_0-1}, v_x)|v_{n_0-1} \in C'_{n_0-1} \wedge$
$\quad\ v_x \in C_x\}$

$-\ E_3 = \{(v_i, v'_i)|v_i \in C_i \wedge v'_i \in C'_{n_0-1-i}\}$, where $i \in \{1, 2, \ldots, n_0 - 2\}$

We can observe that G_{n_0+1} resembles G_{n_0} if n_0 is replaced with $n_0 + 1$. Thus, through this construction, we have proved that there always exists a asymmetric directed graph G with $\mathcal{A}_G = (a_1 < a_2 < \ldots < a_n)$, of order $a_{n-1} + a_n + 1$. Hence, the minimum order $\mu_A(D) \leq a_{n-1} + a_n + 1$. □

We now identify a condition that is sufficient in order to achieve the lower bound in theorem 10.

Lemma 2. *If* $D = \{a_1 < a_2 < \ldots < a_n\}, n \geq 2$ *is a set of positive integers which satisfies the following condition:* $a_i + a_{n+1-i} = a_j + a_{n+1-j} \forall i < j$ *then,* $\mu_A(D) = a_1 + a_n + 1$.

Proof. From Theorem 10, we know that $\mu_A(D) \geq a_1 + a_n + 1$. So, we only have to show that, if the given condition is satisfied, $\mu_A(D) \leq a_1 + a_n + 1$. To do this, we will come up with a construction of a directed graph G with $\mathcal{A}_G = \{a_1 < a_2 < \ldots < a_n\}$ and order $a_1 + a_n + 1$.

D satisfies the given condition. We shall construct a directed graph with order $a_1 + a_n + 1$ by induction on n.

For $n = 2$, $a_1 + a_n + 1 = a_{n-1} + a_n + 1$. Therefore, by Theorem 10, we can always construct a directed graph for $n = 2$ with order $a_1 + a_2 + 1$.

Now take $n = 3$, define G to be the directed graph with vertex set $V(G) = \{v_1, v_2, \ldots, v_{a_1+a_3+1}\}$ and $E(G) = \{(v_1, v_j)|2 \leq j \leq a_1 + 1\} \cup \{(v_j, v_1)|a_1 + 2 \leq j \leq a_1 + a_3 + 1\} \cup \{(v_{a_1+a_3+1}, v_j)|2 \leq j \leq a_3\} \cup \{(v_j, v_{a_1+a_3+1})|a_3 + 1 \leq j \leq a_1 + a_3\}$. G has $a_1 + a_3 - 1$ vertices of indegree and outdegree 1. Since we know that the given condition is satisfied, $a_1 + a_3 = 2a_2$ and $a_1 + a_3 - 1 = 2(a_2 - 1) + 1$. From Lemma 1 1, we can construct a directed graph G_1 of order $a_1 + a_3 - 1$ with $\mathcal{A}_{G_1} = \{a_2 - 1\}$. Superimposing G_1 on the vertices with outdegree and indegree 1 in G, we get a directed graph for $n = 3$ with order $a_1 + a_3 + 1$.

Now, let us assume that such a construction is possible for $n = m$. We will try to construct a graph of order $a_1 + a_n + 1$ for $n = m + 2$. Define G to be the directed graph with $V(G) = \{v_1, v_2, \ldots, v_{a_1+a_n+1}\}$ and $E(G) = \{(v_1, v_j)|2 \leq j \leq a_1 + 1\} \cup \{(v_j, v_1)|a_1 + 2 \leq j \leq a_1 + a_n + 1\} \cup \{(v_{a_1+a_n+1}, v_j)|2 \leq j \leq a_n\} \cup \{(v_j, v_{a_1+a_n+1})|a_n + 1 \leq j \leq a_1 + a_n\}$. G has $a_1 + a_n - 1$ vertices of indegree and outdegree 1. Since we know that the required condition is satisfied, $a_1 + a_n = a_2 + a_{n-1}$ and $a_1 + a_n - 1 = (a_2 - 1) + (a_{n-1} - 1) + 1$.

From our induction assumption, we can construct a graph G_1 of order $a_1 + a_n - 1$ with $\mathcal{A}_{G_1} = \{a_2 - 1, a_3 - 1, \ldots, a_{n-1} - 1\}$ (because G_1 has m number of vertices. Superimposing G_1 on the vertices with outdegree and indegree 1 in G, we get the desired graph for $n = m + 2$. This completes the construction and the proof. □

We are able to prove exact bounds for a special case of the degree set.

Lemma 3. *If* $D = \{0, a_2\}$, *then* $\mu_A(D) = 2a_2$.

Proof. Consider a directed asymmetric graph G for which $\mathcal{A}_G = \{0, a_2\}$. We know that G has at least one vertex, say v_1, with outdegree equal to a_2. Its indegree can be equal to either 0 or a_2. Consider the case in which its indegree is a_2. Since the graph is asymmetric, v_1 connects to $2a_2$ distinct points. Thus the order of G in this case would be at least $2a_2 + 1$. Now, consider the case where $d^-(v_1) = 0$. Here, v_1 connects to a_2 vertices (say $v_2, v_3, \ldots, v_{a_2+1}$), whose indegrees now cannot be equal to 0, and so are all equal to a_2. So, v_2 has edges coming in from $a_2 - 1$ vertices apart from v_1. If any of these vertices are one of $v_2, v_3, \ldots, v_{a_2+1}$, then that particular vertex would have both indegree and outdegree equal to a_2, realizing our earlier case and thus making the order of G at least $2a_2 + 1$. However, if v_2 does not connect to any of $v_2, v_3, \ldots, v_{a_2+1}$, then it connects to $a_2 - 1$ new vertices ($v_{a_2+2}, v_{a_2+3}, \ldots, v_{2a_2}$). Thus the order of G would be at least $2a_2$. From the above cases, we can see that the order of the directed graph must be at least $2a_2$, i.e. $\mu_A(\{0, a_2\}) \geq 2a_2$. To complete the proof, we need to prove that $\mu_A(\{0, a_2\}) \leq 2a_2$. To do this, we will come up with a construction of a directed graph with $\mathcal{A}_G = \{0, a_2\}$ and order $2a_2$.

Define G to be the directed graph with $V(G) = \{v_1, v_2, \ldots, v_{2a_2}\}$ and $E(G) = \{(v_i, v_j) | 1 \leq i \leq a_2 \text{ and } a_2 + 1 \leq j \leq 2a_2\}$. Then G is asymmetric with order $2a_2$ and $\mathcal{A}_G = \{0, a_2\}$. Hence, the proof.   □

6   Complexity Results on Tree Extension Problem

We argue complexity results on the following algorithmic problems related to degree set realizations of trees. We define the problems formally first.

TREE EXTENSION PROBLEM(TEP) : Given a degree set D and an integer r, test if there is a tree having $\mu_T(D) + r$ vertices that realizes the degree set D.

UNARY TREE EXTENSION PROBLEM (UTEP) : Given a tree T on ℓ vertices and a string 1^r, test if there is another tree T' having exactly $\ell + r$ vertices and the degree set same as that of T.

One important ingredient of our arguments about complexity of the above stated problems is the following combinatorial connection first proved by Gupta *et al*[7] between realizability and the well-studied Frobenius problem. We state it differently here, but the proof can be derived from the proof of Theorem 3 in [7]. However, we also give an alternative proof for the forward direction.

Lemma 4 ([7]). *If the degree set $D = \{a_1 = 1 < a_2 < \ldots < a_n\}$ is realized by a tree $T(V, E)$ then we can get another tree realization $T_1 = (V_1, E_1)$ where $|V_1| = |V| + r, r$ is a positive integer, if and only if r is a linear combination of $(a_i - 1)$, i.e.*

$$r = \sum_{i=2}^{n} k_i(a_i - 1) \tag{1}$$

where k_i's are non-negative integers.

Proof. Without loss of generality, we fix $T(V, E)$ as the tree with minimum order realizing D. Let m_i be the multiplicity of vertices with degree a_i in T. Hence, $m_i = 1$, for each $2 \leq i \leq n$ and $m_1 = \sum_{i=1}^{n} a_i - 2n + 3$ which is also the minimum multiplicity of pendant vertices in any tree realization. Now we add r vertices so that exactly k_i vertices are produced with degree a_i, where k_i's are non-negative integers, to get $T_1 = (V_1, E_1)$ and hence $r = \sum_{i=1}^{n} k_i$. So $\{1^{\sum_{i=1}^{n} a_i - 2n + 3 + k_1}, a_2^{1+k_2}, \ldots, a_i^{1+k_i}, \ldots, a_n^{1+k_n}\}$ is the degree sequence of T_1. Consider two following cases:

Case 1 : $a_2 \geq 3$. By the bounds from [1], $\sum_{i=1}^{n} a_i - 2n + 3 + k_1 = \sum_{i=2}^{n}(k_i + 1)(a_i - 2)$ From this we get $k_1 = \sum_{i=2}^{n} k_i(a_i - 2)$. Hence $r = \sum_{i=1}^{n} k_i = \sum_{i=1}^{n} k_i(a_i - 1)$.

Case 2 : $a_2 = 2$. Since $(a_2 - 2) = 0$ so $\sum_{i=3}^{n}(k_i + 1)(a_i - 2) = \sum_{i=2}^{n}(k_i + 1)(a_i - 2)$. Hence we will get the same result. This completes the proof. $\square$

Using the above Lemma, we show the following theorem:

Theorem 11. UNARY TREE EXTENSION PROBLEM *can be solved in log-space.*

Proof. We prove the theorem by reducing the problem to unary subset sum problem which can be solved in log-space. The unary subset sum problem is defined as follows. Given a (multi)-set S of m integers $b_1, b_2, \ldots b_m$ and a value c (all inputs in unary) test if there is a subset S' of these integers such that $\sum_{i \in S'} b_i = c$. The reduction runs in log-space as follows. For $1 < i \leq n$, let $t_i = \lceil \frac{r}{a_i - 1} \rceil$. Given a tree T and r in unary, write down the following set S and r in unary, choose $c = r$ and define:

$$S = \bigcup_{i=2, j=1}^{i=n, j=t_i} \{(a_i - 1)j\}$$

Indeed, if there is a subset of S that sums up to r, then it is clear that this choice of the j's satisfies equation 1. Any solution for the k_i's in equation 1, it must be that $k_i \leq t_i$ for all i. Hence the corresponding terms $k_i(a_i - 1)$ will appear in the set S as well. Choosing these terms in S' ensures $\sum_{i \in S'} b_i = r = c$. To argue the complexity of the reduction, notice that we can compute a_i's each time on the fly by enumerating the degree up to the maximum degree. This can be done in log-space. $\square$

The idea in the above proof can be adapted to argue that TREE EXTENSION PROBLEM is equivalent to INTEGER KNAPSACK PROBLEM(IKP) which can be stated as follows : Given non-negative integers $c_1, \ldots, c_k$, and a value d - the problem asks if there are non-negative integers $d_1, d_2, \ldots, d_k$ such that $\sum_i c_i d_i = d$. Given a degree set D, consider the IKP instance with $k = |D| - 1$ and $c_i = a_{i+1} - 1$ for all $1 \leq i \leq k$. Choose $d = r$. In the reverse direction, given non-negative integers $c_1, \ldots, c_k$, and a value d, consider the degree set $D = \{1, c_1, \ldots, c_k\}$ and $r = d$. The correctness of the reductions follow from Lemma 4 directly. This discussion gives us the following proposition.

Proposition 3. TREE EXTENSION PROBLEM *is equivalent to* INTEGER KNAP-
SACK PROBLEM.

We consider two natural parameterizations of the unary version of the TREE
EXTENSION PROBLEM and argue theorem 4.

Parameterizing with Respect to $|D|$ When r Is Given in Unary : In this
setting, we give a reduction to VARIETY SUBSET SUM PROBLEM. The variety
subset-sum problem : given a multiset A, and a target sum b, the problem asks
if there is a sub(multi)set of A that adds up to exactly b. To do the reduction,
we will list down the number $(a_i - 1)$ where $a_i \in A$, exactly r number of times
in the subset. Since r is given in unary we can, in polynomial time, write out
these numbers. There will be exactly nr of them. The correctness and resource
bounds of the reduction follow easily. The VARIETY SUBSET SUM PROBLEM is
known[4] to be fixed-parameter tractable with respect to the number of distinct
elements in A as the parameter. As we can see in the above case, this is precisely
$|D|-1$. Hence TREE REALIZABILITY PROBLEM is fixed-parameter tractable with
respect to $|D|$ as the parameter.

**Parameterizing with Respect to r as the Parameter, When r Is Given
in Unary** : We first notice that VARIETY SUBSET SUM PROBLEM reduces
to MAXIMUM KNAPSACK PROBLEM. We define the problem first. Given a set
$\{x_1, x_2, \ldots x_n\}$ with sizes $s_1, s_2, \ldots s_n$ and profits $p_1, p_2, \ldots p_n$ respectively, and
two values knapsack capacity b and profit threshold k - test if there exists a
subset $S \subseteq [n]$ such that : $\Sigma_{i \in S} s_i \leq b$ and $\Sigma_{i \in S} p_i \geq b$. To reduce VARIETY
SUBSET SUM PROBLEM, given $A = \{a_1, a_2, \ldots a_n\}$ and target sum t, produce
x_i's such that $s_i = p_i = a_i$ and $b = p = t$. The inequalities ensures that the Maximum Knapsack Problem has a solution if and only if there is a subset $A' \subseteq A$
which adds up to exactly t. Fernau[5] has shown that the MAXIMUM KNAPSACK
PROBLEM is fixed parameter tractable with respect to the parameter b. Since
our reduction maps the parameter t to exactly p, this shows that the TREE EX-
TENSION PROBLEM is fixed parameter tractable with respect to the parameter r
when r is given in unary.

6.1   Tree Extension Problem for Directed Trees

In this section we address similar computational problem for directed trees under
the $\wedge$-realizability and the $\vee$-realizability. Surprisingly in both cases, it turns out
to be the case that for every non-negative integer r, there are directed trees with
$\ell + r$ vertices $\wedge$-realizing and $\vee$-realizing (where ℓ takes appropriate values from
Theorem 8 and Theorem 9 respectively). We prove these two results now.

Theorem 12. *For the degree set $D = \{0, 1, a_2, \ldots, a_n\}$ if we have a directed tree
realization $T_d(V_d, E_d)$ then we can have another tree realization[5] $T_{d_1} = (V_{d_1}, E_{d_1})$
where $|V_{d_1}| = |V_d| + r$ for each non-negative integer r.*

---
[5] For $\vee$-realizability, we do not require 0 to be in D.

Proof. Without loss of generality, we fix $T_d(V_d, E_d)$ as the directed tree with minimum order realizing D. We now consider two cases depending on the number of pendant vertices, say V_p, in T_d:

Case 1: when $r \leq V_p$

Add r number of pendant vertices to any r number of already existing pendant vertices in T_d so that if $d^-(p) = 1$, make p adjacent to newly added vertex by an outgoing edge and similarly if $d^+(p) = 1$, make p adjacent to newly added vertex by an incoming edge. Since $0, 1 \in D$, degree set remains unchanged and we get another tree T_{d_1} with k vertices more than T_d.

Case 2 : when $r \geq V_p$, let $r = l * V_p + r_0$ where l is a positive integer ≥ 1 and r_0 is another non-negative integer $\leq V_p - 1$.

First add V_p number of pendant vertices to T_d in the way described in case 1 and repeat the same procedure $(l-1)$ times more with directed tree obtained from the previous iteration and in the process degree set also does not change as explained above. In last iteration, we will do the same for remaining r vertices. This completes the proof.

$\square$

References

1. Arikati, R., Srinavasa, Maheshwari, A.: Realizing degrees sequences in parallel. SIAM Journal of Discrete Mathematics 9, 317–338 (1996)
2. Chartrand, G., Lesniak, L., Roberts, J.: Degree sets for digraphs. Periodica Mathematica Hungarica 7, 77–85 (1976)
3. Erdös, P., Gallai, T.: Graphs wiyh prescribed degrees of vertices. Mat. Lapok 11, 264–274 (1960)
4. Fellows, M.R., Gaspers, S., Rosamond, F.A.: Parameterizing by the Number of Numbers. In: Raman, V., Saurabh, S. (eds.) IPEC 2010. LNCS, vol. 6478, pp. 123–134. Springer, Heidelberg (2010)
5. Fernau, H.: Parameterized algorithms: A graph-theoretic approach. Technical report, Universität Tübingen, Tübingen, Germany (2005)
6. Flum, J., Grohe, M.: Parameterized Complexity Theory. Texts in Theoretical Computer Science. An EATCS Series. Springer-Verlag New York, Inc., Secaucus (2006)
7. Gupta, G., Joshi, P., Tripathi, A.: Graphic sequences of trees and a problem of frobenius. Czechoslovak Mathematical Journal 57, 49–52 (2007)
8. Guy, R.K.: Unsolved Problems in Number Theory, Unsolved Problems in Intuitive Mathematics, 3rd edn., vol. I. Springer, New York (2004)
9. Louis Hakimi, S.: On realizability of a set of integers as degrees of the vertices of a linear graph i. SIAM Journal of Discrete Mathematics 10, 496–506 (1962)
10. Havel, V.: Eine bemerkung über die existenz der endlichen graphen. Časopis Pěst. Mat. 80, 477–480 (1955)
11. Kapoor, S.F., Polimeni, A.D., Wall, C.E.: Degree sets for graphs. Fundamental Mathematics 95, 189–194 (1977)

An FPT Algorithm for TREE DELETION SET

Venkatesh Raman[1], Saket Saurabh[1], and Ondřej Suchý[2],[*]

[1] The Institute of Mathematical Sciences, Chennai, India
{vraman,saket}@imsc.res.in
[2] Faculty of Information Technology, Czech Technical University
Prague, Czech Republic
ondrej.suchy@fit.cvut.cz

Abstract. We give a $5^k n^{O(1)}$ fixed-parameter algorithm for determining whether a given undirected graph on n vertices has a subset of at most k vertices whose deletion results in a tree. Such a subset is a restricted form of a feedback vertex set. While parameterized complexity of feedback vertex set problem and several of its variations have been well studied, to the best of our knowledge, this is the first fixed-parameter algorithm for this version of feedback vertex set.

1 Introduction

The goal of parameterized complexity is to find ways of solving NP-hard problems more efficiently than brute force: our aim is to restrict the combinatorial explosion to a parameter that is hopefully much smaller than the input size. Formally, a *parameterization* of a problem is assigning an integer k to each input instance and we say that a parameterized problem is *fixed-parameter tractable* (FPT) if there is an algorithm that solves the problem in time $f(k) \cdot n^{O(1)}$, where n is the size of the input and f is an arbitrary computable function depending on the parameter k only. There is a long list of NP-hard problems that are FPT under various parameterizations: finding a vertex cover of size k, finding a cycle of length k, finding a maximum independent set in a graph of treewidth at most k, etc. For more background, the reader is referred to the monographs [6,7,19].

One of the most well studied direction in parameterized complexity is to "delete vertices of the input graph such that the resulting graph satisfies some interesting properties". More precisely, a natural optimization problem associated with a graph class $\mathcal{G}$ is the following: given a graph G, what is the minimum number of vertices to be deleted from G to obtain a graph in $\mathcal{G}$? For example, when $\mathcal{G}$ is the class of empty graphs, forests or bipartite graphs, the corresponding problems are VERTEX COVER, FEEDBACK VERTEX SET and ODD CYCLE TRANSVERSAL, respectively. In the parameterized setting, the parameter for

---

[*] Part of the work was done while with the Universität des Saarlandes, Saarbrücken, supported by the DFG Cluster of Excellence on Multimodal Computing and Interaction (MMCI) and DFG project DARE (GU 1023/1-2), and while visiting IMSc Chennai, supported by the Indo-German Max Planck Center for Computer Science (IMPECS).

S.K. Ghosh and T. Tokuyama (Eds.): WALCOM 2013, LNCS 7748, pp. 286–297, 2013.
© Springer-Verlag Berlin Heidelberg 2013

vertex-deletion problems is the solution size, that is, the number of vertices to be deleted so that the resulting graph belongs to the given graph class. This line of research has been at the forefront of research in parameterized complexity and various new results have been obtained in the last few years. For examples, an improved algorithm for ODD CYCLE TRANSVERSAL [13,18], meta theorems for class of deletion problems [9,12], PROPER INTERVAL VERTEX DELETION [21], DIRECTED/UNDIRECTED SUBSET FEEDBACK VERTEX SET [4,5].

In this paper we consider the following variant of the classical FEEDBACK VERTEX SET problem in the realm of parameterized complexity:

WEIGHTED TREE DELETION SET (WTDS)

Input:　　　An undirected graph $G = (V, E)$, a weight function $w : V \to \mathbb{N}^+$ on vertices, and an integer $k \in \mathbb{N}$.

Parameter:　k

Question:　Is there a set $S \subseteq V$ of total weight $\sum_{v \in S} w(v)$ at most k, such that $G[V \setminus S]$ is a tree?

If $w(v) = 1$ for every $v \in V$, then we speak simply about TREE DELETION SET (TDS). We also refer the subset S as a *tree deletion set* .

TDS is a special case of WTDS, but on the other hand, if $k = n^{O(1)}$, where $n = |V|$, then WTDS is polynomial time reducible to TDS, by adding to each vertex v of the graph $min\{k, w(v) - 1\}$ pendant vertices. Clearly the resulting unweighted graph has a TDS of size at most k if and only if the original graph has a WTDS of weight at most k. This is because if an original vertex is in the tree deletion set, then all the pendant vertices adjacent to it must also be in that set.

If we simply want to find a subset S of vertices such that $G[V \setminus S]$ is a forest, then S is a feedback vertex set. Finding a (size at most k or minimum) feedback vertex set is a well known NP-complete problem and has been well studied in the paradigms of parameterized complexity [2,3,20], approximation [1] and exact algorithms [8]. As a tree deletion set is also a feedback vertex set, it is clear that the size of the minimum tree deletion set is at least the size of the minimum feedback vertex set. However the minimum tree deletion set can be arbitrarily large compared to the size of the minimum feedback vertex set. Consider a graph which simply has a cycle on three vertices, with each vertex attached to a large number of pendant vertices. Any of the vertices of the cycle forms a feedback vertex set (of size 1), but the minimum tree deletion set must contain all the pendant vertices attached to that vertex. Furthermore, standard preprocessing rules like deleting degree one vertices and 'short circuiting' degree two vertices no longer work for tree deletion sets. We would like to point out that TREE DELETION SET has been considered before in the realm of approximation algorithm and has been shown to be hard to approximate within $O(n^{1-\epsilon})$ for any

$\epsilon > 0$ unless P=NP [22]. This is in sharp contrast to the fact that FEEDBACK VERTEX SET has a factor 2-approximation algorithm [1].

Variations of FEEDBACK VERTEX SET and several other problems like DOMINATING SET, VERTEX COVER, when S is required to induce an independent set or a connected graph or some other hereditary properties have also been well studied [15,16,17,14]. To the best of our knowledge, this is the first paper that studies the variation of a problem where the demand of connectivity is on $G[V \setminus S]$ rather than the solution. That is, in our problem we want $G[V \setminus S]$ to induce a connected graph (i.e. it is a tree).

We first show that the problem in NP-complete, and then show that it is fixed-parameter tractable when parameterized by k, the solution size by giving an $O^*(5^k)$ time[1] fixed-parameter tractable algorithm. This is in contrast to, and comes reasonably close to the $O^*(3.83^k)$ bound known for the general feedback vertex set problem [2].

The next section proves the problem NP-complete. Section 3 is the main section that gives the fixed-parameter algorithm for the problem. Finally in Section 4, we conclude with open problems.

2   NP-Completeness

Proposition 1. TREE DELETION SET *is* NP-*complete.*

Proof. The problem is obviously in NP. For the hardness we reduce VERTEX COVER (VC), which is well known to be NP-complete [11]. An instance of VC consists of a graph G and a positive integer k and the question is whether there is a set S of at most k vertices (*vertex cover*) such that $G \setminus S$ contains no edges. Given an instance (G, k) of VC we obtain an equivalent instance (G', k) of TDS as follows. G' is obtained from G by introducing a new universal vertex u (i.e. u is adjacent to all vertices of G) and attaching $k + 1$ new pendant vertices to it. Now if S is a vertex cover in G, then $G \setminus S$ is a star. On the other hand, if S is a tree deletion set in G', then $u \notin S$, as otherwise there would be at least two of the newly added pendant vertices left in $G \setminus S$ and they would become disconnected. But then, as u is adjacent to all vertices of G, there must be no edge in $G \setminus S$ in order for $G' \setminus S$ to be a tree, which implies that S is a vertex cover for G.   □

3   FPT Algorithm

The main result of this section is the following:

Theorem 1. WEIGHTED TREE DELETION SET *can be solved in time* $O^*(5^k)$.

The rest of this section is devoted to the proof of this theorem.

---

[1] O^* notation ignores polynomial factors

3.1 Reduction Rules

We begin with some reduction rules which simplify the input instance. These rules modify the graph G, the weight function w, and the parameter k. For the purpose of the analysis, we denote the original value of the parameter k given on input by k_0. We say that a reduction rule is safe if the instance obtained by application of the rule is a yes-instance if and only if the original instance was. The following two rules formalize obvious constraints to the solvability of the instance.

Reduction Rule 1. *If $k < 0$, then answer NO.*

Reduction Rule 2. *Let N' be the set of vertices which have weight more than k. If $G[N']$ contains a cycle, then answer NO.*

Lemma 1. *Reduction Rule 2 is safe.*

Proof. As no vertex of weight more than k can be included in any set of total weight at most k, no set of total weight at most k forms a tree deletion set. □

While the structure of the vertices of weight more than k is fixed, the following rule helps to simplify the neighborhood of such vertices.

Reduction Rule 3. *Let N' be the set of vertices which have weight more than k. If there is a vertex v in $V(G) \setminus N'$ which has two neighbors in the same connected component of $G[N']$ then delete v and decrease k by $w(v)$.*

Lemma 2. *Reduction Rule 3 is safe.*

Proof. The vertex v must be included in any tree deletion set of total weight at most k, as otherwise it would form a cycle together with the vertices in a connected component of N'. □

The following rule helps us to deal with isolated vertices and the case when the graph is disconnected.

Reduction Rule 4. *If the input graph is disconnected, then delete all vertices in connected components of weight less than $\left(\sum_{v \in V} w(v)\right) - k$ and decrease k by the weight of the deleted vertices.*

Lemma 3. *Reduction Rule 4 is safe.*

Proof. If the vertices of this connected component were not taken into the constructed tree deletion set, then all vertices outside this connected component have to be taken, as the resulting graph must have only one component. But this would mean that the constructed tree deletion set would contain vertices of total weight more than $\left(\sum_{v \in V} w(v)\right) - \left[\left(\sum_{v \in V} w(v)\right) - k\right] = k$ — a contradiction. □

Remark 1. If $\left(\sum_{v \in V} w(v)\right) > 2k$ or there is a vertex of weight more than k, then after the application of Reduction Rule 4 the graph has at most one connected component.

The following rule deals with vertices of degree 1 in the graph.

Reduction Rule 5. *If v is of degree 1 and u is its only neighbor, then delete v and set $w(u) = w(u) + w(v)$.*

Lemma 4. *Reduction Rule 5 is safe.*

Proof. Let G, w, k be the instance before the application of the rule and G', w', k the instance after the application of the rule. Let us first assume, that S is a tree deletion set in G with $w(S) \le k$. If S does not contain u, then $S \setminus \{v\}$ is also a tree deletion set for G of lower total weight and it is also a tree deletion set in G' of the same weight. If S contains u, but not v, then v is the only vertex not in S, $S \setminus \{u\}$ is also a tree deletion set for G of lower total weight and it is also a tree deletion set in G' of the same weight. Finally, if S contains both u and v, then $S \setminus \{v\}$ is a tree deletion set in G' of the same weight.

Assume now that S' is a tree deletion set in G'. If S' does not contain u, then S' is also a tree deletion set in G of the same weight. If S' contains u, then $S' \cup \{v\}$ is a tree deletion set in G of the same weight. □

Finally, the following rule is aimed on reducing the number of degree two vertices by shortening of long paths.

Reduction Rule 6. *If $v_0, v_1, \ldots, v_l, v_{l+1}$ is a path in the input graph, such that $l \ge 3$ and $\deg(v_i) = 2$ for every $i \in \{1, \ldots, l\}$, then (a) replace the vertices $v_1, \ldots, v_l$ by two vertices u_1 and u_2 with edges $\{v_0, u_1\}$, $\{u_1, u_2\}$, and $\{u_2, v_{l+1}\}$ and with $w(u_1) = \min\{w(v_i) \mid 1 \le i \le l\}$ and $w(u_2) = \left(\sum_{i=1}^{l} w(v_i)\right) - w(u_1)$. Moreover, if $l \ge 2$ and $w(v_0) > k$ or $w(v_{l+1}) > k$, then apply (a) and then (b) delete u_2 and connect u_1 directly to v_{l+1}.*

Lemma 5. *Reduction Rule 6 is safe.*

Proof. Let G, w, k be the instance before the application of the rule and G', w', k the instance after the application of the rule. Let us first assume, that S is a tree deletion set in G with $w(S) \le k$. We show that there is a tree deletion set S' for G' with $w'(S') \le w(S)$. We distinguish three cases.

- S *contains both* v_0 *and* v_{l+1}. Note that this can only happen when only the reduction (a) was applied. In this case v_1 is disconnected from $G \setminus \{v_0, \ldots, v_{l+1}\}$ and, therefore, either $\{v_1, \ldots, v_l\} \subseteq S$ or $(V(G) \setminus \{v_1, \ldots, v_l\}) \subseteq S$. In the former case $S' = (S \setminus \{v_1, \ldots, v_l\}) \cup \{u_1, u_2\}$ is a tree deletion set for G' with $w'(S') = w(S)$ while in the latter case, for $S' = S$ and we have $G \setminus S'$ is a path and $w'(S') = w(S)$.

- *S contains exactly one of v_0 and v_{l+1}.* As the situation is symmetric, we can assume that v_{l+1} is in S. As $\{v_1, \ldots, v_l\}$ induces a path in G, $\{v_1, \ldots, v_l\} \setminus S$ induces a path in $G \setminus S$ and $G \setminus (S \cup \{v_1, \ldots, v_l\})$ is also a tree. Attaching at a node of this tree a path, we obtain again a tree. Hence, $S' = S \setminus \{v_1, \ldots, v_l\}$ is also a tree deletion set in G. Since $G \setminus (S \cup \{v_1, \ldots, v_l\}) = G' \setminus (S \cup \{v_1, \ldots, v_l\})$, it follows that S' is also a tree deletion set in G' with $w'(S') = w'(S') \leq w(S)$. This is true both in case (a) and (b).

- *S contains none of v_0 and v_{l+1}.* If $S \cap \{v_1, \ldots, v_l\} = \emptyset$, then S is also a tree deletion set in G'. Otherwise, $\{v_1, \ldots, v_l\} \setminus S$ induces two paths each attached to two different nodes in the tree $G \setminus S$. Assume that we have $w(v_r) = \min\{w(v_i) \mid 1 \leq i \leq l\}$. We first show that $S' = (S \setminus \{v_1, \ldots, v_l\}) \cup \{v_r\}$ is also a tree deletion set for G. This is true, as attaching the two pending paths $v_1, \ldots, v_{r-1}$ and $v_{r+1}, \ldots, v_l$ to the tree $G \setminus (S \cup \{v_1, \ldots, v_l\}) = G' \setminus (S \cup \{v_1, \ldots, v_l\})$ again creates a tree. By the same reason $S'' = (S' \setminus \{v_r\}) \cup \{u_1\}$ is a tree deletion set in G'. Also $w(S') \leq w(S)$ as v_r is the vertex of the minimum weight and $w'(S'') = w(S')$.

Now assume that S' is a tree deletion set in G' and we have $w(v_r) = \min\{w(v_i) \mid 1 \leq i \leq l\}$. First observe, that if $u_2 \in S'$ and $v_0 \in S'$, then also u_1 in S' or u_1 is the only vertex of $G' \setminus S'$, as otherwise the graph would be disconnected. If $u_2 \in S'$ but $v_0 \notin S'$ then $G' \setminus ((S' \setminus \{u_2\}) \cup \{u_1\})$ is also a tree. Since we have $w'(u_2) > w'(u_1)$ as $l \geq 3$, we can assume that whenever u_2 is in S', then also u_1 is in S'.

Now let us distinguish, which of the vertices u_1, u_2 the set S' contains. If S' contains neither u_1 neither u_2, then S' is also a tree deletion set in G and $w(S') = w'(S')$.

If only the part (a) was applied, and S' contains u_1 but not u_2 then we also distinguish, which of the vertices v_0, v_{l+1} is contained in S'. If S' contains v_0, then $S = S' \setminus \{u_1\}$ is also tree deletion set in G' and also in G. If S' contains v_{l+1} then it must also contain v_0 as otherwise u_2 would be disconnected from v_0, and the previous case applies. Finally, if S' contains neither v_0 nor v_{l+1}, then we let $S = (S' \setminus \{u_1\}) \cup \{v_r\})$. Then, $G \setminus S$ is a tree, as $\{v_1, \ldots, v_l\}$ induces two pending paths $v_1, \ldots, v_{r-1}$ and $v_{r+1}, \ldots, v_l$ in it. We also have $w(S) = w'(S')$, as $w(v_r) = w'(u_1)$.

If also part (b) was applied and S' contains u_1, then we distinguish two cases. Note that S' contains at most one of v_0, v_{l+1}. If it contains exactly one of them, then $S = S' \setminus \{u_1\}$ is also tree deletion set in G' and also in G. If S' contains neither v_0 nor v_{l+1}, then $S = (S' \setminus \{u_1\}) \cup \{v_r\})$ is a tree deletion set in G of the same weight.

Finally, if S' contains both u_2 and u_1 then we let $S = (S' \setminus \{u_1, u_2\}) \cup \{v_1, \ldots, v_l\}$. We have $w(S) = w'(S')$, $G \setminus S = G' \setminus S'$ and, hence, it is a tree.   □

3.2 Branching Steps

Our FPT algorithm is based on a branching strategy similar to the one applied in [15]. First we use the algorithm of Cao et al. [2] to determine whether G has a feedback vertex set of size at most k in $O^*(3.83^k)$ time. As a tree deletion set of weight at most k is also a feedback vertex set of size at most k, if the algorithm answers NO, we can also answer NO. Otherwise let F be the feedback vertex set for G found by the algorithm.

Our algorithm now branches into several cases and it returns YES if and only if at least one of the branches answers YES. In a search for a tree deletion set X we first guess its intersection Y with the known feedback vertex set F. This means that we branch into $2^{|F|}$ branches, each corresponding to one subset $Y \subseteq F$. From now on, we assume that our guess was correct and therefore limit our search to the tree deletion sets X with $X \cap F = Y$.

As the vertices of Y are included in the tree deletion set constructed, we remove them and decrease k by $\sum_{v \in Y} w(v)$. Since the guess was correct, we know that none of the vertices in $N = F \setminus Y$ takes part in the sought tree deletion set. Thus we assign them weight $k + 1$. We also know that $G \setminus (Y \cup N)$ is a forest, as $F = (Y \cup N)$.

Now we are ready to describe the branching part of the algorithm. It modifies G, w, k and N. In between any two branching we apply Reduction Rules 1 to 5 and we only apply Reduction Rule 6 if none of $\{v_1, \ldots, v_{l-1}\}$ is in N. We always assume that the graph is reduced with respect to these reduction rules.

Let $H = V \setminus N$. Our branching step picks a vertex v from H, and branches by picking v into the tree deletion set or by not picking it (and hence adding it to N) and recursively solving the resulting problem. When we pick v into the solution, k drops by 1. The key observation in [3] for undirected feedback vertex set was that if v is adjacent to two connected components of N, then when v is added to N, the number of connected components of N decreases by at least one resulting in some progress. However, unlike in the undirected feedback vertex case, we are not guaranteed such vertices (as we couldn't apply the standard preprocessing rules for undirected feedback vertex set) here.

Let us call a vertex *useful* if it is in H, have exactly two neighbors in G and both these neighbors are in N. To bound the depth of the recursion, we use a measure $\mu = k + c - u$ where

k is the budget - the weight of the vertices we can still add to the tree deletion set being constructed. Initially, we have $k \le k_0 - \sum_{v \in Y} w(Y)$.

c is the number of components in $G[N]$. Initially, we have $N = F \setminus Y$ and therefore $c \le k - |Y|$.

u is the number of useful vertices in H.

First we show that the reduction rules do not increase the measure. Note also that none of the rules introduces a cycle to $G[H]$ and therefore $G[H]$ is still a forest.

Lemma 6. *No reduction rule increases μ.*

Proof. None of the rules increases k. The only reduction rule which could potentially increase the number of components of $G[N]$ is Reduction Rule 3, but then the deleted vertex must be in N, have a weight more than k and the branch can be rejected by Reduction Rule 2. While Reduction rules 3 and 4 delete some useful vertices, in such cases, k is decreased by the weight of the deleted vertices and therefore by at least the number of deleted useful vertices.   □

After applying the reduction rules, we distinguish three cases. If μ is negative, we return NO, which is justified by the lemma below.

Lemma 7. *If the measure μ becomes negative, then there is no tree deletion set for the current branch.*

Proof. Suppose there is a tree deletion set X of total weight at most k for the graph G and that U' is the set of useful vertices deleted by this tree deletion set. Consider the subgraph of G induced by N and the set U of remaining useful vertices and contract each connected components of $G[N]$ to a single vertex. Let us call the set of vertices created this way $\tilde{N}$. This does not create parallel edges, because the graph is reduced with respect to Reduction Rule 3. Moreover, the subgraph G' induced by $\tilde{N} \cup U$ is still a tree, because contracting edges of a tree cannot make it disconnected or create a cycle. Therefore, G' has $|\tilde{N}| + |U| - 1$ edges. On the other hand, each vertex in U has exactly two neighbors in $\tilde{N}$ and therefore the graph has $2|U|$ edges. It follows that $|U| = |\tilde{N}| - 1$. Since $c = |\tilde{N}|$, $u = |U| + |U'|$ and $|U'| \leq k$, we have $\mu = k + c - u \geq k + |\tilde{N}| - (|\tilde{N}| - 1 + k) = 1$ — a contradiction.   □

If μ is nonnegative, and if there is a vertex v in H which satisfies at least one of the following conditions:

(i) it has total degree at least three in G and at least two neighbors in N;
(ii) it has a neighbor in N and a neighbor which is a leaf in $G[H]$; or
(iii) it has at least two neighbors in H, which are both leaves in $G[H]$;

then we branch on this vertex v. More precisely, for such a vertex v we consider two cases:

- v is a part of the tree deletion set constructed — then we delete v from the graph and decrease k by $w(v)$;
- v is not in the sought tree deletion set— then we set the weight of v to $k+1$ and add v to N.

In both cases the procedure is called recursively on the modified G, w, k, N and the procedure returns YES if in at least one of the branches the recursive call returns YES. If there are several vertices satisfying the conditions, then we select a vertex which satisfies condition (i) if such a vertex is available. We only select other vertices if there is no vertex satisfying the condition (i).

In this 'two way branch' we show that in each such recursive call, the value of μ is at least one less than that in the current call.

Lemma 8. *If the vertex we branch on satisfies at least one of the conditions (i) to (iii), then the measure decreases by at least one in each branch.*

Proof. Let us first consider the case that we delete the vertex we branch on. Since it is not in N, deleting it cannot increase the number of connected components in $G[N]$. Moreover, since it has degree at least three in case (i) and neighbors in H in cases (ii) and (iii), it is not a useful vertex and therefore the number of useful vertices remains the same. Since k is decreased by the weight of the vertex deleted, the measure drops by at least one.

Consider now the case that we add the vertex v to N and suppose it satisfies the condition (i). Then it has at least two neighbors in N and since the graph is reduced with respect to Reduction Rule 3, these neighbors are in different connected components of $G[N]$. Therefore c is decreased, k remains the same and u is not decreased, which means that the measure drops.

If the vertex satisfies the condition (ii), then adding it to N does not increase the number of components in $G[N]$ as it already has a neighbor in N. On the other hand, its neighbor in H is a leaf in $G[H]$ and since it does not satisfy condition (i), it has exactly one other neighbor, which is in N. Hence, it becomes useful and the measure is decreased as k remains the same.

Finally, if the vertex satisfies condition (iii), then adding it to N may increase the number of components in $G[N]$ by one, but both its neighbors, which are leaves in $G[H]$ and do not satisfy condition (i) become useful. Therefore the measure decreases also in this case.   □

Finally, if μ is nonnegative, but there is no vertex in H satisfying the conditions, then either N is empty or every vertex in H is useful as we argue below.

Lemma 9. *If no vertex satisfies any of the conditions* (*i*) *to* (*iii*) *and* N *is nonempty, then every vertex in* H *is useful.*

Proof. We show that if there is a vertex $v' \in H$ which is not useful, then there is a vertex in H which satisfies some of the conditions (*i*) to (*iii*).

If v' is isolated in $G[H]$, then it must have at least three neighbors in N, as the graph is reduced with respect to Reduction Rules 4, 5, and 3, v' is not useful, and there are no isolated vertices by Remark 1 as N is nonempty. But then it satisfies condition (i).

Recall that $G[H]$ is a forest. If v' is non-isolated in $G[H]$, then consider a leaf v in the same connected component of $G[H]$, which is the furthest apart from v'. We know, that v has degree at least two in G, as the graph is reduced with respect to Reduction Rule 5. If it has degree at least three, then it satisfies the condition (i). Otherwise consider its neighbor u in H. Since the graph is reduced with respect to Reduction Rule 6, u has degree at least three in G. If it has a neighbor in N, then it satisfies the condition (ii).

If u has degree three in $G[H]$, then consider a neighbor w of u which is not on the unique path between v and v' in $G[H]$. If w is a leaf in $G[H]$ then it either satisfies the condition (i), or has degree two in G and therefore u satisfies condition (iii). If w is not a leaf in $G[H]$ then any leaf in the subtree of $G[H]$ rooted in w which does not contain u is further apart from v' than v, which contradicts the way we selected v.   □

If all vertices in H are useful vertices, we proceed as follows. Note that the graph is formed by the vertices in N and useful vertices adjacent to them. We contract each connected component of $G[N]$ to a single vertex. Let us again call the set of vertices created this way $\tilde{N}$. Recall that there is no cycle in $G[N]$ as the graph is reduced with respect to Reduction Rule 2. As we only search for a tree deletion set X among vertices in H, it is easy to verify that $X \subseteq H$ is a tree deletion set in the graph after contraction if and only if it was a tree deletion set in the original graph. Note that this does not create parallel edges, because the graph is reduced with respect to Reduction Rule 3 and, hence, there is no vertex in H with both its neighbors in the same connected component of $G[N]$.

Now if there are two vertices in H with the same neighbors in $\tilde{N}$, then we delete the one with the lower weight and decrease k by its weight. Clearly at least one of them must be in the constructed tree deletion set and if only one of them is in the tree deletion set, then we can assume it is the one with lower weight. Next we construct an auxiliary graph $\overline{G}$ with vertex set $\tilde{N}$ and a weighted edge between a pair of vertices if there is a vertex v in H with this pair of vertices as its neighbors in G. The weight of the edge equals the weight of v. It is easy to see, that a minimum tree deletion set in G corresponds to the edge complement of a maximum spanning tree in $\overline{G}$ and vice versa. More precisely if $T = (\tilde{N}, E')$ is a spanning tree, then the set X of vertices v of H such that the edge corresponding to $N(v)$ in $\overline{G}$ is not in E' is a tree deletion set for G. Similarly, if $X \subseteq H$ is a tree deletion set in G, then $T = (\tilde{N}, S)$, where $S = \{N(v) \mid v \in H \setminus X\}$ is a spanning tree of $\overline{G}$. The weight of S is always $\sum_{v \in H \setminus X} w(v)$

Hence, we use the standard algorithm [10] to find a maximum spanning tree $T = (\tilde{N}, S)$ of $\overline{G}$, and answer YES if and only if $(\sum_{v \in H} w(v)) - w'(S)$ is at most k, where $w'(S)$ denotes the weight of the tree T.

If N is empty, then the graph consists of isolated vertices, since $G[H]$ is a forest and it is reduced with respect to Reduction Rule 5. Therefore it is enough to delete all vertices but the one with the largest weight and answer YES if and only if the weight of the deleted vertices is at most k. This finishes the description of the algorithm.

The correctness of the algorithm has been already argued within its description, here we argue about the running time of the algorithm. First note that an application of any of the reduction rules can be recognized as well as applied in linear time. Since the Reduction Rules 1 and 2 only apply once, while the Reduction Rules 3–6 reduce the number of vertices, the rules can be exhaustively applied in $O(nm)$ time. To check the value of the measure and to find a vertex to branch on takes a linear time. Finally, one can contract the connected components in $G[N]$ in $O(mn)$ time and find the maximum spanning tree in $\overline{G}$ in $O(m)$ time. Hence the time spent in each node of the search tree is $O(mn)$.

It remains to count the number of nodes in the search tree. We first branch into at most $2^{|F|}$ branches, each corresponding to one subset Y of the feedback vertex set F. Then we keep branching into two branches, each time reducing the measure μ by at least one. Since at the beginning we have $\mu = k + c - u \leq 2k_0 - 2|Y|$, this part of the search tree has at most $2 \cdot 2^{2k_0 - 2|Y|}$ nodes. Summing this according to $y = |Y|$, we have that the total number of nodes in the search tree is at most $\sum_{y=0}^{k_0} \binom{k_0}{y} 2 \cdot 2^{2k_0 - 2y} = 2 \cdot (1 + 4)^{k_0} = 2 \cdot 5^{k_0}$. Recall that the first step in our algorithm is to determine whether G has a feedback vertex set of size at most k. This step is done in $O^*(3.83^k)$ time using the algorithm of Cao et al. [2]. Therefore the whole algorithm runs in $O^*(5^k)$ time. This completes the proof.

4   Conclusions and Open Problems

We have shown that the WEIGHTED TREE DELETION SET problem is fixed-parameter tractable. Improving the running time of our algorithm is a natural open problem. Another direction, which has attracted a lot of attention in parameterized complexity recently, is to study the *kernelization complexity* of the problem. Our fixed-parameter algorithm immediately implies an exponential kernel for the problem, but the natural open question is whether the problem has a polynomial size kernel. That is, is there a polynomial time algorithm that reduces the given input (G, k) to an equivalent graph with polynomial in k many vertices and edges? While the related feedback vertex set problem has an $O(k^2)$ sized kernel [20], we conjecture that the TREE DELETION SET problem does not admit a polynomial sized kernel under standard complexity theoretic assumptions.

References

1. Bafna, V., Berman, P., Fujito, T.: A 2-approximation algorithm for the undirected feedback vertex set problem. SIAM J. Discrete Math. 12, 289–297 (1999)
2. Cao, Y., Chen, J., Liu, Y.: On Feedback Vertex Set New Measure and New Structures. In: Kaplan, H. (ed.) SWAT 2010. LNCS, vol. 6139, pp. 93–104. Springer, Heidelberg (2010)
3. Chen, J., Fomin, F.V., Liu, Y., Lu, S., Villanger, Y.: Improved algorithms for feedback vertex set problems. J. Comput. Syst. Sci. 74, 1188–1198 (2008)
4. Chitnis, R., Cygan, M., Hajiaghayi, M., Marx, D.: Directed Subset Feedback Vertex Set Is Fixed-Parameter Tractable. In: Czumaj, A., Mehlhorn, K., Pitts, A., Wattenhofer, R. (eds.) ICALP 2012, Part I. LNCS, vol. 7391, pp. 230–241. Springer, Heidelberg (2012)
5. Cygan, M., Pilipczuk, M., Pilipczuk, M., Wojtaszczyk, J.O.: Subset Feedback Vertex Set Is Fixed-Parameter Tractable. In: Aceto, L., Henzinger, M., Sgall, J. (eds.) ICALP 2011, Part I. LNCS, vol. 6755, pp. 449–461. Springer, Heidelberg (2011)
6. Downey, R.G., Fellows, M.R.: Parameterized Complexity. Monographs in Computer Science. Springer, New York (1999)
7. Flum, J., Grohe, M.: Parameterized Complexity Theory. Springer, Berlin (2006)
8. Fomin, F.V., Gaspers, S., Pyatkin, A.V., Razgon, I.: On the minimum feedback vertex set problem: Exact and enumeration algorithms. Algorithmica 52, 293–307 (2008)
9. Fomin, F.V., Lokshtanov, D., Misra, N., Saurabh, S.: Planar F-deletion: Approximation, kernelization and optimal fpt algorithms. To appear in FOCS 2012 (2012)

10. Fredman, M.L., Willard, D.E.: Trans-dichotomous algorithms for minimum spanning trees and shortest paths. Journal of Computer and System Sciences 48, 533–551 (1994)
11. Garey, M.R., Johnson, D.S.: Computers and Intractability: A Guide to the Theory of NP-Completeness. W. H. Freeman and Company (1979)
12. Kim, E.J., Langer, A., Paul, C., Reidl, F., Rossmanith, P., Sau, I., Sikdar, S.: Linear kernels and single-exponential algorithms via protrusion decompositions. CoRR, abs/1207.0835 (2012)
13. Lokshtanov, D., Narayanaswamy, N.S., Raman, V., Ramanujan, M.S., Saurabh, S.: Faster parameterized algorithms using linear programming. CoRR, abs/1203.0833 (2012)
14. Marx, D., O'Sullivan, B., Razgon, I.: Treewidth reduction for constrained separation and bipartization problems. In: Marion, J.-Y., Schwentick, T. (eds.) STACS. LIPIcs, vol. 5, pp. 561–572. Schloss Dagstuhl - Leibniz-Zentrum fuer Informatik (2010)
15. Misra, N., Philip, G., Raman, V., Saurabh, S.: On Parameterized Independent Feedback Vertex Set. In: Fu, B., Du, D.-Z. (eds.) COCOON 2011. LNCS, vol. 6842, pp. 98–109. Springer, Heidelberg (2011)
16. Misra, N., Philip, G., Raman, V., Saurabh, S., Sikdar, S.: FPT algorithms for connected feedback vertex set. J. Comb. Optim. 24, 131–146 (2012)
17. Mölle, D., Richter, S., Rossmanith, P.: Enumerate and expand: Improved algorithms for connected vertex cover and tree cover. Theory Comput. Syst. 43, 234–253 (2008)
18. Narayanaswamy, N.S., Raman, V., Ramanujan, M.S., Saurabh, S.: LP can be a cure for parameterized problems. In: STACS, pp. 338–349 (2012)
19. Niedermeier, R.: Invitation to Fixed Parameter Algorithms. Oxford Lecture Series in Mathematics and Its Applications. Oxford University Press, USA (2006)
20. Thomassé, S.: A $4k^2$ kernel for feedback vertex set. ACM Transactions on Algorithms 6 (2010)
21. Villanger, Y.: Proper Interval Vertex Deletion. In: Raman, V., Saurabh, S. (eds.) IPEC 2010. LNCS, vol. 6478, pp. 228–238. Springer, Heidelberg (2010)
22. Yannakakis, M.: The effect of a connectivity requirement on the complexity of maximum subgraph problems. J. ACM 26, 618–630 (1979)

Circular Graph Drawings
with Large Crossing Angles

Hooman Reisi Dehkordi, Quan Nguyen, Peter Eades, and Seok-Hee Hong

School of Information Technologies, University of Sydney
{hooman.dehkordi,quan.nguyen,peter.eades,seokhee.hong}@sydney.edu.au

Abstract. This paper is motivated by empirical research that has shown that increasing the angle of edge crossings reduces the negative effect of crossings on human readability. We investigate circular graph drawings (where each vertex lies on a circle) with large crossing angles. In particular, we consider the case of *right angle crossing (RAC)* drawings, where each crossing angle is $\pi/2$.

We characterize *circular RAC graphs* that admit a circular RAC drawing, and present a linear-time algorithm for constructing such a drawing, if it exists. We also describe a quadratic programming approach to construct circular drawings that maximise crossing angles. This method significantly increases crossing angles compared to the traditional equal-spacing algorithm.

Keywords: Graph drawing, Right Angle Crossing (RAC) drawing, Large Angle Crossing (LAC) drawing, Circular graph drawing.

1 Introduction

Since the late 1970s, researchers have studied automatic Graph Drawing, that is, algorithms to find "good" drawings of graphs. This research is motivated by the need to visualize biological networks, social networks, computer networks, and large software structures. One of the most popular graph drawings is a *circular drawing*: all the vertices are located on a circle and edges are drawn as straight-lines. Consequently, many commercial graph drawing softwares (for example, TomSawyer software and yworks) produce circular drawings.

Purchase et al. [13] show that human understanding is negatively correlated to the number of crossings in a drawing. This motivates a large number of algorithms to draw graphs with a small number of crossings. For example, Bauer and Brandes [3] presented methods for reducing the number of crossings in circular drawings by computing a good ordering of the vertices on the circle.

Huang et al. [11] show that human understanding is enhanced if angles at which edges cross is large. A number of investigations on the combinatorics and algorithmics of drawing graphs with large crossing angles followed Huang's work; see, for example, [2,7,8,10]. These papers have concentrated on properties of *right angle crossing (RAC)* graph drawings, where each crossing angle is $\frac{\pi}{2}$.

S.K. Ghosh and T. Tokuyama (Eds.): WALCOM 2013, LNCS 7748, pp. 298–309, 2013.

In this paper, we study circular drawings with large crossing angles. In particular, we concentrate on the right angle crossing case. To construct a circular graph drawing, we need two steps: (1) order the vertices around the circle to reduce the number of crossings, and (2) choose locations for the vertices to increase the crossing angles. For the first step, we can use existing methods such as the Bauer-Brandes [3] algorithm. This paper is concerned with the second step.

We characterize *circular RAC graphs* that admit a circular RAC drawing, and present a linear-time algorithm for constructing such a drawing, if it exists. The input of the algorithms described in this paper is the clockwise sequence S of the vertices around the circle as well as the sequence of edges incident with each vertex u, in clockwise order around u. Once such an input is provided, it is straightforward to maintain data structures so that we can test in constant time whether two edges cross each other. Further, we can easily identify and safely ignore edges between consecutive vertices around the circle.

Note that not every topological graph admits a circular RAC drawing. We thus study the problem of circular LAC drawings, i.e., circular drawing with large crossing angles. We present a quadratic programming approach to construct circular drawings that maximise crossing angles. Experimental results suggest that this method significantly increases crossing angles compared to the traditional equal-spacing drawings.

The next Section defines some useful terminology. Section 3 gives a combinatorial characterization of circular RAC graphs, and Section 4 gives a linear-time algorithm to compute a circular RAC drawing. Section 5 presents a quadratic programming approach to construct circular drawings that maximise crossing angles. Section 6 concludes with an open problem.

2 Terminology

We start by reviewing the basic terminology of graph drawing. For further background, see [5].

A *drawing* of a simple graph $G = (V, E)$ is a representation D of G in the plane, where each vertex is a point and each edge is a closed Jordan arc between the points representing its endpoints.

Two edges *cross*, if they share a point other than their endpoints. A *crossing* $X = \{(a, b), (c, d)\}$ is a set of two distinct edges (a, b) and (c, d) such that (a, b) crosses (c, d). We define the *crossing point* x of the crossing X to be the shared point of (a, b) and (c, d).

In this paper, we assume that graph drawings are not degenerate, in that they satisfy the following conditions: (1) an edge does not contain a vertex other than its endpoints; (2) no edge crosses itself; (3) edges must not meet tangentially, that is, they must either properly cross or not cross at all; and (4) no three edges share a crossing point.

A *topological embedding* of a graph G is an equivalence class of drawings of G under homeomorphism. A topological graph G is an *outer topological graph*, if and only if all of its vertices are on the same face; we normally assume this face to be the outer face.

A *straight-line drawing* of a graph G is a drawing of G in which each edge is represented by a straight-line segment. A straight-line drawing of a graph G is a *right angle crossing (RAC) drawing*, if edges cross each other at an angle of $\frac{\pi}{2}$. A RAC drawing of a graph G is called a *circular RAC drawing*, if each vertex of G lies on a circle. We define a *circular RAC graph* G to be a topological graph that admits a circular RAC drawing.

The *crossing graph* G_c of topological graph G is a graph where each vertex of G_c corresponds to an edge of G and two vertices of G_c are adjacent if and only if their corresponding edges in G cross each other.

Two crossings $X_1 = \{(a,b),(c,d)\}$ and $X_2 = \{(u,v),(w,x)\}$ are *edge disjoint* if and only if $X_1 \cap X_2 = \emptyset$. Two edge disjoint crossings X_1 and X_2 *cross* each other if an edge of X_1 crosses an edge of X_2.

In a connected outer topological graph, vertex v *follows* vertex u, if and only if u and v are two consecutive vertices in the sequence of the vertices on the outer face. We denote such a relation with $u \sim v$ or $v \sim u$. We use $u \sim^+ v$ (respectively, $u \sim^- v$) to denote that u is followed by v in the clockwise (respectively, counter clockwise) order.

Let a and b be two vertices on a cycle C such that the edge (a,b) is a chord of C. Let u and v be two vertices of the cycle other than a and b. Then we define u and v to be on *same side* of (a,b), if and only if it is possible to reach v from u by traversing the vertices of C in order without passing through a or b. Otherwise, we say that u and v are on *different sides* of (a,b).

In this paper, in order to distinguish between the two sides of an edge e, we shall call them the *left-hand side* and the *right-hand side* of e. For example, if u and v are on different sides of an edge e and u is on the left-hand side of it, then v is on the right-hand side of that edge.

We define an edge $e' = (a,b)$ to be at *one side* (left-hand side or the right-hand side) of another edge $e = (u,v)$, if and only if the two endpoints of it, a and b, are at the same side of e. Similarly, we define a crossing $X_1 = \{(a,b),(c,d)\}$ to be at *one side* (left-hand side or the right-hand side) of an edge $e = (u,v)$, if and only if its edges, (a,b) and (c,d), are at the same side of e.

Suppose that a and b are two points on a circle with center o. Then $\overset{\frown}{ab}$ denotes the clockwise angle $\angle aob$.

3   A Characterization of Circular RAC Graphs

We now state the main Theorem of this paper.

Theorem 1. *An outer topological graph G is a circular RAC graph, if and only if (1) its corresponding crossing graph is bipartite; and (2) any two edge-disjoint crossings in G cross each other.*

An immediate consequence of Theorem 1 is that the crossing graph of a circular RAC graph has only one non-trivial connected component. Since the bipartition of a connected bipartite is unique, we can deduce the following two important Corollaries of Theorem 1.

Corollary 1. *A circular RAC drawing can be rotated such that the crossed edges are either vertical or horizontal.*

For the remainder of this paper, we assume that the blue edges are horizontal and the green edges are vertical.

Corollary 2. *Each vertex in a circular RAC graph drawing is incident to at most two crossed edges. If two crossed edges share an endpoint, then they are perpendicular to each other.*

The next two subsections prove the necessity and sufficiency of the two conditions in Theorem 1.

3.1 Necessity

The necessity of condition (1) of Theorem 1 follows from fundamental results on RAC graphs in [6].

For condition (2) of Theorem 1, consider Fig. 1. Intuitively, as the minimum crossing angles at x_1 and x_2 increase toward $\frac{\pi}{2}$, a becomes close to a' and d becomes closer to d'; if both crossing angles are precisely $\frac{\pi}{2}$, then either (a, c) crosses (a', c'), or (b, d) crosses (b', d'). For a formal proof, see [14].

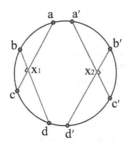

Fig. 1. Crossing X_1 does not cross crossing X_2

3.2 Sufficiency

Throughout this section, $G = (V, E)$ denotes an outer topological graph with a bipartite crossing graph such that any two edge-disjoint crossings in G cross each other. We shall prove that any graph with such properties has a circular RAC drawing and hence is a circular RAC graph.

Following [9], we can color the edges of G as follows. An edge is *red* if it is not crossed by any other edges. The bipartition of the remaining edges allows to color them *blue* and *green*, such that a blue edge is only crossed by green edges and a green edge is only crossed by blue edges. We denote the sets of *red, green* and *blue edges* of graph G by E_r, E_g and E_b respectively.

We first show that the red edges can be eliminated from consideration.

Lemma 1. *Suppose that $G = (V, E)$ is an outer topological graph with a bipartite crossing graph such that any two edge-disjoint crossings in G cross each other. Let E_r be the set of the red edges of G, and let V_r denote the set of vertices that for which all incident edges are red. If there exists a circular RAC drawing D' of $G' = (V \setminus V_r, E \setminus E_r)$, then there exists a circular RAC drawing D of G.*

Proof. The proof of this Lemma is straightforward (see [14]).

From Lemma 1, we now assume that every edge is either blue or green.

We define two isochromatic edges $e_1 = (u_1, v_1)$ and $e_2 = (u_2, v_2)$ of graph G to be *adjoining*, if and only if $u_1 \sim u_2$ and $v_1 \sim v_2$ on the outer face of G (see Figure 2a). We define a blue edge (b, b') and a green edge (g, g') to be *adjoining*, if and only if $g \sim^- b$ and $g \sim^+ b'$ (see Figure 2b) or $g' \sim^- b'$ and $g' \sim^+ b$ (see Figure 2c).

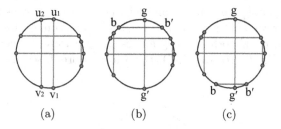

Fig. 2. (a) Two adjoining green edges (u_1, v_1) and (u_2, v_2), (b) A blue edge (b, b') which is adjoining to a green edge (g, g') at vertex g and (c) A blue edge (b, b') which is adjoining to a green edge (g, g') at vertex g'

The most important part of the proof of sufficiency in Theorem 1 is the existence of "major" edges. A *major green edge* of a circular RAC graph drawing is a green edge which is crossed by all the blue edges. Similarly, we define a *major blue edge*. The next Lemma proves the existence of such edges.

Lemma 2. *Suppose that $G = (V, E)$ is an outer topological graph with a nonempty bipartite crossing graph such that any two edge-disjoint crossings in G cross each other. Then G contains a major green edge and a major blue edge.*

Proof. Let g_1 be a green edge with a maximum number of crossings; we show that g_1 crosses every blue edge. On the contrary let b_1 be a blue edge that does not cross g_1. But as b_1 is blue, it is crossed by a green edge g_2 with $g_2 \neq g_1$. As g_1 has at least as many crossings as g_2, there is a blue edge b_2 that crosses g_1 but does not cross g_2. Therefore there are two edge-disjoint crossings $X_1 = \{g_1, b_2\}$ and $X_2 = \{g_2, b_1\}$ that do not cross each other; this contradicts our assumption. The Lemma follows. □

The next Lemma exploits the existence of the major green edge and provides a fundamental base of the drawing algorithm presented later in this section.

Lemma 3. *Let e_g be a major green edge of G. Then there exists either a blue edge adjoining to e_g or a green edge adjoining to e_g.*

Proof. See [14].

In a circular RAC drawing D, we define a *divider* to be the diameter of the outer circle such that it is parallel to the green edges. The next Theorem provides a recursive algorithm to construct a circular RAC drawing of G.

Note that the drawing algorithm preserves the given ordering of its vertices on the outer face. The input of the drawing algorithm also includes a major edge. An algorithm to find a major edge will be presented in the next Section.

Theorem 2. *Let e_g be a major green edge of G. Then there exists two different circular RAC drawings D and D' of G such that:*

- *D and D' differ only by the location of e_g; that is, if we delete e_g from both D and D', then D is the same as D'.*
- *e_g is located on the left side of the divider in D, and on the right side of the divider in D'.*
- *e_g is the closest green edge to the center o of the circle Σ in both D and D' (o is the origin of the coordinate system).*

Proof. This proof is by induction on the number of crossing edges of G. Suppose that G has only one green edge e_g and one blue edge e_b. Then it is easy to construct D and D' such that e_g is located on the left side of the divider in D and on the right side of it in D'.

Now suppose that G has $k+1$ crossing edges. By Lemma 3, there exists either a blue edge e_b adjoining to the major edge e_g, or a green edge e'_g adjoining to e_g. We have the following two cases.

Case 1. There exists a blue edge $e_b = (b_1, b_2)$ adjoining to $e_g = (g_1, g_2)$. Let $G^- = (V \setminus \{b_1, b_2\}, E \setminus \{e_b\})$ be the graph induced by removing e_b and its endpoints from G. First we show that G^- has the necessary properties of circular RAC graphs, and it does not contain any red edges. Since G has a bipartite crossing graph, G^- has a bipartite subgraph. As e_b was adjoining to e_g, after removing e_b, all the edges of G^- are still either green or blue. Also, note that since all the crossings in G cross each other, any two crossings in G^- cross each other as well. Therefore, G^- has the necessary properties of circular RAC graphs, and it does not contain any red edges.

Now by induction hypothesis, there exist two circular RAC drawings D^- and $D^{-'}$ of G^- such that all the vertices of G^- are located on circle Σ, and

$e_g = (g_1, g_2)$ is the closest green (vertical) edge to o from the left in D^- and from the right in $D^{-'}$. Since e_g is the closest edge to o in both drawings, g_1 and g_2 have the biggest and smallest y-coordinates among the endpoints of the green edges. Hence it is easy to add the blue edge e_b such that it is adjoining to e_g, and it crosses no other green edge. See Figure 3.

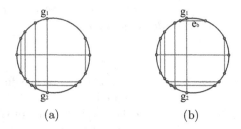

(a)   (b)

Fig. 3. (a) G^- without the blue edge (b) G after adding the blue edge

Case 2. There exists a green edge $e'_g = (g_3, g_4)$ adjoining to e_g.
Let $G^- = (V \setminus \{g_1, g_2\}, E \setminus \{e_g\})$ be the graph obtained by removing e_g and its endpoints from G. Since we have handled the blue edges adjoining to e_g in Case 1, we now assume that no more blue edges are adjoining to e_g. Therefore, by the removing green edge e_g, all the edges of G^- are still colored in blue or green. Also by an argument similar to Case 1, G^- has a bipartite crossing graph and all the crossings in G^- cross each other. Therefore, it satisfies the necessary conditions of circular RAC graphs, and it does not contain any red edges.

Now by the induction hypothesis, the Theorem holds for G^- with k crossed edges. That is, G^- has two circular RAC drawing D^- and $D^{-'}$, such that all the vertices of G^- are located on a circle Σ and $e'_g = (g_3, g_4)$ is the closest green (vertical) edge to o from the left in D^- and from the right in $D^{-'}$. Next we shall show how to add e_g to D^- on the right side of e'_g or to $D^{-'}$ on the left side of e'_g, to construct a circular RAC drawing of G.

Without loss of generality, suppose that e'_g is located on the left side of e_g in G. That is, we shall add e_g to D^- on the right side of e'_g to construct two drawings D and D' of G. Let e'_g be located at $x = 0 - \epsilon^-$ in D^-, where ϵ^- is a positive value smaller than the radius of circle Σ. We add e_g to D^- at two x-coordinates x_1 and x_2 in order to construct D and D' respectively, such that $0 - \epsilon^- < x_1 < 0$ and $0 < x_2 < 0 + \epsilon^-$. That is, e_g is located on the left side of the divider in D and on the right side of the divider in D', such that it is closer to o than e'_g and hence any other green edge of the two drawings.

Since e'_g is the closest green edge to o in D^- and all the horizontal blue edges in D^- cross e'_g, no vertex has an x-coordinate between $0 - \epsilon^-$ and $0 + \epsilon^-$. Now since e'_g is located on the left side of the divider in D^-, by adding e_g at an x-coordinate between $0 - \epsilon'$ and $0 + \epsilon'$, we can ensure that $g_3 \sim g_1$ and $g_2 \sim g_4$. Therefore, we construct D and D' such that in both drawings, e_g is located on the right side of e'_g.

As e_g is closer to o than any other green edges, the y-coordinate of its upper endpoint is bigger than the y-coordinate of the upper endpoints of the other green edges, and the y-coordinate of its lower endpoint is smaller than the y-coordinate of all the other lower endpoints of other green edge. Hence, it crosses all the existing blue edges.

Note that in both cases, we construct D and D', by adding e_b and e_g to D^- and $D^{-'}$, such that they are adjoining to e_g and e'_g respectively. Therefore, we preserve the ordering of the vertices on the outer face of G. The Theorem follows.   □

The algorithm described in the proof of Theorem 2 runs in linear time, since by Corollary 2, there are at most a linear number of blue and green edges in G. Next Section describes how to find a major edge in linear time.

4   A Linear-Time Circular RAC Drawing Algorithm

In the previous Section, we prove that any outer topological graph G is a circular RAC graph, if and only if it has a bipartite crossing graph and any two edge-disjoint crossings in G cross each other. Although the proof is constructive, it requires a major edge as an input of the algorithm. In this Section, we complete the drawing algorithm by providing a linear-time algorithm to find a major edge.

Theorem 3. *There is a linear-time algorithm that tests circular RAC graphs and constructs a circular RAC drawings if it exists.*

Using a double-ended stack, we can eliminate red edges in linear time; for details, see [14]. Henceforth, we assume that the graph has no red edges.

To complete the proof of the Theorem, we need to show that the major edges can be found in linear time. We shall assume that G is a circular RAC graph without any red edges.

To find a major edge, we first find an edge e, which is crossed a small number of times. Then we search for a major edge among the edges crossing e. We define a *green side edge* of a circular RAC graph G to be a green edge of G such that it has only one green adjoining edge. Similarly, we define a *blue side edge*. Intuitively, a green (respectively, blue) side edge is an edge where it has no other green (respectively, blue) edges on at least one side of it. Figure 4a shows the side edges of a circular RAC graph.

We also define an edge e' to be *parallel* with an edge e, if and only if e and e' do not cross each other. That is, in a circular RAC graph G, e' is parallel to e, if both of the endpoints of e' are on the same side of e.

It is useful to note that we can test whether two edges cross or are parallel in constant time, just by checking the indices of their endpoints in the sequence of the vertices on the outer face. The next Lemma shows how to find side edges in linear time.

Lemma 4. *Let G be a circular RAC graph with no red edges. Given a sequence of vertices on the boundary of G, there is a linear-time algorithm to find a side edge e of G.*

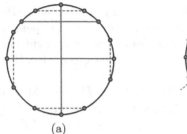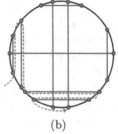

(a)   (b)

Fig. 4. (a)The dashed edges indicate the side edges, (b) the traversal path to find a side edge.

Proof. In order to find a side edge in linear time, we sweep the parallel edges of graph G such that no edge is visited twice. Suppose that during the sweep, we encounter the parallel edges in the following order, $e_1, e_2, \ldots, e_k$. If, for each i, the i-th edge is on the left-hand side (or right-hand side) of the $(i-1)$-th edge, then no edge is visited twice. We describe this sweeping process more rigourously next.

Choose a direction d to be either clockwise or counter clockwise. Pick an arbitrary vertex u to initiate the sweep. Follow the crossed edge e_1 incident to u, in order to reach its other endpoint u'. Traverse the circular boundary from u' in direction d, until we reach a vertex v of an edge e_2, that is parallel to edge e_1 (the new edge might not be of the same color of the first one). Then start searching for the next parallel edge by traversing the circle from the other endpoint of e_2, namely v', in a direction d', opposite to d. Repeat this process, until we reach an edge e_k such that there are no more parallel edges on one side of it. This sweeping process is shown in Figure 4b.

The "clockwise/counter clockwise" alternation in the algorithm is to make sure that we are advancing towards one side of the cycle. Hence, no vertex is visited more than once, and therefore it guarantees that the algorithm runs in linear time. □

Now we have found a side edge. Next, we describe an algorithm for finding a major edge. From Lemma 2, there is a major edge that crosses the side edge. Further note that all the edges crossing the side edge are isochromatic.

Lemma 5. *Let G be a circular RAC graph with no red edges. Given a sequence S of vertices on the outer face of G and a side edge (u, v) as input, there exists a linear-time algorithm to find a major edge m of G.*

Proof. Without loss of generality, let (u, v) be a green edge. Since (u, v) is a side edge, there are no parallel edges on one side of it. Let u' be the consecutive vertex to u on that side of (u, v), where there are no parallel edges. Similarly, let u'' be the consecutive vertex to v on that side of (u, v), where there are no parallel edges (see Figure 5a). Name the other endpoints of the edges incident to u' and u'' as v' and v'' correspondingly. Obviously, (u', v') and (u'', v'') are blue edges. Let d be the direction of which we traverse S to reach u' from u. We start traversing the cycle from v' in the opposite direction of d, until we reach v'' or two endpoints m and n of an edge (m, n), such that (m, n) is not among the set $\mathcal{E}$ of the edges crossing (u, v).

Then we have the following three cases:

1. We reach v'' and there is a vertex m (but not a vertex n) between v' and v'', such that m does not belong to any of the edges in $\mathcal{E}$: Then, either (u', v') or (u'', v'') is a major edge.
2. We reach v'' and all the vertices between v' and v'' are incident to edges of $\mathcal{E}$: Then, all of these edges are major edges.
3. We reach n: Then, those edges of $\mathcal{E}$ which cross (m, n) are major edges.

Case 1. We reach v'' and there is a vertex m (but not a vertex n) between v' and v'' (see Figure 5b).
Suppose that m is a vertex between v' and v'', which is not an endpoint of the edges in $\mathcal{E}$. Let us call the other endpoint of the edge incident to m as n. Since n is not between v' and v'' on S, it is either located between v'' and v or between u and v' in the clockwise order. Without loss of generality assume that n is located between u and v', and hence it crosses (u', v'). Now, note that no green edge e is incident to two vertices between m and v (or between n and u), since otherwise no blue edge can cross e, (u, v) and (m, n) simultaneously, which contradicts Lemma 2. Hence, (u', v') crosses all the blue edges and is a major edge.

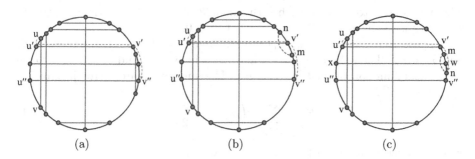

Fig. 5. The traversal path to find a major edge

Case 2. We reach v'' and all the vertices between v' and v'' are incident to edges of $\mathcal{E}$.
Clearly, by Lemma 2, at least one of the edges in $\mathcal{E}$ is a major edge. Since there exists no vertex m or n between the endpoints of these edges, none of them has more crossings than the others. Hence, all of them are major edges.

Case 3. We reach n (see Figure 5c).
Obviously, the major edges of G cross both edges (u, v) and (m, n). Since no vertex is located between the endpoints of the edges crossing (u, v) and (m, n), none of them have more crossings than the others. Hence, all of them are major edges.

The Lemma follows.   □

Lemma 5 completes the proof of Theorem 3.

The overall drawing algorithm can be described as follows:

1. Identify and delete the red edges.
2. Find a side edge, using the algorithm in the proof of Lemma 4
3. Find a major edge, using the algorithm in the proof of Lemma 5.
4. Using the recursive algorithm in the proof of Theorem 2, draw the blue and green edges.
5. Draw the red edges.

5   Circular Drawings with Large Angle Crossings

The results of the previous Sections are useful when a RAC drawing is possible. However, in many cases, such a drawing is not possible. In this Section, we describe a practical quadratic programming approach for increasing crossing angles in a circular drawing.

Again, we assume that the ordering $(u_0, u_2, \ldots, u_{n-1})$ of the vertices around the circle is computed by a crossing reduction algorithm in [3]. The quadratic programming approach was first reported in [12] as a poster.

As noted in Section 4, red edges can be eliminated. The partition of the remaining edges into blue and green is straightforward using the circular ordering given in the input.

We assume that the vertices are drawn on a unit circle, centered at the origin. The next step is to find, for each i, the angle θ_i that the line from the origin to u_i makes with the x axis. We assume that $\theta_0 = 0$.

Suppose that edge $e = (u_i, u_j)$ crosses edge $e' = (u_{i'}, u_{j'})$, where $\theta_i < \theta_j$ and $\theta_{i'} < \theta_{j'}$. Then the crossing angle $\alpha(e, e')$ between e and e' can be written as a linear function of $\theta_i, \theta_j, \theta_{i'}$ and $\theta_{j'}$. The sum of squares error in crossing angles is:

$$F = \sum_{(e,e') \in \Omega} \left(\alpha_{(e,e')} - \frac{\pi}{2} \right)^2, \tag{1}$$

where Ω denote the set of pairs of crossing edges.

The ordering of vertices around the circle induces a set of linear constraints for the θ_i. In practice, we need further constraints. The *vertex resolution* of a drawing is the minimum distance between two vertices, given a fixed size enclosing rectangle for the drawing. To ensure reasonable vertex resolution, we enforce an angular *gap* g between vertices, that is, $\theta_i + g < \theta_{i+1}$ for $i < n - 1$ and $\theta_{n-1} \leq 2\pi$. We can minimise the function F in equation (1) subject to these gap constraints using standard quadratic programming methods.

Using the Rome data set [1], CPLEX, and a relatively old CPU, we tested this method. The execution time varied around 10ms, which is very little overhead on the Bauer-Brandes algorithm. We found a significant improvement in crossing angles over the simple approach of equally spacing vertices around the circle. The improvement varied, depending on the chosen value of the gap g, but averaged around 60%.

6   Conclusion

In this paper, we have characterized circular RAC graphs and provided a linear-time algorithm to construct a circular RAC drawing. We also presented a practical quadratic programming approach for crossing angle maximisation.

In future, we would like to extend our work to find a characterization of *outer RAC graphs*, that is, graphs which have a RAC drawing in which every vertex appears on the outside face. Further preliminary work toward this goal is presented in [4].

References

1. The Rome graphs (2010), http://www.graphdrawing.org/data/index.html
2. Argyriou, E.N., Bekos, M.A., Symvonis, A.: The Straight-Line RAC Drawing Problem Is NP-Hard. In: Černá, I., Gyimóthy, T., Hromkovič, J., Jefferey, K., Královič, R., Vukolić, M., Wolf, S. (eds.) SOFSEM 2011. LNCS, vol. 6543, pp. 74–85. Springer, Heidelberg (2011)
3. Baur, M., Brandes, U.: Crossing Reduction in Circular Layouts. In: Hromkovič, J., Nagl, M., Westfechtel, B. (eds.) WG 2004. LNCS, vol. 3353, pp. 332–343. Springer, Heidelberg (2004)
4. Reisi Dehkordi, H., Eades, P.: Every outer-1-plane graph has a right angle crossing drawing. International Journal of Computational Geometry and Applications (to appear, 2012)
5. Di Battista, G., Eades, P., Tamassia, R., Tollis, I.G.: Graph Drawing: Algorithms for the Visualization of Graphs. Prentice Hall (July 1998)
6. Didimo, W., Eades, P., Liotta, G.: Drawing Graphs with Right Angle Crossings. In: Dehne, F., Gavrilova, M., Sack, J.-R., Tóth, C.D. (eds.) WADS 2009. LNCS, vol. 5664, pp. 206–217. Springer, Heidelberg (2009)
7. Didimo, W., Eades, P., Liotta, G.: Drawing graphs with right angle crossings. Theoretical Computer Science 412(39), 5156–5166 (2011)
8. Dujmovic, V., Gudmundsson, J., Morin, P., Wolle, T.: Notes on large angle crossing graphs. In: Computing: The Australasian Theory Symposium (2010)
9. Eades, P., Liotta, G.: Right Angle Crossing Graphs and 1-Planarity. In: van Kreveld, M., Speckmann, B. (eds.) GD 2011. LNCS, vol. 7034, pp. 148–153. Springer, Heidelberg (2012)
10. Di Giacomo, E., Didimo, W., Eades, P., Liotta, G.: 2-Layer Right Angle Crossing Drawings. In: Iliopoulos, C.S., Smyth, W.F. (eds.) IWOCA 2011. LNCS, vol. 7056, pp. 156–169. Springer, Heidelberg (2011)
11. Huang, W., Hong, S.H., Eades, P.: Effects of crossing angles. In: Proceedings of PacificVis, pp. 41–46 (2008)
12. Nguyen, Q., Eades, P., Hong, S.-H., Huang, W.: Large Crossing Angles in Circular Layouts. In: Brandes, U., Cornelsen, S. (eds.) GD 2010. LNCS, vol. 6502, pp. 397–399. Springer, Heidelberg (2011)
13. Purchase, H.C., Cohen, R.F., James, M.: Validating Graph Drawing Aesthetics. In: Brandenburg, F.J. (ed.) GD 1995. LNCS, vol. 1027, pp. 435–446. Springer, Heidelberg (1996)
14. Reisi Dehkordi, H., Nguyen, Q., Eades, P., Hong, S.H.: Circular graph drawings with large crossing angles. Technical Report TR691, School of Information Technologies, University of Sydney

On Graphs That Are Not PCGs

Stephane Durocher[1,*], Debajyoti Mondal[1,**], and Md. Saidur Rahman[2,***]

[1] Department of Computer Science, University of Manitoba
[2] Department of Computer Science and Engineering,
Graph Drawing & Information Visualization Laboratory,
Bangladesh University of Engineering and Technology
{durocher,jyoti}@cs.umanitoba.ca, saidurrahman@cse.buet.ac.bd

Abstract. Let T be an edge weighted tree and let d_{min}, d_{max} be two nonnegative real numbers. Then the pairwise compatibility graph (PCG) of T is a graph G such that each vertex of G corresponds to a distinct leaf of T and two vertices are adjacent in G if and only if the weighted distance between their corresponding leaves in T is in the interval $[d_{min}, d_{max}]$. Similarly, a given graph G is a PCG if there exist suitable T, d_{min}, d_{max}, such that G is a PCG of T. Yanhaona, Bayzid and Rahman proved that there exists a graph with 15 vertices that is not a PCG. On the other hand, Calamoneri, Frascaria and Sinaimeri proved that every graph with at most seven vertices is a PCG. In this paper we construct a graph of eight vertices that is not a PCG, which strengthens the result of Yanhaona, Bayzid and Rahman, and implies optimality of the result of Calamoneri, Frascaria and Sinaimeri. We then construct a planar graph with sixteen vertices that is not a PCG. Finally, we prove a variant of the PCG recognition problem to be NP-complete.

1 Introduction

Let T be an edge weighted tree and let d_{min}, d_{max} be two nonnegative real numbers. Then the *pairwise compatibility graph (PCG)* of T is a graph G such that each vertex of G corresponds to a distinct leaf of T and two vertices are adjacent in G if and only if the weighted distance between their corresponding leaves in T is in the interval $[d_{min}, d_{max}]$. Similarly, a given graph G is a PCG if there exist suitable T, d_{min}, d_{max}, such that G is a PCG of T. Figure 1(a) illustrates an edge weighted tree T, and Figure 1(b) shows the corresponding PCG G, where $d_{min} = 2$ and $d_{max} = 3.5$. Figure 1(c) shows another edge weighted tree T' such that G is a PCG of T' when $d_{min} = 1.5$ and $d_{max} = 2$.

In 2003, Kearney et al. [7] introduced the concept of PCG and showed how to use them to model evolutionary relationships among a set of organisms. Moreover, they proved that the problem of finding a maximal clique can be solved

* Work of the author is supported in part by the Natural Sciences and Engineering Research Council of Canada (NSERC).
** Work of the author is supported in part by a University of Manitoba Graduate Fellowship.
*** Work of the author is supported in part by the Ministry of Science and Information & Communication Technology, Government of Bangladesh.

S.K. Ghosh and T. Tokuyama (Eds.): WALCOM 2013, LNCS 7748, pp. 310–321, 2013.
© Springer-Verlag Berlin Heidelberg 2013

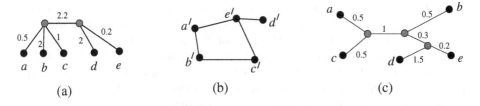

Fig. 1. (a) An edge weighted tree T. (b) A PCG G of T, where $d_{min} = 2, d_{max} = 3.5$. (c) Another edge weighted tree T' such that G is a PCG of T' when $d_{min} = 1.5, d_{max} = 2$.

in polynomial time for pairwise compatibility graphs if one can find their corresponding edge weighted trees in polynomial time. They hoped that every graph is a PCG, but later, Yanhaona et al. [12] constructed a 15-vertex graph that is not a PCG. Several researchers have attempted to characterize pairwise compatibility graphs. Yanhaona et al. [13] proved that graphs having cycles as its maximal biconnected components are PCGs. Salma and Rahman [10] proved that every triangle free maximum degree three outerplanar graph is a PCG. Calamoneri et al. [5] gave some sufficient conditions for a split matrogenic graph to be a PCG, and examined the graph classes that arise from using the intervals $[0, d_{max}]$ (LPG) and $[d_{min}, \infty]$ (mLPG). They proved that the intersection of these classes is not empty, and neither of them is contained in the other. The graph classes LPG, mLPG and PCG are similar to the leaf powers and their variants, which have been extensively studied in the literature [1–3, 6, 8, 9]. For example, the complement of PCG and the graph class LPG are closely related to the exact k-leaf powers, (k, l)-leaf powers and k-leaf powers, respectively.

Finding a pairwise compatibility tree of a given graph appeared to be difficult, even for graphs with few vertices. Kearney et al. [7] showed that every graph with at most five vertices is a PCG. The smallest graph known not to be a PCG is a 15-vertex graph constructed by Yanhaona et al. [12]. This is a bipartite graph with partite sets A and B, where $|A| = 5$ and $|B| = 10$, and each subset of three vertices of A is adjacent to a distinct vertex of B. Recently, Calamoneri et al. [4] proved that every graph with at most seven vertices is a PCG.

In this paper we construct a graph of eight vertices that is not a PCG, which strengthens the result of Yanhaona et al. [12], and implies optimality of the result of Calamoneri et al. [4]. We then construct a planar graph with sixteen vertices that is not a PCG; this is the first planar graph known not to be a PCG. Finally, we prove a variant of the PCG recognition problem to be NP-complete.

2 Preliminaries

In this section we introduce some definitions and review some relevant results.

Let $G = (V, E)$ be a graph with vertex set V and edge set E. The complement graph $\overline{G}$ of G is the graph with vertex set V and edge set $\overline{E}$, where $\overline{E}$ consists of the edges that are determined by the non-adjacent pairs of vertices of G.

Let T be an edge weighted tree. Let u and v two leaves of T. By P_{uv} we denote the unique path between u and v in T. By $d_T(u, v)$ we denote the *weighted distance* between u and v, i.e., the sum of the weights of the edges on P_{uv}. Let d_{min}, d_{max} be two nonnegative real numbers. Then by $PCG(T, d_{min}, d_{max})$ we denote the PCG of T that respects the interval $[d_{min}, d_{max}]$. By $T_{x_1 x_2 ... x_t}$ we denote the subgraph of T induced by the paths $P_{x_i x_j}$, where $1 \leq i, j \leq t$. Figures 2(a)–(b) illustrate an example of such a subgraph.

Lemma 1 (Yanhaona et al. [12]). *Let T be an edge weighted tree, and let u, v and w be three leaves of T such that P_{uv} is the longest path in T_{uvw}. Let x be a leaf of T other than u, v and w. Then $d_T(w, x) \leq d_T(u, x)$, or $d_T(w, x) \leq d_T(v, x)$.*

Let $G = PCG(T, d_{min}, d_{max})$. Then by u' we denote the vertex of G that corresponds to the leaf u of T. The following lemma illustrates a relationship between a PCG and its corresponding edge weighted tree, which holds based on the proof of [12, Lemma 3.3].

Lemma 2. *Let $G = PCG(T, d_{min}, d_{max})$. Let a, b, c, d, e be five leaves of T and a', b', c', d', e' be the corresponding vertices of G, respectively. Let P_{ae} and P_{bd} be the longest path in T_{abcde} and T_{bcd}, respectively. Then any vertex x' in G that is adjacent to a', c', e' must be adjacent to b' or d'.*

The rest of the paper is organized as follows. In Section 3 we construct a graph G_1 with nine vertices that is not a PCG. In Section 4 we prove that the graph obtained by deleting a vertex of degree three from G_1 is not a PCG. In Section 5 we construct a planar graph that is not a PCG. In Section 6 we prove the NP-hardness result. Finally, Section 7 concludes the paper.

3   Not All 9-Vertex Graphs Are PCGs

In this section we construct a graph G_1 of nine vertices that is not a PCG. Here we describe an outline of the construction.

We use three lemmas to construct G_1. In Lemma 3 we prove that for a cycle a', b', c', d' of four vertices, $d_T(a, c)$ and $d_T(b, d)$ cannot be both greater than d_{max}. We then construct a graph H with six vertices a', b', c', d', i', j' such that each pair of vertices in H are adjacent except the pairs $(a', c'), (b', d'), (i', d')$, $(j', b'), (i', j')$, as shown in Figure 2(c). Using Lemma 3 we prove in Lemma 4 that at least one of $d_T(a, c), d_T(b, d), d_T(i, d), d_T(j, b), d_T(i, j)$ must be greater than d_{max}. In Lemma 5 we prove that any PCG that contains H as an induced subgraph must satisfy the inequality $d_T(a, c) < d_{min}$, where a' and c' are the only vertices of degree four in H. We add three vertices k', u', v' to H to construct G_1, as shown in Figure 2(d). In Theorem 1 we show that for every non-adjacent pair (x', y') in H, the graph G_1 contains an induced subgraph isomorphic to H that contains x' and y' as its degree four vertices. By Lemma 5, $d_T(x, y) < d_{min}$. Observe that this contradicts Lemma 4. Consequently, G cannot be a PCG.

The following lemma proves that for a cycle a', b', c', d' of four vertices, $d_T(a, c)$ and $d_T(b, d)$ cannot be both greater than d_{max}. We omit its proof due to space constraints.

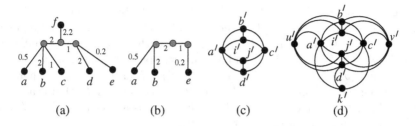

Fig. 2. (a) An edge weighted tree T. (b) T_{abe}. (c) Illustration for H. (d) G_1.

Lemma 3. *Let* $G = PCG(T, d_{min}, d_{max})$, *which is a cycle* a', b', c', d' *of four vertices. Let* a, b, c, d *be the leaves of* T *that correspond to the vertices* a', b', c', d' *of* G, *respectively. Then* $d_T(a, c)$ *and* $d_T(b, d)$ *cannot be both greater than* d_{max}.

We now construct a graph H with six vertices a', b', c', d', i', j' such that each pair of vertices in H are adjacent except the pairs $(a', c'), (b', d'), (i', d')$, $(j', b'), (i', j')$, as shown in Figure 2(c). The following lemma proves that at least one of $d_T(a, c), d_T(b, d), d_T(i, d), d_T(j, b), d_T(i, j)$ must be greater than d_{max}.

Lemma 4. *Let* $H = PCG(T, d_{min}, d_{max})$. *Let* a, b, c, d, i, j *be the leaves of* T *that correspond to the vertices* a', b', c', d', i', j' *of* H. *Then at least one of* $d_T(a, c)$, $d_T(b, d), d_T(i, d), d_T(j, b), d_T(i, j)$ *must be greater than* d_{max}.

Proof. For each pair $(x', y') \in \{(a', c'), (b', d'), (i', d'), (j', b'), (i', j')\}$, x' and y' are non-adjacent in H. Therefore, either $d_T(x, y) < d_{min}$ or $d_T(x, y) > d_{max}$.

If one of $d_T(a, c), d_T(b, d), d_T(i, d), d_T(j, b)$ is greater than d_{max}, then the lemma holds irrespective of whether $d_T(i, j) < d_{min}$ or $d_T(i, j) > d_{max}$. We thus assume that each of $d_T(a, c), d_T(b, d), d_T(i, d), d_T(j, b)$ is less than d_{min}, and then prove that $d_T(i, j)$ must be greater than d_{max}.

Suppose for a contradiction that $d_T(i, j) < d_{min}$. Recall that we assumed $d_T(j, b) < d_{min}$. Consequently, since i' and b' are adjacent in H, the path P_{ib} must be the longest path T_{ijb}. By Lemma 1, $d_T(j, d) \leq d_T(i, d)$ or $d_T(j, d) \leq d_T(b, d)$. Since we assumed that $d_T(i, d) < d_{min}$ and $d_T(b, d) < d_{min}$, the inequality $d_T(j, d) < d_{min}$ holds. But this contradicts that j', d' are adjacent in G. Therefore, $d_T(i, j)$ must be greater than d_{max}. □

In the following lemma we prove that any PCG that contains H as an induced subgraph must satisfy the inequality $d_T(a, c) < d_{min}$, where a' and c' are the only vertices of degree four in H.

Lemma 5. *Let* $G = PCG(T, d_{min}, d_{max})$ *be a graph that contains an induced subgraph* G' *isomorphic to* H. *Let* a, b, c, d, i, j *be the leaves of* T *that correspond to the vertices* a', b', c', d', i', j' *of* G'. *Let* a' *and* c' *be the vertices of degree four in* G'. *Then* $d_T(a, c)$ *must be less than* d_{min}.

Proof. Since a', c' are non-adjacent in G', either $d_T(a, c) < d_{min}$ or $d_T(a, c) > d_{max}$. Suppose for a contradiction that $d_T(a, c) > d_{max}$.

Since the subgraph induced by a', b', c', d' is a cycle, by Lemma 3, $d_T(b', d') < d_{min}$. Again, since the subgraph induced by a', i', c', d' is a cycle, by Lemma 3, $d_T(i', d') < d_{min}$. Consequently, P_{bi} is the longest path in T_{ibd}. Observe that we assumed $d_T(a, c) > d_{max}$. On the other hand, for each pair $(x', y') \in \{(a', b'), (a', d'), (a', i'), (b', d'), (b', c'), (b', i'), (c', d'), (c', i'), (d', i')\}$, $d_T(x, y) \le d_{max}$. Therefore, P_{ac} is the longest path in T_{abcdi}. By Lemma 2, any vertex j' in G' that is adjacent to a', c', d' must be adjacent to i' or b'. Although j' is adjacent to a', c', d' in G, neither i' nor b' is adjacent to j', a contradiction.   □

We now add three vertices k', u', v' to H to construct G_1, as shown in Figures 3(a)–(b). In the following theorem we show that G_1 is not a PCG.

Theorem 1. G_1 *is not a PCG.*

Proof. For every non-adjacent pair (x', y') in H, the graph G_1 contains an induced subgraph isomorphic to H that contains x' and y' as its degree four vertices, as shown in Figures 3(c)–(g). By Lemma 5, $d_T(x, y) < d_{min}$. This contradicts Lemma 4 that says there exists a non-adjacent pair (x', y') in H such that $d_T(x, y) > d_{max}$. Consequently, G cannot be a PCG.   □

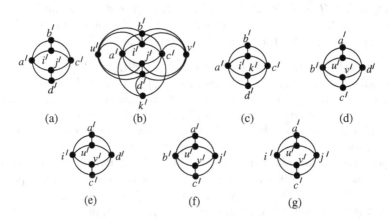

Fig. 3. (a) H. (b) G_1. (c)–(g) Five induced subgraphs of G, when (c) $d_T(a, c) > d_{max}$, (d) $d_T(b, d) > d_{max}$, (e) $d_T(i, d) > d_{max}$, (f) $d_T(j, b) > d_{max}$, (g) $d_T(i, j) > d_{max}$.

4   Not All 8-Vertex Graphs Are PCGs

In this section we analyze the structure of the graph G_1, and modify it to obtain a graph of eight vertices that is not a PCG.

We refer the reader to Figure 3. Observe that G_1 has only one vertex of degree three, i.e., vertex k'. The proof of Theorem 1 refers to vertex k' only in the case when $d_T(a, c) > d_{max}$, as shown in Figure 3(c). This observation inspired us to examine whether the graph $G_1 \setminus k'$ is a PCG or not. In this section we denote the graph $G_1 \setminus k'$, shown in Figure 4(a), by G_2 and prove that G_2 is not a PCG. The following lemma will be useful to prove the main result.

Lemma 6. *Let $G = PCG(T, d_{min}, d_{max})$ be a graph of four vertices a', b', c', d' and two edges (a', b') and $(c'd')$. Let a, b, c, d be the leaves of T that correspond to the vertices a', b', c', d' of G, respectively. Then at least one of $d_T(a, d), d_T(b, d), d_T(b, c), d_T(a, c)$ must be greater than d_{max}.*

Proof. Since every pair of vertices among $(a', d'), (b', d'), (b', c'), (a', c')$ are non-adjacent in G, each of $d_T(a, d), d_T(b, d), d_T(b, c), d_T(a, c)$ is either greater than d_{max} or less than d_{min}. Suppose for a contradiction that $d_T(a, d), d_T(b, d), d_T(b, c), d_T(a, c)$ are less than d_{min}.

Since a' and b' are adjacent and $d_T(a, c), d_T(b, c)$ are less than d_{min}, P_{ab} must be the longest path in T_{abc}. By Lemma 1, $d_T(c, d) \leq d_T(a, d)$ or $d_T(c, d) \leq d_T(b, d)$. By assumption, both $d_T(a, d)$ and $d_T(b, d)$ are less than d_{min}. Therefore, $d_T(c, d) < d_{min}$, which contradicts that c' and d' are adjacent in G.   □

We now use Lemma 6 to obtain the following corollary.

Corollary 1. *Let $G_2 = PCG(T, d_{min}, d_{max})$ and let a, b, c, d, i, j, u, v be the leaves of T that correspond to the vertices $a', b', c', d', i', j', u', v'$ of G_2. Then (a) at least one of $d_T(u, v), d_T(a, v), d_T(a, c), d_T(u, c)$ must be greater than d_{max}, and (b) at least one of $d_T(b, j), d_T(b, d), d_T(i, d), d_T(i, j)$ must be greater than d_{max}.*

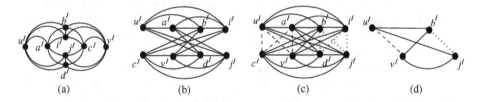

Fig. 4. (a) G_2. (b) Another drawing of G_2. (c) Illustration for $((w', x'), (y', z'))$, where (w', x') and (y', z') are shown in dashed lines and dotted lines, respectively. (d) $((w', x'), (y', z')) = ((u', v'), (b', j'))$.

Theorem 2. *G_2 is not a PCG.*

Proof. Suppose for a contradiction that $G_2 = PCG(T, d_{min}, d_{max})$, where $a, b, c,$ d, i, j, u, v are the leaves of T that correspond to the vertices $a', b', c', d', i', j', u', v'$ of G_2. Observe that for any $((w', x'), (y', z'))$, where $(w', x') \in \{(u', v'), (a', v'),$ $(a', c'), (u', c')\}$ and $(y', z') \in \{(b', j'), (b', d'), (i', d'), (i', j')\}$, the vertices $\{w', x', y', z'\}$ induce a cycle C such that w', x' and y', z' are non-adjacent in C. Figures 4(b)–(d) illustrate this scenario. By Corollary 1, for some $((w', x'), (y', z'))$, both $d_T(w, x)$ and $d_T(y, z)$ are greater than d_{max}. This contradicts Lemma 3 since the vertices $\{w', x', y', z'\}$ induce a cycle.   □

5   Not All Planar Graphs Are PCGs

In this section we prove that the planar graph G_p, shown in Figure 5(a), is not a PCG.

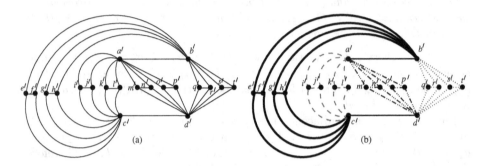

Fig. 5. (a) G_p. (b) Illustration for the proof of Theorem 3. The graphs isomorphic to H are shown in bold lines ($d_T(b,c) > d_{max}$), regular dashed lines ($d_T(a,c) > d_{max}$), regular dotted lines ($d_T(b,d) > d_{max}$) and irregular dashed lines ($d_T(a,d) > d_{max}$).

Theorem 3. G_p *is not a PCG.*

Proof. Suppose for a contradiction that $G_p = PCG(T, d_{min}, d_{max})$, where $a, b, \ldots, s, t$ are the leaves of T that correspond to the vertices $a', b', \ldots, s', t'$ of G_p.

Since the subgraph induced by a', b', c', d' consists of exactly two edges (a', b') and (c', d'), by Lemma 6, at least one of $d_T(a,d), d_T(b,d), d_T(b,c), d_T(a,c)$ must be greater than d_{max}. For any pair $(x', y') \in \{(a', d'), (b', d'), (b', c'), (a', c')\}$, there exists an induced subgraph in G_p that is isomorphic to H (i.e., the graph of Figure 3(c)) that contains x' and y' as its degree four vertices. By Lemma 5, $d_T(x, y) < d_{min}$, which contradicts that at least one of $d_T(a,d), d_T(b,d), d_T(b,c), d_T(a,c)$ must be greater than d_{max}. Consequently, G_p cannot be a PCG.   □

Observe that G_p has twenty vertices. However, the proof of Theorem 3 holds even for the planar graph obtained from G_p by merging the pair of vertices $(e', t'), (h', i'), (l', m'), (p', q')$ and then removing the resulting multi-edges. Therefore, there exists a planar graph with sixteen vertices that is not a PCG. We omit the details due to space constraints.

6   NP-Hardness

In this section we examine a generalized PCG recognition problem that given a graph G and a set $S \subseteq \overline{E}^1$, asks to determine a PCG $G' = (T, d_{min}, d_{max})$ that contains G as a subgraph[2] but does not contain any edge of S. Observe that if

---

[1] $\overline{E}$ is the set of edges of the complement graph of G.

[2] Not necessarily an induced subgraph.

$S = \overline{E}$, then the problem asks to decide whether G is a PCG. We prove that the generalized PCG recognition problem is NP-hard if we require the maximum number of edges of S to have weighted tree distance greater than d_{max} between their corresponding leaves. A decision version of the problem is as follows.

Problem : MAX-GENERALIZED-PCG-RECOGNITION
Instance : A graph G, a subset S of the edges of its complement graph, and a positive integer k.
Question : Is there a PCG $G' = PCG(T, d_{min}, d_{max})$ such that G' contains G as a subgraph[2], but does not contain any edge of S; and at least k edges of S have distance greater than d_{max} between their corresponding leaves in T?

We prove the NP-hardness of MAX-GENERALIZED-PCG-RECOGNITION by reduction form MONOTONE-ONE-IN-THREE-3-SAT [11].

Problem : MONOTONE-ONE-IN-THREE-3-SAT
Instance : A set U of variables and a collection C of clauses over U such that each clause consists of exactly three non-negated literals.
Question : Is there a satisfying truth assignment for U such that each clause in C contains exactly one true literal?

Given an instance $I(U, C)$ of MONOTONE-ONE-IN-THREE-3-SAT, we construct an instance $I(G, S, k)$ of MAX-GENERALIZED-PCG-RECOGNITION such that $I(U, C)$ has an affirmative answer if and only if $I(G, S, k)$ has an affirmative answer. The idea of the reduction is as follows. Given an edge weighted tree T with n leaves, $d_{min} = 0$ and $d_{max} = +\infty$, the corresponding PCG is a complete graph K_n of n vertices. Observe that as the interval $[d_{min}, d_{max}]$ begins to shrink, more and more edges of K_n disappear. Some edges disappear due to the increase of d_{min} and some other edges disappear due to the decrease of d_{max}. We use these two events to set the truth values of the literals.

Let G_{not} be the graph of Figure 6(a). The following lemma shows how to use this graph as a NOT gate.

Lemma 7. *Assume that $G_{not} = PCG(T, d_{min}, d_{max})$, where $a, b, \ldots, q$ are the leaves of T that correspond to the vertices $a', b', \ldots, q'$ of G_{not}. Then $d_T(a, b) < d_{min}$ if and only if $d_T(c, d) > d_{max}$.*

Proof. By Lemma 6, one of $d_T(e, g), d_T(e, h), d_T(f, g), d_T(f, h)$ must be greater than d_{max}. Observe that for any pair $(x, y) \in \{(e', g'), (e', h'), (f', g'), (f', h')\}$, the vertices b', x', d', y' form an induced cycle. Therefore, by Lemma 3, $d_T(b, d) < d_{min}$. Similarly, we can prove that $d_T(a, q) < d_{min}$ and $d_T(c, q) < d_{min}$. Since a', c', b', q', d' induce a cycle of five vertices, one of $d_T(a, b), d_T(c, d), d_T(a, q), d_T(c, q), d_T(b, d)$ is greater than d_{max} [5, Lemma 2]. Since $d_T(a, q), d_T(c, q), d_T(b, d)$ are less than d_{min}, one of or both $d_T(a, b)$ and $d_T(c, d)$ are greater than d_{max}.

Without loss of generality assume that $d_T(a, b) > d_{max}$. Then by Lemma 1, $d_T(c, d) \leq d_T(a, d)$ or $d_T(c, d) \leq d_T(b, d)$. Since $d_T(a, d) \leq d_{max}$ and $d_T(b, d) <$

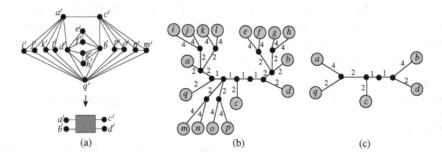

Fig. 6. (a) G_{not}, and its hypothetical representation. (b) $G_{not} = PCG(T, 7, 11)$. (c) Simplified representation of T.

$d_{min} \leq d_{max}$, $d_T(c, d)$ must be less than d_{min}. Similarly, we can prove that if $d_T(c, d) > d_{max}$, then $d_T(a, b) < d_{min}$.   □

Properties of G_{not}. The vertices a, b and c, d play the role of the input and output of a NOT gate, respectively. Figure 6(b) illustrates a pairwise compatibility tree T, where $G_{not} = PCG(T, 7, 11)$ and $d_T(a, b) > d_{max}$. Observe that once we construct the tree T_{abqcd}, it becomes straightforward to add the trees T_{efgh}, T_{ijkl} and T_{mnop}. Therefore, in the rest of this section we only consider the simplified representation for T, as shown in Figure 6(c). We can cascade several NOT gates to duplicate or invert the input, but we omit the details due to space constraints.

In the reduction, all the edges of $\overline{G_{not}}$ will belong to S. Every G_{not} has 101 non-adjacent pairs, and by construction, in any pairwise compatibility tree T' of G_{not}, $d_{T'}(a, q), d_{T'}(c, q), d_{T'}(b, d)$ and one of $d_{T'}(a, b), d_{T'}(c, d)$ must be less than d_{min}. Therefore, at most 97 edges of $\overline{G_{not}}$ can have distance greater than d_{max} between their corresponding leaves in T'. Since the tree T, shown in Figure 6(b), determines 97 such edges, it maximizes the number of edges of $\overline{G_{not}}$ that have distance greater than d_{max} between their corresponding leaves.

Gadget. Each literal gadget consists of a pair of non-adjacent vertices. Every edge determined by these two vertices, belongs to S. We say that *a literal (or, any non-adjacent pair of vertices) (a', b') is true* if and only if $d_T(a, b) > d_{max}$; otherwise, it is *false*.

Every clause gadget G_{clause}, as shown in Figure 7(a), corresponds to a logic circuit L that is consistent if and only if at most one of its three inputs is true. The three pairs of vertices $(a', b'), (c', d')$, and (e', f') of G_{clause} play the role of the inputs. For each pair of inputs, e.g., $((a', b'), (c', d'))$, G_{clause} contains a G_{not} such that the ports o'_1, o'_2 of G_{not} form a cycle with a', b', and the ports o'_3, o'_4 of G_{not} form a cycle with c', d'. In the following we show that L is consistent if and only if at most one input is true.

Suppose for a contradiction that at least two of the three inputs, without loss of generality (a', b') and (c', d'), are true. Since (a', b') is true, by Lemma 3,

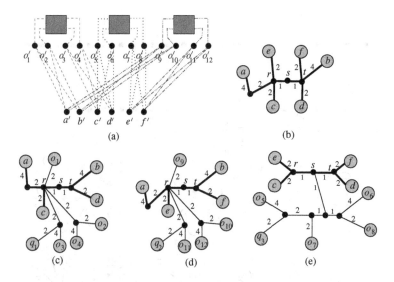

Fig. 7. (a) A clause gadget G_{clause}. (b) Simplified representation of a pairwise compatibility tree T that determines the truth values of its literals. Here, $(a', b'), (c', d')$ and (e', f') correspond to the values true, false and false, respectively. (c)–(e) Subtrees of T that correspond to a G_{not} and its associated literal gadgets.

(o'_1, o'_2) must be false. Consequently, (o'_3, o'_4) must be true. Since c', o'_3, d', o'_4 induce a cycle, by Lemma 3, (c', d') must be false, a contradiction.

Assume now that at most one of the three inputs is true. In this case, we show how to construct a pairwise compatibility tree such that the corresponding PCG G'_{clause} contains G_{clause} as a subgraph. Without loss of generality assume that (a', b') is true. (The construction when when all the inputs are false are similar.) Construct an edge weighted tree T as illustrated in Figure 7(b). Observe that $d_T(c, d) < d_{min}, d_T(e, f) < d_{min}$ and $d_T(a, b) > d_{max}$, which implies that $(c', d'), (e', f')$ are false and (a', b') is true. We call r, s, t the *medial path* of T. Figure 7(c)–(e) illustrates how to add the subtrees (shown in thin lines) that correspond to the G_{not}s to T. These trees not only realize the G_{not}s, but also determine the cycles that are incident to the inputs of the clause gadget.

We now have the following theorem. We omit the details due to space constraints.

Theorem 4. MAX-GENERALIZED-PCG-RECOGNITION *is NP-hard.*

Proof (Outline). Given an instance $I(U, C)$ of MONOTONE-ONE-IN-THREE-3-SAT, we construct a corresponding instance $I(G, S, k)$ of MAX-GENERALIZED-PCG-RECOGNITION in polynomial time by constructing a clause gadget for each clause, and duplicating the literals that occurs in multiple clauses by cascading NOT gates, as illustrated in Figure 8(a). The set S consists of the edges of $\overline{G_{not}}$s and the edges that are determined by the literal gadgets. Let N and t' be the number of NOT gates and clauses, respectively. We set $k = 97N + t'$.

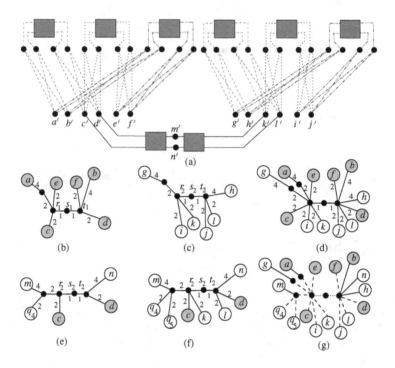

Fig. 8. (a) The graph G that correspond to the instance $I(U, C) = (x_1 \lor x_2 \lor x_3) \land (x_4 \lor x_2 \lor x_5)$, where x_1, x_2, x_3, x_4, x_5 correspond to $(a', b'), (c', d'), (e', f'), (g', h'), (i', j')$, respectively. (b)–(c) Compatibility trees for the clauses, where the literals except x_1 and x_2 are false. (d) Merging the medial paths. (e)–(f) Compatibility trees for G_{not}s that propagate the truth value from (c', d') to (k', l'). (g) A compatibility tree of G'. The edges with weights $1, 2$ and 4 are shown in dotted, dashed and solid lines, respectively.

Assume first that $I(U, C)$ has an affirmative answer. For each clause, we construct an edge weighted tree as shown in Figure 7(a). We then merge the medial paths of these trees, as shown in Figures 8(b)–(d). Finally, we add the subtrees that correspond to the G_{not}s that we used to duplicate (or, propagate) the input values, as depicted in Figures 8(e)–(g). Let G' be the PCG of the resulting tree. G' contains G as a subgraph since we constructed T using the trees for the basic gadgets. G' does not contain any edge of S since every redundant edge of G' lie between different G_{not}s, or different literal gadgets, or between a G_{not} and a literal gadget. Finally, there are 97 edges in each $\overline{G_{not}}$ that contribute to k, and t' true literals, one from each clause, that contribute to k.

Assume now that $I(U, C)$ does not have any affirmative answer. Since each $\overline{G_{not}}$ can have at most 97 edges that contribute to k, at least t' edges that contribute to k must come from the literal gadgets. Since no two literal gadget that lie in the same clause can be true, each clause must have at least one true literal, which contradicts that $I(U, C)$ does not have any affirmative answer.   □

7 Conclusion

We have constructed a nonplanar graph with eight vertices that is not a PCG, but the graph is not split matrogenic. Therefore, the question of Calamoneri et al. [5] of whether every split matrogenic is a PCG remains open. We also construct a planar graph that is not a PCG, but the graph is not outerplanar. Since every triangle-free outerplanar graph with degree at most three is a PCG [10], an interesting question is whether there exists any outerplanar graph that is not a PCG. An important open problem that remains is to determine the complexity of the (original, or generalized) PCG recognition problem.

References

1. Brandstädt, A., Hundt, C., Mancini, F., Wagner, P.: Rooted directed path graphs are leaf powers. Discrete Mathematics 310(4), 897–910 (2010)
2. Brandstädt, A., Le, V.B., Rautenbach, D.: Exact leaf powers. Theoretical Computer Science 411(31-33), 2968–2977 (2010)
3. Brandstädt, A., Wagner, P.: Characterising (k, l)-leaf powers. Discrete Applied Mathematics 158(2), 110–122 (2010)
4. Calamoneri, T., Frascaria, D., Sinaimeri, B.: All graphs with at most seven vertices are pairwise compatibility graphs. The Computer Journal (to appear, 2012), http://arxiv.org/abs/1202.4631
5. Calamoneri, T., Petreschi, R., Sinaimeri, B.: On Relaxing the Constraints in Pairwise Compatibility Graphs. In: Rahman, M. S., Nakano, S.-I. (eds.) WALCOM 2012. LNCS, vol. 7157, pp. 124–135. Springer, Heidelberg (2012)
6. Fellows, M.R., Meister, D., Rosamond, F.A., Sritharan, R., Telle, J.A.: Leaf Powers and Their Properties: Using the Trees. In: Hong, S.-H., Nagamochi, H., Fukunaga, T. (eds.) ISAAC 2008. LNCS, vol. 5369, pp. 402–413. Springer, Heidelberg (2008)
7. Kearney, P.E., Munro, J.I., Phillips, D.: Efficient Generation of Uniform Samples from Phylogenetic Trees. In: Benson, G., Page, R.D.M. (eds.) WABI 2003. LNCS (LNBI), vol. 2812, pp. 177–189. Springer, Heidelberg (2003)
8. Kennedy, W.S., Lin, G., Yan, G.: Strictly chordal graphs are leaf powers. Journal of Discrete Algorithms 4(4), 511–525 (2006)
9. Nishimura, N., Ragde, P., Thilikos, D.M.: On graph powers for leaf-labeled trees. Journal of Algorithms 42(1), 69–108 (2002)
10. Salma, S.A., Rahman, M.S.: Triangle-Free Outerplanar 3-Graphs Are Pairwise Compatibility Graphs. In: Rahman, M. S., Nakano, S.-I. (eds.) WALCOM 2012. LNCS, vol. 7157, pp. 112–123. Springer, Heidelberg (2012)
11. Schaefer, T.J.: The complexity of satisfiability problems. In: Proceedings of Symposium on Theory of Computing, STOC 1978, pp. 216–226 (1978)
12. Yanhaona, M.N., Bayzid, M.S., Rahman, M.S.: Discovering pairwise compatibility graphs. Discrete Mathematics, Algorithms and Applications 2(4), 607–623 (2010)
13. Yanhaona, M.N., Hossain, K.S.M.T., Rahman, M.S.: Pairwise compatibility graphs. Journal of Applied Mathematics and Computing 30(1-2), 479–503 (2009)

On Embedding of Certain Recursive Trees and Stars into Hypercube

Indhumathi Raman

School of Information Technology and Engineering
VIT University, Vellore - 632014, India
indhumathi.r@vit.ac.in

Abstract. In this paper, a rooted binary tree T_h of height h is called a *recursive tree* if T_h can be defined in terms of T_{h-k} for certain values of k, $1 \leq k \leq h$. We prove that certain recursive trees are subtrees of the hypercube graph whose dimension is close to optimal. We also embed the stars $K_{1,2^n 3^m}$ and $K_{1,2^n 5^m}$ into hypercube with dilation $n + 2m$ and $n + 3m$ respectively.

Keywords: Embedding, hypercube, recursive trees, stars.

1 Introduction

The processors in a parallel computer communicate by the exchange of messages. A key element in the design of a parallel computer is the interconnection network of processors, which must be made to operate as efficient as possible. An interconnection network N can be represented by a graph $G(N)$ as follows: Each vertex of $G(N)$ represents a processor and each edge represents the link between the two processors. On the other hand, the data processing by a parallel algorithm A can be represented by a graph $G(A)$ as follows: Each vertex of $G(A)$ represents a data allocated to the local memory of the processor and each edge represents a computation between two data sets. An important goal of a parallel algorithm designer is to map the algorithm graph $G(A)$ into the corresponding graph $G(N)$ of the target machine's interconnection network N. This goal leads to the following definition.

Definition 1. *An embedding f of a (guest) graph $G = (V_G, E_G)$ into a (host) graph $H = (V_H, E_H)$ is an injection $f : V_G \rightarrow V_H$ (not necessarily onto) such that every edge of G is mapped onto a path of H. ie., if $(v_1, v_2) \in E_G$, then $f(v_1)$ and $f(v_2)$ are connected by a path in H.*

Some of the parameters used to measure the quality of an embedding f are dilation and expansion.

1. $\text{Dilation}(f) := \max\{d_H(f(u), f(v)) : (u, v) \in E_G\}$ where $d_H(x, y)$ denotes the length of a shortest path between x and y in H.
2. $\text{Expansion}(f) := \frac{|V_H|}{|V_G|}$

S.K. Ghosh and T. Tokuyama (Eds.): WALCOM 2013, LNCS 7748, pp. 322–333, 2013.
© Springer-Verlag Berlin Heidelberg 2013

These parameters represent respectively the maximum time required to route a message in the host graph (computation latency) and the relative number of unutilized processors in the host graph. For most embedding problems, it is very difficult to obtain an embedding that minimizes both the parameters simultaneously. Therefore, some tradeoffs among these parameters must be made (Leighton [8]). If $dilation(f) = 1$, then G is isomorphic to a subgraph of H and we write as $G \subseteq H$.

Among various interconnection networks, the hypercube network simulates many other network topologies like mesh, pyramids and tree-related networks (Kemal efe [7]; Wu [10]). This important property of hypercube makes it more popular in the topic of interconnection networks and hence is chosen as the host graph of the embedding in this paper.

Definition 2. *An n-dimension hypercube, Q_n, has 2^n vertices each labelled with a binary string of length n. Two vertices are adjacent iff their labels differ in exactly one position. Figure 1 shows hypercubes of dimension 1, 2, 3 and 4.*

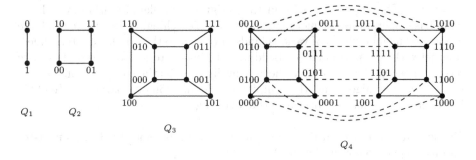

Fig. 1. Hypercubes of dimension 1, 2, 3, 4: Q_4 is constructed using two copies of Q_3

The hypercube Q_n also admits a recursive construction as follows.

Definition 3 (Recursive Construction). *$Q_1 := K_2$ and for $n \geq 2$, Q_n is recursively constructed by taking two copies of Q_{n-1}, denoted by Q_{n-1}^0 and Q_{n-1}^1 and adding 2^{n-1} edges as follows: Let $V(Q_{n-1}^0) = \{0U = 0u_2u_3\ldots u_n : u_i = 0 \text{ or } 1\}$ and $V(Q_{n-1}^1) = \{1V = 1v_2v_3\ldots v_n : v_i = 0 \text{ or } 1\}$. A vertex $0U = 0u_2\ldots u_n$ of Q_{n-1}^0 is joined to a vertex $1V = 1v_2\ldots v_n$ of Q_{n-1}^1 iff $u_i = v_i$ for every i, $2 \leq i \leq n$.*

Figure 1 shows a recursive construction of Q_4 from two copies of Q_3 by adding the 2^{n-1} edges as defined above (shown in dashed lines).

In the above construction of Q_n, if we delete the set of edges $\{(0U, 1V) : U \in V(Q_{n-1}^0) \text{ and } V \in V(Q_{n-1}^1)\}$, then we get two disjoint copies Q_{n-1}^0 and

Q_{n-1}^1 each isomorphic with Q_{n-1} from Q_n. This process of obtaining two copies of Q_{n-1} from Q_n is called as the **canonical decomposition** of Q_n and is denoted by (Q_{n-1}^0, Q_{n-1}^1).

One of the most interesting properties of the hypercube Q_n is that it is a vertex-symmetric, edge-symmetric and P_3-symmetric graph. (Here, P_k is a path on k vertices and a graph G is called a P_k-symmetric graph if for any two paths $(u_1, u_2, \ldots u_k)$ and $(v_1, v_2, \ldots v_k)$, there exists an automorphism α of G such that $\alpha(u_i) = v_i$, for every $i = 1, 2, \ldots k$. P_1-symmetric and P_2-symmetric graphs are commonly called vertex- and edge-symmetric graphs.)

In an embedding of a graph G into Q_n, it is important to minimize the size n of the cube since in a parallel computer the most expensive parts are processor nodes. So, a natural question is to ask for a hypercube with least number of nodes so that $G \subseteq Q_n$. This enquiry leads to the following definition.

Definition 4. *If G is a graph such that $2^{n-1} < |V(G)| \leq 2^n$, then Q_n is called the optimal hypercube of G, Q_{n+1} is called the next-to-optimal hypercube and n is called the optimal dimension of G (denoted by $dim(G)$).*

From a computing perspective, trees form an important class of computational structures. They naturally arise in the design of parallel algorithms which require basic operations like merging, sorting and searching. Hence, there is a large literature on embeddings of various kinds of trees into the graphs of interconnection networks. In particular, embeddings of binary trees into hypercubes have received special attention since they naturally arise as the computational structures of parallel algorithms that employ divide and conquer paradigm. In 1985, Bhatt et al. ([1]) have conjectured the following.

Conjecture (Bhatt et al [1]): Any binary tree can be embedded in its next-to-optimal hypercube with dilation 1.

Several partial results have been proved supporting the conjecture (Bier et al [2]; Chen et al [3]; Wagner [9]). In this paper, we settle the conjecture in affirmative for certain binary trees called recursive trees. Moreover, we prove that these recursive trees are embeddable in a hypercube of dimension close to the optimal (and hence less than next-to-optimal) dimension with dilation 1.

A natural reason to consider such a class of recursive trees is that the host graph, hypercube also admits a recursive definition (see Definition 3). A trivial technique for embedding a recursive tree T into hypercube is mathematical induction. In this technique, if the induction is applied on the height h of T, then the dimension of the hypercube increases linearly with the increase in h. However we aim at reducing the dimension of the hypercube. To overcome the difficulty of the linear growth of the hypercube dimension, we do some case-analysis in addition to mathematical induction. The case-analysis depends on the difference in the height of T and its smaller (left and right) trees.

2   Recursive Trees and Their Embeddings

Algorithms that solve decision problems usually have binary trees as their representation. A rooted tree represents a data structure with a hierarchical relationship among its various elements. We call a rooted binary tree T_h of height h, a *recursive tree* if T_h can be defined in terms of T_{h-k} for certain values of k, $1 \le k \le h$. A trivial classical example of recursive trees is the complete binary tree CT_h of height h since it can be defined using two copies of CT_{h-1}. In the following definitions, we define three classes of recursive trees.

Definition 5. *Let $X_0 := K_1$. For every even $h \ge 2$, X_h is obtained by taking three copies of X_{h-2} with roots r_1, r_2, r_3 (say), two new vertices R, S and adding the edges $(R, S), (S, r_1), (S, r_2)$ and (R, r_3). We designate R as the root of X_h; see Figure 2(a).*

Definition 6. *Let $Y_0 := K_1$. For every $h > 0$ and $h \equiv 0 \pmod 3$, Y_h is obtained by taking five copies of Y_{h-3} with roots r_1, r_2, r_3, r_4, r_5 (say), four new vertices P, Q, R, S and adding the edges $(R, P), (R, Q), (P, S), (Q, r_4), (Q, r_5), (S, r_1),$ (S, r_2) and (P, r_3). We designate R as the root of Y_h; see Figure 2(b).*

Definition 7. *Let $T_0, T_1 \in \{S_1, S_2, S_3\}$ where $S_1 := K_1$, $S_2 := K_2$, $S_3 = K_{1,2}$. $\mathbb{T}_0(T_0, T_1) := T_0, \mathbb{T}_1(T_0, T_1) := T_1$. For every $h \ge 2$, $\mathbb{T}_h(T_0, T_1)$ is obtained by taking a copy of $\mathbb{T}_{h-1}(T_0, T_1)$, a copy of $\mathbb{T}_{h-2}(T_0, T_1)$, a new vertex R and joining R to the roots of $\mathbb{T}_{h-1}(T_0, T_1)$ and $\mathbb{T}_{h-2}(T_0, T_1)$.*

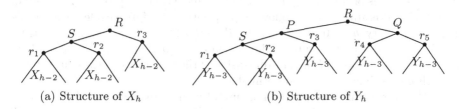

(a) Structure of X_h (b) Structure of Y_h

Fig. 2. Examples of recursive trees

One may observe that since $T_0, T_1 \in \{S_1, S_2, S_3\}$, there are nine possibilities for $\mathbb{T}_h(T_0, T_1)$. They are (1) $\mathbb{T}_h(S_1, S_1)$, (2) $\mathbb{T}_h(S_1, S_2)$, (3) $\mathbb{T}_h(S_1, S_3)$, (4) $\mathbb{T}_h(S_2, S_1)$, (5) $\mathbb{T}_h(S_2, S_2)$, (6) $\mathbb{T}_h(S_2, S_3)$, (7) $\mathbb{T}_h(S_3, S_1)$, (8) $\mathbb{T}_h(S_3, S_2)$, (9) $\mathbb{T}_h(S_3, S_3)$. In the next section, we present an embedding which embeds every tree in each of the nine classes into the hypercube whose dimension is close to the optimal dimension of the tree.

From Definitions 5, 6 and 7 we can easily deduce the following respective observations on the number of vertices of the trees.

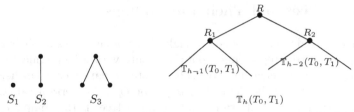

(a) Trees of height 0   (b) A recursive construction of $\mathbb{T}_h(T_0, T_1)$ where
     and 1 $T_0, T_1 \in \{S_1, S_2, S_3\}$

Fig. 3. Structure of the recursive tree $\mathbb{T}_h(T_0, T_1)$

1. $|V(X_h)| = 3|V(X_{h-2})| + 2$ with $|V(X_0)| = 1$. On solving this recurrence relation we have $|V(X_h)| = 2(3)^{\frac{h}{2}} - 1$ and the optimal dimension of $dim(X_h) = \lceil \frac{h}{2} \log_2 3 \rceil + 1$.
2. $|V(Y_h)| = 5|V(Y_{h-3})| + 4$ with $|V(Y_0)| = 1$. On solving this recurrence relation we have $|V(Y_h)| = 2(5)^{\frac{h}{3}} - 1$ and the optimal dimension of $dim(X_h) = \lceil \frac{h}{3} \log_2 5 \rceil + 1$.
3. $|V(\mathbb{T}_h(T_0, T_1))| = |V(\mathbb{T}_{h-1}(T_0, T_1))| + |V(\mathbb{T}_{h-2}(T_0, T_1))| + 1$. This recursive formula is similar to the Fibonacci sequence $f(n) = f(n-1) + f(n-2)$ with $f(0) = 0$ and $f(1) = 1$. If $f(i)$ represents the ith Fibonacci number, then $f(i)$ is approximately $2^{0.6932i}$. Using this fact, one can deduce that the optimal dimension $dim(\mathbb{T}_h(T_0, T_1)) = O(0.6932h)$.

The embeddings of X_h and Y_h into the hypercube have been proved by Choudum et al ([6]) and is stated in the following propositions for further reference. In this paper, for every $h \geq 0$ and every $T_0, T_1 \in \{S_1, S_2, S_3\}$, we embed the trees $\mathbb{T}_h(T_0, T_1)$ into the hypercube of dimension $d_h = \lfloor 0.25(3h + a + b + 3) \rfloor$ where $a = |V(T_0)|$ and $b = |V(T_1)|$. The dimension of the hypercube in the embeddings, though not optimal, is close to optimal.

In the above embeddings, we make use of a small extension in the structure of $T \in \{X_h, Y_h\}$ to embed it into hypercube. We denote by T^* a supertree of T whose root is R, formed by adding two new vertices A, B and two new edges (A, B) and (B, R). We call T^* the *auxiliary tree* of T and call the path (A, B, R) the *auxiliary path* of T^*. An injective embedding $L_h : \mathbb{T}_h(T_0, T_1) \to Q_{d_h}$ is a labelling of the vertices of $\mathbb{T}_h(T_0, T_1)$ with binary strings of length d_h. Clearly, $\mathbb{T}_h(T_0, T_1) \subseteq Q_{d_h}$ iff $L_h(u)$ and $L_h(v)$ differ in exactly one position whenever (u, v) is an edge in $\mathbb{T}_h(T_0, T_1)$. Figure 4(a) shows a labelling of $\mathbb{T}_4^*(S_3, S_3)$ using binary strings of length 5 which proves that $\mathbb{T}_4^*(S_3, S_3) \subseteq Q_5$. This labelling technique is widely used in the proof of embedding $\mathbb{T}_h(T_0, T_1)$ into hypercube. If A is a set of binary strings, we denote by $0A$ the set $\{0X : X \in A\}$, where $0X$ refers to prefixing X by 0. Similarly $A0 := \{X0 : X \in A\}$.

Proposition 1. *([6]) For every even integer $h \geq 2$, $X_h^* \subseteq Q_{d_h}$ where $d_h = \lceil \frac{4h}{5} \rceil + 1$.* $\qquad\square$

Proposition 2. *([6]) For every positive integer* $m \geq 1$, $Y_{3m}^* \subseteq Q_{d_m}$ *where* $d_m = \lceil \frac{12m}{5} \rceil + 1$. $\qquad\square$

Theorem 1. *Let* $T_0, T_1 \in \{S_1, S_2, S_3\}$. *Let* $a = |V(T_0)|$ *and* $b = |V(T_1)|$. *Then for every* $h \geq 0$ *and every* $T_0, T_1 \in \{S_1, S_2, S_3\}$, $\mathbb{T}_h(T_0, T_1)$ *is a subtree of* Q_{d_h} *where* $d_h = \lfloor 0.25(3h + a + b + 3) \rfloor$.

Proof: For $0 \leq h \leq 3$, one can easily and explicitly label $\mathbb{T}_h(T_0, T_1)$ for each of the nine possibilities. For $h \geq 4$, we prove the following claim.

Claim : $\mathbb{T}_h^*(T_0, T_1) \subseteq Q_{d_h}$ where $d_h = \lfloor 0.25(3h + a + b + 3) \rfloor$ such that the auxiliary path (A_h, B_h, R_h) of $\mathbb{T}_h^*(T_0, T_1)$ is labelled $(0^{d_h-2}11, 0^{d_h-1}1, 0^{d_h})$.

It is clear that if the claim holds, then the theorem holds. We call the condition on the auxiliary path as the auxiliary condition. We prove the claim by induction on h.

Proof of the claim: As a basic case, we explicitly label the vertices of $\mathbb{T}_h^*(T_0, T_1)$ for every h, $4 \leq h \leq M$ where $M = \begin{cases} a + b + 4 \text{ if } a + b \leq 3 \\ a + b \text{ if } a + b > 3 \end{cases}$ and every $T_0, T_1 \in \{S_1, S_2, S_3\}$ satisfying the auxiliary condition. For the sake of crispness of the current paper, we present only the labellings of $\mathbb{T}_4^*(S_3, S_3)$, $\mathbb{T}_5^*(S_3, S_3)$ and $\mathbb{T}_6^*(S_3, S_3)$ in Figures 4(a), 4(b) and 4(c) respectively. By such a labelling, we note that $\mathbb{T}_4^*(S_3, S_3) \subseteq Q_{d_h}$; $d_h = \lfloor 0.25(3 * 4 + 3 + 3 + 3) \rfloor = 5$. Similarly, $\mathbb{T}_4^*(S_3, S_3) \subseteq Q_6$ and $\mathbb{T}_4^*(S_3, S_3) \subseteq Q_6$.

We proceed to the next step in the induction. For $h > M$, we assume that the claim is true for $\mathbb{T}_k^*(T_0, T_1)$ for all $M \leq k \leq h - 1$ and we prove the theorem for $\mathbb{T}_h^*(T_0, T_1)$ in 3 cases. The following proof technique holds for every $T_0, T_1 \in \{S_1, S_2, S_3\}$ and hence we use $\mathbb{T}_h$ in analogous to $\mathbb{T}_h(T_0, T_1)$.

<u>Case 1</u>: $h \equiv a + b + 2 \pmod{4}$ **or** $h \equiv a + b + 3 \pmod{4}$
In this case, $d_h - d_{h-1} = 1$ and $d_h - d_{h-2} = 2$. By induction hypothesis, we are given embeddings $L^1 : \mathbb{T}_{h-1}^* \to Q_{d_{h-1}}$ and $L^2 : \mathbb{T}_{h-2}^* \to Q_{d_{h-2}}$ satisfying the auxiliary condition. For notational convenience, we denote $L^1(\mathbb{T}_{h-1}^*)$ by $\mathbb{T}_{h-1}^1$ and $L^2(\mathbb{T}_{h-2}^*)$ by $\mathbb{T}_{h-2}^2$. We extend these two labellings to a labelling L_h of $\mathbb{T}_h^*$ by suffixing the labels of all the vertices of $\mathbb{T}_{h-1}^1$ by 0 and the labels of all the vertices of $\mathbb{T}_{h-2}^2$ by 11; see Figure 5(a). By such a labelling, every label of $\mathbb{T}_h^*$ is of length d_h since $d_h - d_{h-1} = 1$ and $d_h - d_{h-2} = 2$. Clearly, L_h is an injection and hence a required embedding. On applying the automorphism $\psi_{12} : V(Q_{d_h}) \to V(Q_{d_h})$ defined by

$$\psi_{12}(x_1 x_2 \ldots x_{d_h}) = x_1 x_2 \ldots x_{d_h-3} \overline{x_{d_h-1}} x_{d_h} x_{d_h-2}$$

to the labelling L_h, we obtain a labelling of $\mathbb{T}_h^*$ satisfying the auxiliary condition.

<u>Case 2</u>: $h \equiv a + b + 1 \pmod{4}$
In this case, $d_h - d_{h-1} = d_h - d_{h-2} = 1$. By induction hypothesis, we are given embeddings $L^1 : \mathbb{T}_{h-1}^* \to Q_{d_{h-1}}$ and $L^2 : \mathbb{T}_{h-2}^* \to Q_{d_{h-2}}$ satisfying the auxiliary condition. By applying the automorphism $\alpha : V(Q_{d_{h-2}}) \to V(Q_{d_{h-2}})$ to the labels of $L^2(\mathbb{T}_{h-2}^*)$ defined by

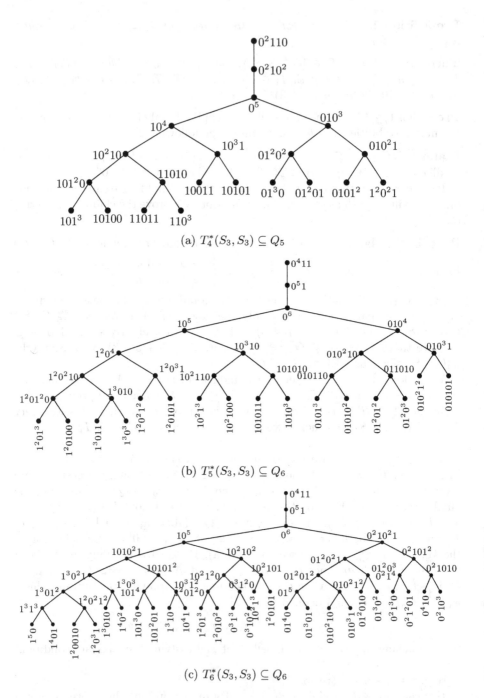

(a) $T_4^*(S_3, S_3) \subseteq Q_5$

(b) $T_5^*(S_3, S_3) \subseteq Q_6$

(c) $T_6^*(S_3, S_3) \subseteq Q_6$

Fig. 4. Labelling of $\mathbb{T}_h^*(S_3, S_3)$ for $4 \leq h \leq 6$

$$\alpha(x_1 x_2 \ldots x_{d_h-2}) = x_1 x_2 \ldots x_{d_h-2-2} \overline{x_{d_h-2}} \; \overline{x_{d_h-2-1}}$$

the labels of the auxiliary path $(A_{h-2}, B_{h-2}, R_{h-2})$ of $\mathbb{T}^*_{h-2}$ is mapped to the path $(R_{h-2}, B_{h-2}, A_{h-2})$. For notational convenience, we denote $L^1(\mathbb{T}^*_{h-1})$ by $\mathbb{T}^1_{h-1}$ and $\alpha \circ L^2(\mathbb{T}^*_{h-2})$ by $\alpha\mathbb{T}^2_{h-2}$. We extend these two labellings to a labelling L_h of $\mathbb{T}^*_h$ by suffixing the labels of all the vertices of $\mathbb{T}^1_{h-1}$ by 0 and the labels of all the vertices of $\alpha\mathbb{T}^2_{h-2}$ by 1; see Figure 5(b). By such a labelling, every label of $\mathbb{T}^*_h$ is of length d_h since $d_h - d_{h-1} = d_h - d_{h-2} = 1$. Clearly, L_h is an injection and hence a required embedding.

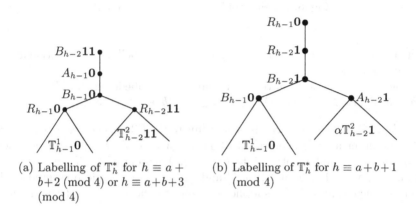

(a) Labelling of $\mathbb{T}^*_h$ for $h \equiv a + b+2 \pmod 4$ or $h \equiv a+b+3 \pmod 4$

(b) Labelling of $\mathbb{T}^*_h$ for $h \equiv a+b+1 \pmod 4$

Fig. 5. Labelling of $\mathbb{T}^*_h$ for $h \equiv a + b + k \pmod 4$, $k \in \{1, 2, 3\}$

On applying the automorphism α (defined above) to the labelling L_h, we obtain a labelling of $\mathbb{T}^*_h$ satisfying the auxiliary condition.

<u>Case 3:</u> $h \equiv a + b \pmod 4$
In this case, $d_h - d_{h-3} = 2$ and $d_h - d_{h-4} = 3$.
By induction hypothesis, we are given embeddings $L^3 : \mathbb{T}^*_{h-3} \to Q_{d_{h-3}}$ and $L^4 : \mathbb{T}^*_{h-4} \to Q_{d_{h-4}}$ satisfying the auxiliary condition. Let $L^3(\mathbb{T}^*_{h-3})$ be denoted by $\mathbb{T}^3_{h-3}$ and $L^4(\mathbb{T}^*_{h-4})$ be denoted by $\mathbb{T}^4_{h-4}$. We note that on applying the automorphism $\phi : V(Q_{d_{h-3}}) \to V(Q_{d_{h-3}})$ to the labels of $\mathbb{T}^3_{h-3}$ defined by

$$\phi(x_1 x_2 \ldots x_{d_h-3}) = x_1 x_2 \ldots x_{d_h-3-2} \overline{x_{d_h-3}} \; \overline{x_{d_h-3-1}}$$

the auxiliary path $(A_{h-3}, B_{h-3}, R_{h-3})$ of $\mathbb{T}^*_{h-3}$ is mapped to the path $(R_{h-3}, B_{h-3}, A_{h-3})$. Let $\phi(\mathbb{T}^3_{h-3})$ be denoted by $\mathbb{T}^5_{h-3}$. We use $\mathbb{T}^3_{h-3}$, $\mathbb{T}^4_{h-4}$, $\mathbb{T}^5_{h-3}$ to obtain a labelling L_h of $\mathbb{T}^*_h$ as shown in Figure 6.
On applying the automorphism $\psi_3 : V(Q_{d_h}) \to V(Q_{d_h})$ defined by

$$\psi_3(x_1 x_2 \ldots x_{d_h}) = x_1 x_2 \ldots x_{d_h-4} x_{d_h} \overline{x_{d_h-2}} x_{d_h-1} x_{d_h-3}$$

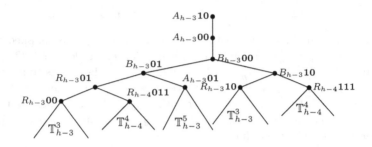

Fig. 6. Labelling of $\mathbb{T}_h^*$ for $h \equiv a + b \pmod 4$

to the labelling L_h shown in Figure 6, we obtain a labelling of $\mathbb{T}_h^*$ satisfying the auxiliary condition.

In each of the cases we have given a labelling (embedding) for $\mathbb{T}_h^*$. Hence for all $h \geq 0$, $\mathbb{T}_h^*(T_0, T_1) \subseteq Q_{d_h}$ where $d_h = \lfloor 0.25(3h + a + b + 3) \rfloor$. $\square$

Remark 1. The tree $\mathbb{T}_h(S_1, S_2)$ is isomorphic to the Fibonacci tree $\mathbb{F}_h$ considered by Choudum et al ([5]). The result of Theorem 1 for $\mathbb{T}_h(S_1, S_2)$ coincides with the result of embedding of $\mathbb{F}_h$ into Q_{d_h} ([5], Theorem 4.1). Hence Theorem 1 is an extension of Theorem 4.1 of [5] and the proof technique presented in this paper is much simpler. The simplicity is due to the presence of the auxiliary path. $\square$

Remark 2. All recursive trees considered in this paper are height-balanced trees (A rooted binary tree T is said to be height-balanced if for every vertex $v \in V(T)$, the heights of the subtrees, rooted at the left and right child of v, differ by at most one). $\square$

3   Embedding of the Stars

As an application to Propositions 1 and 2, we embed the stars $K_{1,2^n 3^m}$ and $K_{1,2^n 5^m}$ into the hypercube.

The star $K_{1,d}$ is a tree of diameter 2 with exactly one vertex of degree d and exactly d vertices of degree 1. Clearly, $K_{1,d} \subseteq Q_d$. In particular, $K_{1,2^n p^m} \subseteq Q_{2^n p^m}$. Our focus is now to reduce the dimension $2^n p^m$. In this section, we embed $K_{1,2^n 3^m}$ into $Q_{\lceil 1.6m \rceil + n + 1}$, however with dilation $n + 2m$. In similar lines, we also embed $K_{1,2^n 5^m}$ into $Q_{\lceil 2.4m \rceil + n + 1}$ with dilation $n + 3m$. The dimensions $\lceil 1.6m \rceil + n + 1$ and $\lceil 2.4m \rceil + n + 1$ are close to the optimal dimension of $K_{1,2^n 3^m}$ and $K_{1,2^n 5^m}$ respectively.

3.1   Embedding of $K_{1,2^n 3^m}$

The proof of embedding $K_{1,2^n 3^m}$ into $Q_{\lceil 1.6m \rceil + n + 1}$ with dilation $n + 2m$ consists of two steps. In the first step, we embed a tree $H_{h,m}$ of height h whose structure is shown in Figure into Q_{d_h} where $d_h = \lceil 1.6m \rceil + h - 2m + 1$ with dilation 1; this is done in the following Theorem 2. In the second step, we embed $K_{1,2^n 3^m}$ into $H_{n+2m,m}$ with dilation $n + 2m$; this is done in Theorem 3. Combining the two results, we get the required embedding.

We note that (1) the induced subgraph of $H_{h,m}$ from level 0 to $h - 2m$ is a complete binary tree of height $h - 2m$ and (2) the induced subgraph of $H_{h,m}$ from level $h - 2m + 1$ to h is the tree X_{2m} of height $2m$ (refer to Definition 5). Hence, $H_{2m,m} \simeq X_{2m}$. The tree $H_{h,m}$ has $(2^{h-2m+1})(3^m) - 1$ vertices and $(2^{h-2m})(3^m)$ leaves and hence optimal dimension of $H(h, m)$ is $\log((2^{h-2m+1})(3^m)) \simeq \lceil 1.585m \rceil + h - 2m + 1$. We next embed $H_{h,m}$ into $Q_{\lceil 1.6m \rceil + h - 2m + 1}$ with dilation 1.

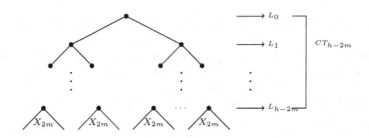

Fig. 7. Structure of $H_{h,m}$; X_{2m} is shown in Figure 2(a)

Theorem 2. *For every integer* $h, k \geq 0$, $H_{h,m} \subseteq Q_{\lceil 1.6m \rceil + h - 2m + 1}$.

Proof: Let $\lceil 1.6m \rceil + 1 = d_m$. Let $v(t, i)$ be the ith vertex from left at level t, $0 \leq t \leq h - 2m, 1 \leq i \leq 2^t$ of $H_{h,m}$ and let $F_{t,i}$ be the subtree of $H_{h,m}$ rooted at $v(t, i)$. We note that $F_{h-2m,i} \simeq X_{2m}$ for all $i, 1 \leq i \leq 2^t$. To prove the theorem, we first prove that $F^*_{h-2m-1,j} \subseteq Q_{d_m+1}$, for every $j, 1 \leq j \leq 2^{h-2m-1}$.

Embedding $F^*_{h-2m-1,j}$ **in** Q_{d_m+1}: Let $(A_1, B_1, R_1 = v(h - 2m, 2j - 1))$ and $(A_2, B_2, R_2 = v(h - 2m, 2j))$ denote the auxiliary paths of $F^*_{h-2m,2j-1}$ and $F^*_{h-2m,2j}$ respectively. Consider a canonical decomposition $(Q^0_{d_m}, Q^1_{d_m})$ of Q_{d_m+1}. Embed $F^*_{h-2m,2j-1} \simeq X^*_{2m}$ in $Q^0_{d_m}$ and $F^*_{h-2m,2j} \simeq X^*_{2m}$ in $Q^1_{d_m}$ (this is possible by Proposition 1). We next add the edges (A_1, B_2) and (B_1, R_2). These edges exist in $Q_{d_m} + 1$ since we can map the edge (B_2, R_2) onto the edge (A_1, B_1) using the edge-symmetric property of $Q^1_{d_m}$. This gives an embedding of $F^*_{h-2m-1,j}$ into Q_{d_m+1} with (B_2, A_1, B_1) as its auxiliary path; see Figure 8.

By repeatedly applying the above procedure of embedding, we have $F^*_{k,j} \subseteq Q_{d_m+h-2m-k}$, for every k, $0 \leq k \leq h - 2m - 2$. In particular, for $k = 0$ we have $F^*_{0,1} \subseteq Q_{d_m+h-2m}$. The theorem follows since $H_{h,m} \subset H^*_{h,m} \simeq F^*_{0,1}$. $\square$

In the next theorem, we embed $K_{1,2^n 3^m}$ into $H_{h,m}$ for $h = n + 2m$.

Theorem 3. *For every* $n \geq 0$ *and every* $m \geq 1$, $K_{1,2^n 3^m}$ *is embeddable in* $H_{n+2m,m}$ *with dilation* $n + 2m$.

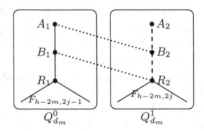

Fig. 8. Embedding of $F^*_{h-2m-1,j}$ in Q_{dm+1}

Proof: The required embedding is given by explicitly defining a map ϕ as follows. Let $\phi : V(K_{1,2^n3^m}) \to V(H_{n+2m,m})$ be such that the leaves of $K_{1,2^n3^m}$ are mapped to the leaves of $H_{n+2m,m}$ (we note that $H_{n+2m,m}$ has 2^n3^m leaves) and the root of $K_{1,2^n3^m}$ is mapped to the root of $H_{n+2m,m}$. Clearly, $dil(\phi)$ is the height of $H_{n+2m,m}$ which is equal to $n + 2m$. $\square$

Combining Theorem 2 and Theorem 3, we have the following theorem.

Theorem 4. *For every $n \geq 0$ and every $m \geq 1$, $K_{1,2^n3^m}$ is embeddable in the hypercube of dimension $\lceil 1.6m \rceil + n + 1$ with dilation $n + 2m$.* $\square$

3.2   Embedding of $K_{1,2^n5^m}$

We next proceed to embed the star $K_{1,2^n5^m}$ into the hypercube of dimension $\lceil 2.4m \rceil + n + 1$ with dilation $n + 3m$. The proof is similar to the embedding of $K_{1,2^n3^m}$. However we use a different intermediate tree $G_{h,m}$ (similar to $H_{h,m}$) which is as follows: (1) the induced subgraph of $G_{h,m}$ from level 0 to $h - 3m$ is a complete binary tree of height $h - 3m$ and (2) the induced subgraph of $G_{h,m}$ from level $h - 3m + 1$ to h is the tree Y_{3m} of height $3m$ (refer to Definition 6). Hence, $G_{3m,m} \simeq Y_{3m}$. The tree $G_{h,m}$ has $(2^{h-3m+1})(5^m) - 1$ vertices and hence optimal dimension of $G_{h,m}$ is $\log((2^{h-3m+1})(5^m)) \simeq \lceil 2.323m \rceil + h - 3m + 1$. Arguing in similar lines of Theorems 2, 3 and using Proposition 2, we have the following theorem.

Theorem 5. *For every $h, m \geq 0$, $G_{h,m} \subseteq Q_{\lceil 2.4m \rceil + h - 3m + 1}$.* $\square$

Theorem 6. *For every $n \geq 0$ and every $m \geq 1$, $K_{1,2^n5^m}$ is embeddable in $G_{n+3m,m}$ with dilation $n + 3m$.* $\square$

Combining the above two theorems, we have the following.

Theorem 7. *For every $n \geq 0$ and every $m \geq 1$, $K_{1,2^n5^m}$ is embeddable in the hypercube of dimension $\lceil 2.4m \rceil + n + 1$ with dilation $n + 3m$.* $\square$

Remark 3. The embeddings of recursive trees X_h and Y_h (stated in Propositions 1 and 2) find application in embedding other trees like $H_{h,m}$, $G_{h,m}$, $K_{1,2^n3^m}$ and $K_{1,2^n5^m}$ into hypercube. $\square$

Remark 4. The recursive trees considered in this paper are not exhaustive. That is, there are recursive trees $\mathbb{H}_n$ as defined below, other than those considered in this paper for embedding purpose. The tree $\mathbb{H}_n$ is obtained by taking (i) a copy of CT_{n-1} and a copy of CT_{n-2} with roots R_1 and R_2; (ii) a new vertex R and (iii) joining R to R_1 and R_2. Choudum et al ([4], Theorem 4.2, see Figure 11) have proved that $\mathbb{H}_n^* \subseteq Q_{n+1}$. $\qquad\qquad\qquad\qquad\qquad\square$

4 Conclusion

In this paper, we have embedded certain recursive trees in the hypercube with unit dilation. The dimension of the hypercube is close to optimal dimension. The embedding result settles a conjecture (stated in page 324) of Bhatt et al. ([1]) in affirmative for the recursive trees. We have also shown how these embedding can be used to embed other trees like stars into hypercube.

In future, one can attempt to recognize more trees that are subgraphs of hypercube or other hypercube-like graphs. One can also attempt to reduce the dilation of embedding stars into hypercube.

References

1. Bhatt, S.N., Ipsen, I.I.F.: How to embed trees in hypercubes, Technical Report YALEU/DCS/RR-443, Yale University (1985)
2. Bier, T., Loe, K.F.: Embedding of binary trees into hypercubes. Journal of Parallel and Distributed Computing 6(3), 679–691 (1989)
3. Chen, W.K., Stallmann, M.F.M.: On Embedding Binary Trees into Hypercubes. Journal of Parallel and Distributed Computing 24(2), 132–138 (1995)
4. Choudum, S.A., Indhumathi, R.: On embedding subclasses of height balanced trees in hypercubes. Information Sciences 179, 1333–1347 (2009)
5. Choudum, S.A., Raman, I.: Embedding height balanced trees and Fibonacci trees in hypercubes. Journal of Applied Mathematics and Computing 30(1-2), 39–52 (2009)
6. Choudum, S.A., Raman, I.: Embedding certain height-balanced trees and complete p^m-ary trees into hypercubes. Journal of Discrete Algorithms (accepted, under revision)
7. Efe, K.: Embedding mesh of trees in the hypercube. Journal of Parallel and Distributed Computing 11(3), 222–230 (1991)
8. Leighton, F.T.: Introduction to Parallel Algorithms and Architectures: Arrays, Trees, Hypercubes. Morgan Kauffmann, San Mateo (1992)
9. Wagner, A.S.: Embedding All Binary Trees in the Hypercube. Journal of Parallel and Distributed Computing 18(1), 33–43 (1993)
10. Wu, A.Y.: Embedding of tree networks in hypercubes. Journal of Parallel and Distributed Computing 2(3), 238–249 (1985)

Box-Rectangular Drawings of Planar Graphs
(Extended Abstract)

Md. Manzurul Hasan[1,2], Md. Saidur Rahman[2], and Md. Rezaul Karim[3]

[1] Alternative Delivery Channel Division, Dutch-Bangla Bank Limited, Bangladesh
[2] Department of Computer Science and Engineering,
Bangladesh University of Engineering and Technology (BUET), Bangladesh
[3] Department of Computer Science and Engineering,
Univeristy of Dhaka, Bangladesh
manzurul@dbbl.com.bd, saidurrahman@cse.buet.ac.bd, rkarim@univdhaka.edu

Abstract. A plane graph is a planar graph with a fixed embedding in
the plane. In a box- rectangular drawing of a plane graph, every vertex is
drawn as a rectangle, called a box, each edge is drawn as either a horizontal line segment or a vertical line segment, and the contour of each face
is drawn as a rectangle. A planar graph is said to have a box-rectangular
drawing if at least one of its plane embeddings has a box-rectangular
drawing. Rahman et al. [8] gave a necessary and sufficient condition for
a plane graph to have a box-rectangular drawing and developed a linear-
time algorithm to draw a box-rectangular drawing of a plane graph if
it exists. Since a planar graph G may have an exponential number of
embeddings, determining whether G has a box-rectangular drawing or
not using the algorithm of Rahman et al. [8] for each embedding of
G takes exponential time. Thus to develop an efficient algorithm to examine whether a planar graph has a box-rectangular drawing or not is
a non-trivial problem. In this paper we give a linear-time algorithm to
determine whether a planar graph G has a box-rectangular drawing or
not, and to find a box-rectangular drawing of G if it exists.

Keywords: Graph drawing, Planar graph, Box-rectangular drawing,
Rectangular drawing, Cyclically 4-edge connected graph.

1 Introduction

For the last two decades automatic drawings of graphs have created intense
interest due to their broad applications, and as a consequence, a number of
drawing styles and corresponding drawing algorithms have emerged [1,3,5,10]. A
plane graph is a planar graph with a fixed embedding in the plane. A *rectangular
drawing* of a plane graph G is a drawing of G, where each vertex is drawn as
a point, each edge is drawn as a horizontal or vertical line segment, and each
face is drawn as a rectangle. On the other hand a *box-rectangular drawing* of a
plane graph G is a drawing of G in which each vertex is drawn as a (possibly
degenerated) rectangle, called a *box*, each edge is drawn as a horizontal line
segment or a vertical line segment, and the contour of each face is drawn as a
rectangle. Figure 1(c) illustrates a box-rectangular drawing of the plane graph

S.K. Ghosh and T. Tokuyama (Eds.): WALCOM 2013, LNCS 7748, pp. 334–345, 2013.
© Springer-Verlag Berlin Heidelberg 2013

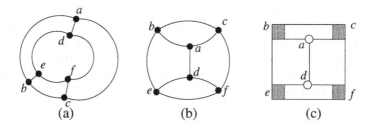

Fig. 1. (a) A planar graph G, (b) a plane embedding Γ of G for which box-rectangular drawing exists, and (c) box-rectangular drawing of the planar graph G

in Figure 1(b). Box-rectangular drawings have practical applications in VLSI floorplanning [8,9] and architectural floorplanning [2,6,7].

All graphs do not have box-rectangular drawings. Rahman et al. [8] gave a necessary and sufficient condition for a plane graph to have a box-rectangular drawing and developed a linear-time algorithm to draw a box-rectangular drawing of a plane graph if it exists. Xin He [4] did the same task for proper box-rectangular drawings of plane graphs. A planar graph is said to have a box-rectangular drawing if at least one of its plane embeddings has a box-rectangular drawing. For the plane embedding in Fig. 1(a) of the planar graph G there is no box-rectangular drawing. But the plane embedding of G in Fig. 1(b) has a box-rectangular drawing as illustrated in Figure 1(c). Thus G has a box-rectangular drawing. Since a planar graph G may have an exponential number of embeddings, determining whether G has a box-rectangular drawing or not using the algorithm of Rahman et al. [8] for each embedding of G takes exponential time. Thus to develop an efficient algorithm to examine whether a planar graph has a box-rectangular drawing or not is a non-trivial problem. In this paper we give a linear-time algorithm to determine whether a planar graph G has a box-rectangular drawing or not, and to find a box-rectangular drawing of G if it exists.

Our approach for finding a box-rectangular drawing of a planar graph is similar to that of Rahman et al. [9] for finding a rectangular drawing of a planar graph. However, our work is not a mere extension of the work of Rahman et al. [9], and we had to face a lot of challenges. In this paper we show that all the plane embeddings of a subdivision of planar 3-connected cubic graph G which is cyclically 4-edge-connected, have box-rectangular drawings, whereas not every such embedding has a rectangular drawing. We denote the maximum degree of a graph G by $\Delta(G)$ or simply by Δ. Rahman et al. [9] deal with planar graphs having $\Delta \leq 3$. But for box-rectangular drawing we deal with planar graphs of the maximum degree 4 or more. We had to face enormous difficulties in dealing with the graphs of maximum degree 4 or more. In [9] Rahman et al. showed that at most four plane embeddings are needed to be checked to determine whether a planar graph has a rectangular drawing. Whereas in case of box-rectangular drawing we showed that at most 81 embeddings are needed to be checked.

The rest of this paper is organized as follows. In section 2, we give some basic terminologies, and some previous results. In Section 3, we describe a necessary and sufficient condition for a planar graph G with $\Delta \leq 3$ to have a box-rectangular drawing and find a drawing if it exists. Section 4 gives a necessary and sufficient condition for a planar graph G with $\Delta \geq 4$ to have a box-rectangular drawing and describes a linear-time algorithm for finding a drawing if it exists. Finally Section 5 concludes the paper.

2   Preliminaries

In this section we give some definitions and present preliminary results.

Throughout the paper we assume that a *graph* G is so called a multigraph which may have *multiple edges*, i.e., edges sharing both ends. If G has no multiple edges, then G is called a *simple* graph. *Subdividing an edge* (u, v) of a graph G is the operation of deleting the edge (u, v) and adding a path $u(= w_0), w_1, w_2, \ldots, w_k, v(= w_{k+1})$ passing through new vertices $w_1, w_2, \ldots, w_k, k \geq 1$, of degree 2. A graph G is called a *subdivision* of a graph G' if G is obtained from G' by subdividing some of the edges of G'.

The *connectivity* $\kappa(G)$ of a graph G is the minimum number of vertices whose removal results in a disconnected graph or a single-vertex graph K_1. We say that G is *k-connected* if $\kappa(G) \geq k$. A graph G is called *cyclically 4-edge-connected* if

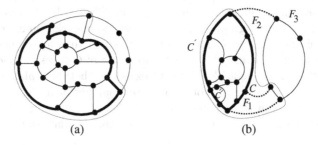

(a)   (b)

Fig. 2. (a) A cyclically 4-edge-connected graph, and (b) a graph which is not cyclically 4-edge-connected

the removal of any three or fewer edges leaves a graph such that exactly one of the components has a cycle. The graph in Fig. 2(a) is cyclically 4-edge-connected. On the other hand, the graph in Fig. 2(b) is not cyclically 4-edge-connected, since the removal of the three edges drawn by thick dotted lines leaves a graph with two connected components each of which has a cycle. We often use the following operation on a planar graph G. Let v be a vertex of degree d in a plane graph Γ of the planar graph G, let $e_1 = vw_1, e_2 = vw_2, \ldots, e_d = vw_d$ be the edges incident to v, and assume that these edges $e_1, e_2, \ldots, e_d$ appear clockwise around v in this order. Replace v with a cycle $v_1, v_1v_2, v_2, v_2v_3, \ldots, v_dv_1, v_1$, and replace the edges vw_i with v_iw_i for $i = 1, 2, \ldots, d$. We call the operation above *replacement of a vertex by a cycle*. The cycle $v_1, v_1v_2, v_2, v_2v_3, \ldots, v_dv_1, v_1$ in the resulting graph is called the *replaced cycle* corresponding to the vertex v of Γ.

Let G be a planar graph, and Γ be an arbitrary plane embedding of G. We denote by $F_o(\Gamma)$ the outer face of Γ. For a cycle C of Γ, we call the plane subgraph of Γ inside C (including C) the *inner subgraph* $\Gamma_I(C)$ for C, and we call the plane subgraph of Γ outside C (including C) the *outer subgraph* $\Gamma_O(C)$ for C. An edge which is incident to exactly one vertex of a cycle C and located outside C is called a *leg* of C. The vertex of C to which a leg is incident is called a *leg-vertex* of C. A cycle C in Γ is called a *k-legged cycle* of Γ if C has exactly k legs and there is no edge which joins two vertices on C and is located outside C. In each of Figs. 2(a) and 2(b), a 3-legged cycle is drawn by thick solid lines. We call a face F of Γ a *peripheral face* for a 3-legged cycle C in Γ if F is in $\Gamma_O(C)$ and the contour of F contains an edge on C. Clearly there are exactly three peripheral faces for any 3-legged cycle in Γ. In Fig. 2(b) F_1, F_2, F_3 are the three peripheral faces for the 3-legged cycle C drawn by thick solid lines. A k-legged cycle C is called a *minimal k-legged cycle* if $\Gamma_I(C)$ does not contain any other k-legged cycle of Γ. The 3-legged cycle C drawn by thick lines in Fig. 2(b) is not minimal, while the 3-legged facial cycle C'' in Fig. 2(b) is minimal. We say that cycles C and C' in Γ are *independent* if $\Gamma_I(C)$ and $\Gamma_I(C')$ have no common vertex. A set S of cycles is *independent* if any pair of cycles in S are independent. A cycle C in Γ is called *regular* if the plane graph $\Gamma - \Gamma_I(C)$ has a cycle. Similarly an edge of Γ which is incident to exactly one vertex of a cycle C in Γ and located inside C is called a *hand* of C. The vertex of C to which a hand is incident is called a *hand-vertex* of C. A cycle C is called a *k-handed cycle* if C has exactly k hands in Γ and there is no edge which joins two vertices on C and is located inside C. We call a k-handed cycle C a *regular k-handed cycle* if $\Gamma - \Gamma_O(C)$ contains a cycle.

We now give some definitions regarding a box-rectangular drawing of a plane graph Γ. We say that a vertex of graph Γ is drawn as a *degenerated box* in a box-rectangular drawing D if the vertex is drawn as a point in D. We often call a degenerated box in D a *point* and call a nondegenerated box a *real box*. We call the rectangle corresponding to $F_o(\Gamma)$ the *outer rectangle*, and we call a corner of the outer rectangle simply a *corner*. A box in D containing at least one corner is called a *corner box*. A corner box may be degenerated. If $n = 1$, that is, Γ has exactly one vertex, then the box-rectangular drawing is trivial: the drawing is just a degenerated box corresponding to the vertex. Thus in the paper, we can assume that $n \geq 2$.

Rahman at al. [8] gave a necessary and sufficient condition for a plane graph Γ to have a box-rectangular drawing, and developed a linear algorithm for finding a drawing of Γ if it exists, as stated in the following lemma.

Lemma 1. *[8] Let G be a connected planar graph with $\Delta \leq 3$, and let Γ be a plane embedding of G. Assume that Γ has neither a vertex of degree 1 nor a 1-legged cycle. Then Γ has a box-rectangular drawing if and only if Γ satisfies the following two conditions:*

(br1) every 2- or 3- legged cycle in Γ contains an edge on $F_o(\Gamma)$; and
(br2) $2c_2 + c_3 \leq 4$ for any independent set ξ of cycles in Γ, where c_2 and c_3 are the numbers of 2- and 3- legged cycles in ξ respectively.

In the problem box-rectangular drawing of a plane graph Γ for $\Delta \geq 4$ Rahman et al. [8] constructed a new plane graph Φ from Γ by replacing each vertex v of degree four or more in Γ by a cycle. Thus $\Delta(\Phi) \leq 3$. The following lemma is the main result for $\Delta \geq 4$.

Lemma 2. *[8] Let Γ be a plane connected graph with $\Delta \geq 4$, and let Φ be the graph transformed from Γ as above. Then Γ has a box-rectangular drawing if and only if Φ has a box-rectangular drawing.*

It is not difficult to derive a characterization of a connected planar graph to have a box-rectangular drawing if we know a characterization of a biconnected planar graph to have a box-rectangular drawing. Throughout this extended abstract we thus consider the planar graph G is biconnected.

3   Box-Rectangular Drawings of Planar Graphs with $\Delta \leq 3$

In this section we give a necessary and sufficient condition for a planar graph G with $\Delta \leq 3$ to have a box-rectangular drawing. We first consider the case where G is a subdivision of a planar 3-connected cubic graph. G has an $O(n)$ number of embeddings, one for each face chosen as outer face. Examining by the linear algorithm in Lemma 1 whether the two conditions (br1) and (br2) hold for each of the $O(n)$ embeddings, one can examine in time $O(n^2)$ whether the planar graph G has a box-rectangular drawing. However, we obtain the following necessary and sufficient condition for G to have a box-rectangular drawing, which leads to a linear-time algorithm to examine whether G has a box-rectangular drawing. We also give a linear-time algorithm to find a drawing if it exists.

Theorem 1. *Let G be a subdivision of a planar 3-connected cubic graph, and let Γ be an arbitrary plane embedding of G.*

(a) Suppose first that G is cyclically 4-edge-connected, that is, Γ has no regular 3-legged cycle. Then the planar graph G has a box-rectangular drawing.

(b) Suppose next that G is not cyclically 4-edge-connected, that is, Γ has a regular 3-legged cycle C. Let F_1, F_2, and F_3 be the three peripheral faces for C, and let Γ_1, Γ_2, and Γ_3 be the plane embeddings of G taking F_1, F_2, and F_3 respectively as the outer face. Then the planar graph G has a box-rectangular drawing if and only if at least one of the three embeddings Γ_1, Γ_2, and Γ_3 has a box-rectangular drawing.

We only prove here Theorem 1(a), the proof of Theorem 1(b) is similar to that of Theorem 3.1(b) in [9]. Before giving a proof of Theorem 1(a), we need the following Lemmas 3 and 4. Lemma 3 is needful to prove the Lemma 4.

Lemma 3. *Let G be a subdivision of planar 3-connected cubic graph, and Γ be an arbitrary plane embedding of G. If G is cyclically 4-edge-connected, then $2c_2 + c_3 \leq 2$ for any independent set ξ of cycles in Γ, where c_2 and c_3 are the numbers of 2- and 3-legged cycles in ξ, respectively.*

Proof. Let G be a subdivision of planar 3-connected cubic graph, and Γ be an arbitrary plane embedding of G. Assume that G is cyclically 4-edge-connected. We first show that Γ does not have two or more independent 2-legged cycles. Assume for a contradiction that Γ has two independent 2-legged cycles, C_1 and C_2. Then removal of the two legs of either C_1 or C_2 leaves a graph with two connected components, each of which has a cycle, contrary to the definition of a cyclically 4-edge-connected graph. Similarly we can prove that Γ can not have two independent 3-legged cycles. Similarly we can also prove that Γ can not have two cycles, one is 2-legged, and another is 3-legged, which are independent. That is, $2c_2 + c_3 \leq 2$ for any independent set ξ of cycles in Γ, where c_2 and c_3 are the numbers of 2- and 3-legged cycles in ξ, respectively.   □

Lemma 4. *Let G be a subdivision of planar 3-connected cubic graph. If G is cyclically 4-edge-connected, then all the plane embeddings of the planar graph G, satisfy (br1) and (br2) of Lemma 1.*

Proof. Let Γ be a plane embedding of G. We first show that Γ satisfies (br1) in Lemma 1. Assume for a contradiction that a 2-legged or a 3-legged cycle C has no edge on the outer face of Γ. Then the removal of the legs of C will result in two connected components having cycles, and G would not be a cyclically 4-edge connected graph, a contradiction. By Lemma 3, Γ satisfies (br2) of Lemma 1.   □

Proof of Theorem 1(a). By Lemma 4, every plane embedding Γ of G satisfies Conditions (br1) and (br2) of Lemma 1; and hence Γ has a box-rectangular drawing by Lemma 1. Therefore the planar graph G has a box-rectangular drawing.   □

We now consider the other case. It can be trivially shown that every biconnected planar graph G having two vertices of degree 3 has a box-rectangular drawing. We may thus assume that G has three or more vertices of degree 3. Then any planar embedding Γ of G has a regular 2-legged cycle; otherwise, G would be a subdivision of a 3-connected cubic graph. In this case we have the following theorem.

Theorem 2. *Let G be a planar biconnected graph with $\Delta \leq 3$ which is not a subdivision of a planar 3-connected cubic graph. Let Γ be a planar embedding of G such that every 2-legged cycle in Γ has leg-vertices on $F_o(\Gamma)$, let Γ have exactly two independent 2-legged cycles, and let C_1 and C_2 be the two minimal 2-legged cycles in Γ. Let $\Gamma_1(=\Gamma)$, Γ_2, Γ_3, and Γ_4 be the four embeddings of G obtained from Γ by flipping $\Gamma_I(C_1)$ or $\Gamma_I(C_2)$ around the the leg vertices of C_1 and C_2. Then G has a box-rectangular drawing if and only if at least one of the four embeddings Γ_1, Γ_2, Γ_3, and Γ_4 has a box-rectangular drawing.*

Using a method similar to that used in the proof of Theorem 3.4 in [9], we can prove Theorem 2. Note that G does not always have such an embedding Γ; if G has no such embedding, then G has no box-rectangular drawing. We omit the detail in this extended abstract.

4   Box-Rectangular Drawings of Planar Graphs with $\Delta \geq 4$

In this section we give a necessary and sufficient condition for a planar graph G with $\Delta \geq 4$ to have a box-rectangular drawing. We also give a linear-time algorithm to find a drawing if it exists. In Subsection 4.1 we consider the case where G is a subdivision of a planar 3-connected graph with $\Delta \geq 4$ and in Subsection 4.2 we consider the other cases.

4.1   Case for a Subdivision of a Planar 3-Connected Graph with $\Delta \geq 4$

Let G be a subdivision of a planar 3-connected graph with $\Delta \geq 4$, and Γ be an arbitrary plane embedding of G. We construct a new planar graph H from Γ by replacing each vertex v of degree four or more in Γ by a cycle.

Figures 3(a), 3(b), and 3(c) illustrate G, Γ, and H respectively. A replaced cycle corresponds to a real box in a box-rectangular drawing of G. We do not replace a vertex of degree 2 or 3 by a cycle since a vertex of degree 3 may be drawn as a point, and a vertex of degree 2 is always drawn as a point. Thus $\Delta(H) \leq 3$. The following theorem is the main result of this subsection.

Theorem 3. *Let G be a subdivision of a planar 3-connected graph with $\Delta \geq 4$, and Γ be an arbitrary plane embedding of G. Let H be the graph transformed from Γ as above. Then G has a box-rectangular drawing if and only if the planar graph H has a box-rectangular drawing.*

We only give a proof for sufficiency of Theorem 3 in this extended abstract; it is rather easy to prove the necessity.

We give some definitions before proving the sufficiency. Figures 3(a) and 3(b) illustrate G and Γ respectively. We replace the vertices of degree 4 or more in Γ by cycles. Each vertex of degree 2 or 3 in Γ has a corresponding vertex of the same degree in H, and we call such a vertex in H an *original vertex*. A vertex on a replaced cycle is called a *replaced vertex*. Now each vertex in H is either a replaced vertex or an original vertex.

Assume that, an arbitrary plane embedding Φ of the planar graph H has a box-rectangular drawing D_Φ. Therefore, Φ satisfies (br1) and (br2) of Lemma 1. We can easily transform D_Φ to a box-rectangular drawing $D_{\Gamma'}$ for any plane embedding Γ' of the planar graph G if only original vertices are drawn as corner boxes in D_Φ, because then each replaced vertex is a point in D_Φ, and each replaced cycle in Φ is a rectangular face in D_Φ, and hence D_Φ can be transformed to $D_{\Gamma'}$ by regarding each replaced cycle as a box. The problem is the

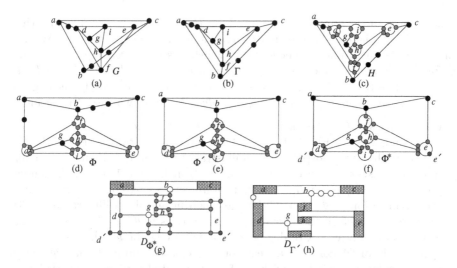

Fig. 3. Illustration for a box-rectangular drawing of a subdivision of a planar 3- connected graph G with $\Delta \geq 4$

case where a replaced vertex is drawn as a corner box in D_Φ. Because such a drawing D_Φ cannot always be transformed to a box-rectangular drawing $D_{\Gamma'}$ of Γ'. However we show that a plane graph Φ^* in Fig. 3(f) obtained from Φ in Fig. 3(d) through an intermediate graph Φ' in Fig. 3(e) with slight modification has a particular box-rectangular drawing D_{Φ^*} which can be easily transformed to a box-rectangular drawing of Γ' as illustrated in Fig. 3(h). Transformation is also not possible when the outer face of Φ is a replaced cycle. However, we are released from the problem by proving the following Lemmas 5 and 6 which are on a planar graph with $\Delta \geq 4$. Lemma 5 is needful to prove the Lemma 6. Proofs of the Lemmas 5 and 6 are omitted in this extended abstract.

Lemma 5. *Let G be a planar graph with $\Delta \geq 4$, and Γ be an arbitrary plane embedding of G. Let H be the transformed graph of Γ by replacing each vertex v of degree four or more in Γ by a cycle, and Φ be an arbitrary plane embedding of the planar graph H. Denote the total number of 2-legged and 3-legged cycles in Γ by l_Γ, and the total number of 2-handed and 3-handed cycles in Γ by h_Γ. Also denote the total number of 2-legged and 3-legged cycles in Φ by l_Φ, and the total number of 2-handed and 3-handed cycles in Φ by h_Φ. If $p_\Gamma = l_\Gamma + h_\Gamma$, and $p_\Phi = l_\Phi + h_\Phi$, then $p_\Gamma = p_\Phi$.*

Lemma 6. *Let G be a planar graph with $\Delta \geq 4$, and Γ be an arbitrary plane embedding of G. Let H be the transformed graph of Γ by replacing each vertex v of degree four or more in Γ by a cycle. Let Φ_R be any arbitrary plane embedding of the planar graph H, such that $F_o(\Phi_R)$ is a replaced cycle in H. Then G is cyclically 4-edge-connected if and only if Φ_R has a box-rectangular drawing.*

We are now ready to prove the sufficiency of Theorem 3.

Sufficiency of Theorem 3. Assume that H has a box-rectangular drawing. Then H has a plane embedding Φ which has a box-rectangular drawing. We now show that G has a plane embedding Γ' which has a box-rectangular drawing. Before entering into the cases we give a definition. The replaced cycle on $F_o(\Phi)$ corresponding to a vertex of degree 4 or more in G contains exactly one edge on $F_o(\Phi)$. We call such an edge in Φ a *green edge*. We have two cases to consider.

Case 1. Φ *does not contain a replaced cycle as outer face.*
Assume that Φ as in Fig. 3(d) has a box-rectangular drawing. Let Φ' be the minimal graph homeomorphic to Φ as illustrated in Fig. 3(e); then Φ' is a cubic graph and satisfies Conditions (br1) and (br2) in Lemma 1. Using the similar approach used in [8], we can designate four vertices as corner vertices after slight modification in Φ'. Let Φ^* be the resulting graph as illustrated in Fig. 3(f). Note that each of the four designated vertices in Φ^* is either an original vertex or a dummy vertex of degree 2 on a green edge of Φ' [8]. Clearly, every 3-legged cycle in Φ^* contains at least one designated vertex. Hence, Φ^* has a box-rectangular drawing D_{Φ}^* with the four designated vertices as corner boxes, as illustrated in Fig. 3(g). Inserting the removed vertices of degree 2 on some vertical and horizontal line segments in D_{Φ}^* and regarding the drawing of each replaced cycle as a box, we immediately obtain a box-rectangular drawing $D_{\Gamma'}$ for a plane embedding Γ' of the planar graph G from D_{Φ}^*, as illustrated in Fig. 3(h).

Case 2. Φ *contains a replaced cycle as outer face.*
In this case $\Phi = \Phi_R$, as in Lemma 6. By Lemma 6, if Φ_R has a box-rectangular drawing, then H is cyclically 4-edge connected. Hence by Theorem 1(a), another plane embedding Γ' of H, whose outer face is not a replaced cycle has also a box-rectangular drawing D. Thus by using the method used in Case 1 we can obtain a box-rectangular drawing of Γ'. □

4.2   The Other Case for a Planar Graph G with $\Delta \geq 4$

It can be trivially shown that every graph G with $\Delta \geq 4$ having two vertices has a box-rectangular drawing. Note that in this case the graph G is a multigraph.

We may thus assume that G is a planar biconnected graph with $\Delta \geq 4$ but not a subdivision of a planar 3-connected graph. In this case the following fact holds.

Fact 7. *Let G be a biconnected planar graph with $\Delta \geq 4$ but not a subdivision of a planar 3-connected graph. Let Γ_1 and Γ_2 be the two arbitrary plane embeddings of G. A minimal 3-legged cycle in Γ_1 is a minimal 3-legged or a maximal 3-handed cycle in Γ_2, and a minimal 2-legged cycle in Γ_1 is a minimal 2-legged or a maximal 2-handed cycle in Γ_2. Similarly a maximal 3-handed cycle in Γ_1 is a maximal 3-handed or a minimal 3-legged cycle in Γ_2, and a maximal 2-handed cycle in Γ_1 is a maximal 2-handed or a minimal 2-legged cycle in Γ_2.*

Let G be a planar biconnected graph with $\Delta \geq 4$ but not a subdivision of a planar 3-connected graph, and Γ be an arbitrary plane embedding of G.

Let $(x_1, y_1), (x_2, y_2), \ldots, (x_l, y_l)$ be all pairs of vertices such that x_i and y_i, $1 \leq i \leq l$, are the leg vertices of a regular 2-legged cycle or the hand-vertices of a regular 2-handed cycle. If there is a plane embedding Γ' of G having a box-rectangular drawing, then the outer face $F_o(\Gamma')$ must contain all vertices $(x_1, y_1), (x_2, y_2), \ldots, (x_l, y_l)$; otherwise, Γ' would have a 2-legged cycle containing no vertex on $F_o(\Gamma')$, and Γ' would not have a box-rectangular drawing. Because after replacing the vertices of degree 4 or more by cycles in Γ', according to Lemma 5, 2-legged cycles will remain same and the total number of 2-legged cycles will also remain same. The graph is denoted by Φ after transformation from Γ'. If Γ' has a 2-legged cycle containing no vertex on $F_o(\Gamma')$, then by Lemma 5, Φ also has a 2-legged cycle containing no vertex on $F_o(\Phi)$. Hence by (br1) of Lemma 1, Φ does not have a box-rectangular drawing, and consequently by Lemma 2, Γ' does not have a box-rectangular drawing. Similarly, if Γ' has a box-rectangular drawing, then by Lemma 2, Φ has a box-rectangular drawing and by [8] exactly two leg vertices of every minimal 3-legged cycle in Φ are on the outer face of the box-rectangular drawing. Thus by Lemma 5, $F_o(\Gamma')$ contains exactly two leg vertices of every minimal 3-legged cycle in Γ'. Hence by Fact 7 exactly two leg vertices of every minimal 3-legged cycle, and exactly two hand vertices of every maximal 3-handed cycle in Γ must be on the outer face $F_o(\Gamma')$.

Let p be the largest integer such that a number p of minimal 2-legged and maximal 2-handed cycles in Γ are independent with each other, and q be the largest integer such that a number q of minimal 3-legged and maximal 3-handed cycles in Γ are independent with each other. If $p > 2$ or $q > 4$, then by [8] Γ' does not have a box-rectangular drawing. Assume the worst case, that is, $p = 2$ and $q = 4$ in Γ. Independent minimal 3-legged and maximal 3-handed cycles in Γ are denoted by C_1, C_2, C_3, and C_4. Let $\{a_k, b_k, c_k\}$ be the set of leg vertices or hand vertices in C_k, for $k = 1, 2, 3$, or 4. We can choose two vertices from each C_1, C_2, C_3, or C_4 in 3 ways. The combinations are $\{(a_k, b_k), (b_k, c_k), \text{ and } (c_k, a_k)\}$, for $k = 1, 2, 3$ or 4. If we want to choose eight vertices from the four cycles, C_1, C_2, C_3, and C_4, two vertices from each C_k, for $k = 1, 2, 3$ and 4, we can choose in $3 \times 3 \times 3 \times 3 = 81$ number of ways. The combinations are $S_1 = \{(a_1, b_1), (a_2, b_2), (a_3, b_3), (a_4, b_4)\}$, $S_2 = \{(a_1, b_1), (a_2, b_2), (a_3, b_3), (b_4, c_4)\}$, $S_3 = \{(a_1, b_1), (a_2, b_2), (a_3, b_3), (c_4, a_4)\}$, $\ldots$, and $S_{81} = \{(c_1, a_1), (c_2, a_2), (c_3, a_3), (c_4, a_4)\}$.

Let G be a planar biconnected graph with maximum degree 4 or more but not a subdivision of a planar 3-connected graph, and Γ be an arbitrary plane embedding of G as in Fig. 4(a). Let $(x_1, y_1), (x_2, y_2), \ldots, (x_l, y_l)$ be all pairs of vertices such that x_i and y_i, $1 \leq i \leq l$, are the leg vertices of a regular 2-legged cycle or the hand-vertices of a regular 2-handed cycle, and $\{a_k, b_k, c_k\}$ be the set of leg vertices or hand vertices in C_k, for $k = 1, 2, 3$ and 4. A dummy vertex z is added in the outer face of Γ. Construct a graph Γ_j^+, for any $j = 1, 2, 3, \ldots$, or 81, by adding dummy edges (x_i, z) and (y_i, z) for all indices $i, 1 \leq i \leq l$, and by adding eight dummy edges from z to all vertices in the set S_j. In this way we can get 81 number of graphs Γ_j^+, for $j = 1, 2, 3, \ldots$, and 81. Γ_1^+ and Γ_2^+ are

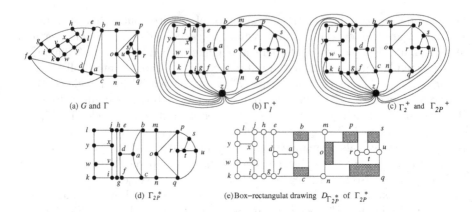

(a) G and Γ (b) Γ_1^+ (c) Γ_2^+ and Γ_{2P}^+

(d) Γ_{2P}^* (e) Box–rectangular drawing $D_{\Gamma_{2P}^*}$ of Γ_{2P}^*

Fig. 4. Illustration for a box-rectangular drawing of a biconnected graph G with $\Delta \geq 4$ but not a subdivision of a 3-connected graph

two such kinds of graphs as illustrated in Fig. 4(b) and in Fig. 4(c) respectively. G may have a box-rectangular drawing, only if, any one of the graphs Γ_j^+, for $j = 1, 2, 3, \ldots$, and 81, has a planar embedding such that z is embedded in the outer face. Γ_{2P}^+ in Fig. 4(c) is such a planar embedding of the graph Γ_2^+, but Γ_1^+ in Fig. 4(b) has no such a planar embedding. That is why, the planar graph G in Fig. 4(a) may have a box-rectangular drawing. Delete the dummy vertex z from Γ_{2P}^+. The graph is then called Γ_{2P}^* as in Fig. 4(d). Lastly by Lemma 2 and by the approach used in [8] we can test whether the plane graph Γ_{2P}^* has a box-rectangular drawing and find a drawing if it exists. $D_{\Gamma_{2P}^*}$ is a box rectangular of the plane graph Γ_{2P}^* as well as of the planar graph G, as illustrated in Fig. 4(e).

We thus have the following theorem.

Theorem 4. *Let G be a planar biconnected graph with $\Delta \geq 4$ which is not a subdivision of a planar 3-connected graph. Then one can determine whether G has a box-rectangular drawing or not by checking at most 81 graphs constructed from G as mentioned above. Furthermore, each of the 81 graphs can be checked in linear time.*

5   Conclusion

In this paper we addressed the problem for finding box-rectangular drawings of planar graphs. We gave a necessary and sufficient condition for a planar graph to have a box-rectangular drawing and developed a linear-time algorithm for finding a drawing if it exists. In this paper we have shown that, at most 81 graphs are required to be checked to take a decision whether a planar biconnected graph with $\Delta \geq 4$ has a box-rectangular drawing or not. In future one may try to minimize the number of graphs required to be checked to take the decision whether the planar biconnected graph with $\Delta \geq 4$ has a box-rectangular drawing or not.

Acknowledgment. This work is done in Graph Drawing & Information Visualization Laboratory of the Department of CSE, BUET, established under the project "Facility Upgradation for Sustainable Research on Graph Drawing & Information Visualization" supported by the Ministry of Science and Information & Communication Technology, Government of Bangladesh.

References

1. Biedl, T.C.: Optimal Orthogonal Drawings of Triconnected Plane Graphs. In: Karlsson, R., Lingas, A. (eds.) SWAT 1996. LNCS, vol. 1097, pp. 333–344. Springer, Heidelberg (1996)
2. Buchbaum, A.L., Gansner, E.R., Procopiuc, C.M., Venkatasubramanian, S.: Rectangular layouts and contact graphs. ACM Transactions on Algorithms 4(1), 8.1–8.28 (2008)
3. Di Battista, G., Eades, P., Tamassia, R., Tollis, I.G.: Graph Drawing. Prentice Hall, Englewood Cliffs (1999)
4. He, X.: A simple linear time algorithm for proper box rectangular drawings of plane graphs. Journal of Algorithms 40(1), 82–101 (2001)
5. Kant, G.: Drawing planar graphs using the canonical ordering. Algorithmica 16, 4–32 (1996)
6. Munemoto, S., Katoh, N., Imamura, G.: Finding an optimal floor layout based on an orthogonal graph drawing algorithm. Journal of Architecture, Planning and Environmental Engineering (Transactions of AIJ) 524, 279–286 (2000)
7. Nishizeki, T., Rahman, M.S.: Planar Graph Drawing. World Scientific, Singapore (2004)
8. Rahman, M.S., Nakano, S., Nishizeki, T.: Box-rectangular drawings of plane graphs. Journal of Algorithms 37, 363–398 (2000)
9. Rahman, M.S., Nishizeki, T., Ghosh, S.: Rectangular drawings of planar graphs. Journal of Algorithms 50, 62–78 (2004)
10. Tamassia, R., Tollis, I.G., Vitter, J.S.: Lower bounds for planar orthogonal drawings of graphs. Information Processing Letters 39, 35–40 (1991)

Author Index

Aoki, Takanori   233
Aschner, Rom   89

Babu, Jasine   17
Bandyapadhyay, Sayan   77
Banik, Aritra   77
Bera, Suman Kalyan   137
Bhattacharya, Binay   217
Biniaz, Ahmad   17
Bourgeois, Nicolas   114
Brunsch, Tobias   182

Chakraborty, Soudipta   217
Chun, Jinhee   53
Claude, Francisco   158
Cornelissen, Kamiel   182
Crespelle, Christophe   126

Damaschke, Peter   245
Das, Ananda Swarup   65
Das, Sandip   77
Dorrigiv, Reza   158
Durocher, Stephane   310

Eades, Peter   298
Eğecioğlu, Ömer   245
El Ouali, Mourad   101

Fekete, Sándor P.   29, 41
Fohlin, Helena   101

Gambette, Philippe   126
Gao, Jiawei   194
Giannakos, Aristotelis   114
Goddard, Wayne   146
Gupta, Prosenjit   65
Gupta, Shalmoli   137

Hasan, Md. Manzurul   334
Hong, Seok-Hee   298
Horiyama, Takashi   53

Iranmanesh, Ehsan   217
Ito, Takehiro   53, 233

Jayalal Sarma, M.N.   274

Kamali, Shahin   158
Kaothanthong, Natsuda   53
Karim, Md. Rezaul   334
Katz, Matthew J.   89
Kawahara, Jun   170
Kishimoto, Akihiro   170
Kloks, Ton   194
Krishnamurti, Ramesh   217
Kröller, Alexander   5
Kumar, Amit   137
Kumar, Prasun   274

López-Ortiz, Alejandro   158
Lucarelli, Giorgio   114

Maheshwari, Anil   17
Manthey, Bodo   182
Milis, Ioannis   114
Minato, Shin-ichi   170
Mishra, Tapas Kumar   257
Moeini, Mahdi   5
Molokov, Leonid   245
Mondal, Debajyoti   310
Morgenstern, Gila   89

Nguyen, Quan   298

Olsen, Martin   206
Ono, Hirotaka   53
Otachi, Yota   53

Pal, Sudebkumar Prasant   257
Panigrahy, Rina   4
Paschos, Vangelis Th.   114
Poon, Sheung-Hung   194
Prałat, Paweł   158

Rahman, Md. Saidur   310, 334
Raman, Indhumathi   322
Raman, Venkatesh   265, 286
Reinhardt, Jan-Marc   41
Reisi Dehkordi, Hooman   298
Revsbæk, Morten   206
Rex, Sophia   29
Röglin, Heiko   182

Romero, Jazmín   158
Roy, Sambuddha   137

Salinger, Alejandro   158
Santoro, Nicola   1
Sarkar, Hirak   77
Saurabh, Saket   286
Sawlani, Saurabh   274
Schmidt, Christiane   5, 29
Schweer, Nils   41
Seco, Diego   158
Shankar, Bal Sri   265
Smid, Michiel   17
Srimani, Pradip K.   146

Srinathan, Kannan   65
Srivastav, Anand   101
Suchý, Ondřej   286

Takeuchi, Shogo   170
Tokuyama, Takeshi   53

Uchizawa, Kei   233
Uehara, Ryuhei   53
Uno, Takeaki   53

Yuditsky, Yelena   89

Zhou, Xiao   233